AUTOMOTIVE AIR-CONDITIONING AND CLIMATE CONTROL SYSTEMS

Automotive Air-conditioning and Climate Control Systems

Steven Daly
BEng, BA (Hons), IEng, Cert Ed, MIMI, LAE, MSAE

ELSEVIER

Amsterdam • Boston • Heidelberg • London • New York • Oxford
Paris • San Diego • San Francisco • Singapore • Sydney • Tokyo
Butterworth-Heinemann is an imprint of Elsevier

Butterworth-Heinemann is an imprint of Elsevier
Linacre House, Jordan Hill, Oxford OX2 8DP
30 Corporate Drive, Suite 400, Burlington, MA 01803

First edition 2006
Reprinted 2007

Notice
No responsibility is assumed by the publisher for any injury and/or damage to persons
or property as a matter of products liability, negligence or otherwise, or from any use
or operation of any methods, products, instructions or ideas contained in the material
herein. Because of rapid advances in the medical sciences, in particular, independent
verification of diagnoses and drug dosages should be made

British Library Cataloguing in Publication Data
A catalogue record for this book is available from the British Library

Library of Congress Cataloging-in-Publication Data
A catalog record for this book is available from the Library of Congress

ISBN: 978-0-7506-6955-9

For information on all Butterworth-Heinemann publications
visit our website at books.elsevier.com

Printed and bound in *Hungary*

07 08 09 10 10 9 8 7 6 5 4 3 2

Working together to grow
libraries in developing countries
www.elsevier.com | www.bookaid.org | www.sabre.org
ELSEVIER BOOK AID International Sabre Foundation

Contents

Preface

Vehicle subsystems are understandably never given the discourse (research) needed to allow the engineer to have a complete understanding of how such technology evolves. The subject of air-conditioning (A/C) is certainly a victim of such negligence within the UK. Textbooks exist for the US market, which contain contributions from US manufacturers like GM, but little literature exists which provides comprehensive coverage for Europe. This problem, combined with the global political pressure on manufacturers to reduce the emission of harmful refrigerant gases (R134a), is providing a catalyst for changes to A/C technology. Research into alternative refrigerants like CO_2 and alternative A/C systems has been ongoing for a number of years. The motor vehicle industry resists such radical moves and wants more of a progressive phasing out of R134a, giving more of a lead time for the replacement technology to be introduced. It is certainly accurate to predict that during the next couple of years A/C technology, which includes systems and procedures and possibly certification to technicians, will radically change.

This book is born out of the current debate between politics and industry and hopes to provide the reader with a thorough up-to-date knowledge of current A/C systems, refrigerants and the new possible replacement systems like CO_2. The book is primarily *technology* focused, providing additional chapters on legislation and the environment. The book also has an unprecedented amount of electronic coverage with some of the very latest sensors and actuators, OBD and EOBD, test procedures using meters, scanners and oscilloscopes and additional information on how to read European wiring diagrams. This information is then applied to three practical case studies based on European manufacturers. It is imperative that A/C engineers have the fundamental understanding of automotive electronic control to enable them to successfully work within the field of automotive Heating, Ventilation and Air-Conditioning (HVAC). This book gives that level of coverage providing the reader with a holistic understanding of the climate control system.

I hope you enjoy reading this book as much as I enjoyed writing it.

Steven Daly

Acknowledgements

This book has been successfully produced due to the contribution of the following companies. They have provided diagrams, information and services in the quest to help provide a comprehensive account of the current and future technological advancement of the A/C industry.

1. Amerigon – Dan Pace
2. Autoclimate – Brian Webster, James Onion
3. Autodata Ltd – Malcolm Rixon
4. Crocodile Clips – Kirsty Gutherie
5. Elsevier, Commissioning editor – Jonathan Simpson
6. Environ – Barry Quested, Scott Mitchell
7. EPA – Kristen Taddonio
8. Fluent CFD – Chris Carey, Helen Rushby
9. Fluke – Simon Worrall
10. Ford – David Grunfeld, Avtar Singh, Alan Jones, Steve Green, George Klinker
11. Rover
12. Sanden UK – Mike Tabb
13. SMMT – Eva de Marchi Taylor
14. Tellurex
15. Toyota UK – Paul Hunt, Lisa Halliday, Heidi Lismore
16. Vauxhall Motor Company – Adam Colins, Tony Rust, Barry James, Paul Usher
17. Visiteon – John Sherringham

All my love to my wife Tina and two children Luke and Jack. Without your support, patience and understanding I could not have completed this book.

Introduction: An overview of the automotive air-conditioning market, training and qualifications

The aim of this section is to:

- Enable the reader to appreciate the growth pattern of the A/C market.
- Enable the reader to appreciate the opportunities available due to the growth and development of the A/C market.

The A/C market can be viewed from various statistical viewpoints, several of which are included below. Ultimately, whatever the perspective, the picture is of tremendous and sustained market growth, both over the last decade and into the coming years.

New registration of cars with A/C

The proportion of A/C registrations (the registration of new vehicles with A/C compared to without A/C) has risen dramatically since the mid-1990s. The pattern is a typical 'S-shaped' growth curve. The fastest rate of increase was between 1995 and 1998, when the penetration of factory-fit A/C tripled.

The global statistics

Figure P.1 provides information on A/C registrations per international region including predictions on future demand. These percentages include vehicles with manual, semi-automatic and Automatic Climate Control (ACC).

The statistics provide evidence of an increased penetration of the ACC system on new vehicles. The ACC system is showing growth in regions where A/C penetration has not increased – NAFTA and Japan. This provides evidence of the increased level of comfort customers expect with the purchase of a new vehicle and of course the competition involved with new vehicle sales.

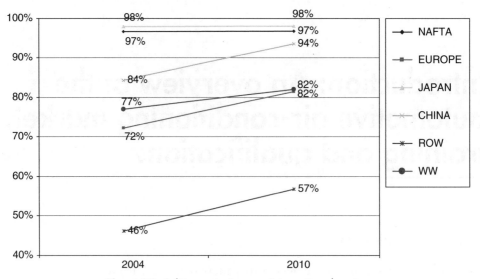

Figure P.1 A/C penetration per international region
Note
NAFTA – Northern American Free Trade Agreement States
ROW – Rest of World
WW – Western World
(courtesy of Valeo)

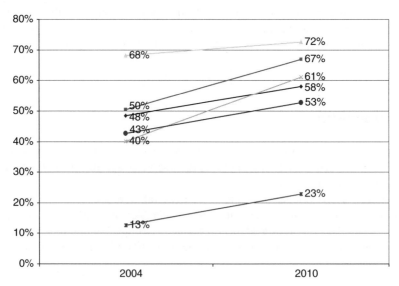

Figure P.2 Climate control penetration
(courtesy of Valeo)

Comparative data

	A/C rate			ACC rate	
	2004	**2010**		**2004**	**2010**
NAFTA	97%	97%	NAFTA	48%	58%
Europe	72%	82%	Europe	50%	67%
Japan	98%	98%	Japan	68%	72%
China	84%	94%	China	40%	61%
Row	46%	57%	Row	13%	23%
WW	77%	82%	WW	43%	53%

Figure P.3
Note
NAFTA – Northern American Free Trade Agreement States
ROW – Rest of World
WW – Western World

The UK figures

A/C penetration on new vehicles in the UK is currently around 75%.

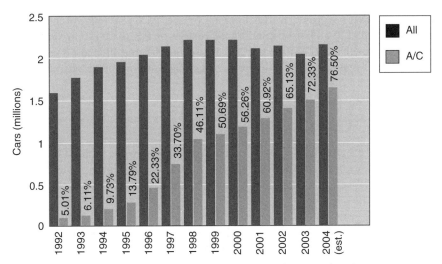

Figure P.4 UK registrations: all cars and cars with A/C
(courtesy of Autoclimate)

Year	A/C registrations	% of all registrations*
1995	267 932	13.79%
1996	451 030	22.33%
1997	726 190	33.70%
1998	1 024 009	46.11%
1999	1 094 381	50.69%
2000	1 168 849	56.26%
2001	1 293 679	60.92%
2002	1 405 092	65.13%
2003	1 500 902	72.33%
2004	1 693 250	76.50%

Figure P.5 The growth in registrations of air-conditioned vehicles
(courtesy of Autoclimate)
*A/C registrations as a percentage of total registrations.

Total numbers of cars with A/C on the road

New registrations only give part of the picture. In fact, the size of the market is more closely related to the total number of cars with A/C on the roads rather than the number of new ones being registered each year, though clearly the latter determines the former. The UK figures are available as an example. The compound figures are shown in Figure P.6.

While year on year growth rate in UK car registrations peaked between 1995 and 1998, the growth rate of the total parc (penetration of vehicles with A/C) has continued to accelerate. Since 1998, the total A/C parc has grown by over 1 million vehicles each year and will continue to do so. There is an estimated 11.95 million air-conditioned vehicles on UK roads (approximately 1 in 2.5). This represents a growth of 422% from the 2.03 million of 1997, and all in just 8 years.

This long period of peak growth is due to two combined factors, namely:

● Very high A/C fitment on new vehicles.
● Very low A/C fitment on vehicles leaving the parc (due to scrapping, accident damage etc.).

The graph in Figure P.7 gives further details on the age range of the estimated 10.58 million air-conditioned cars that will be driving on UK roads in 2004. Each column shows the number of vehicles by year of registration. The bulk of these 10.58 million cars is relatively recently registered vehicles.

Sustained market growth since the mid-1990s and the maturing age profile of vehicles both clearly demonstrate that the A/C service and repair opportunity is broadening. Where franchise dealers previously controlled more than 90% of opportunities, now, the full spectrum of dealers, accident repairers, independent garages and fast fits are seeing enough cars with A/C to justify involvement. In fact, recent years have seen something of a 'rush' to enter this market. The ripple effects caused by this offer training colleges opportunities which are discussed in more detail in Chapter 3.

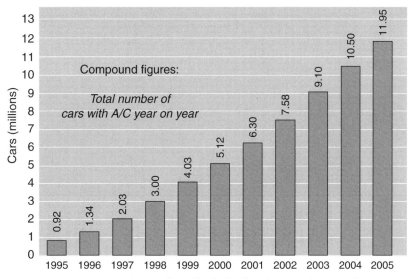

Figure P.6 Increase in the number of vehicles fitted with A/C in UK by 2005
(courtesy of Autoclimate)

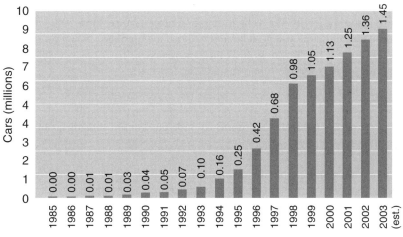

Figure P.7 Age profile of cars with A/C in UK
(courtesy of Autoclimate)

Overview of the global market

The number of automotive businesses working directly on vehicle A/C systems is growing rapidly. A wider trend towards increased reliability (at vehicle, system and component level) and longer service intervals have meant A/C service and repair have received substantial attention as a rare aftermarket growth area. A/C has become almost standard-fit and it requires service and maintenance, offering forward thinking garages a profit opportunity. Statistics also show an increased percentage in the number of ACC systems fitted, which means that the systems will be more complex including a greater level of electronic control.

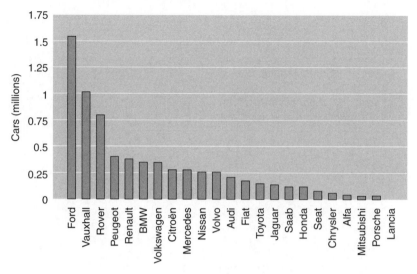

Figure P.8 Total number of cars with A/C by manufacturer in UK in 2003 (courtesy of Autoclimate)

It is now no longer just the franchised dealers that are involved. Each passing year sees more and more companies moving into A/C and the job of selling the service and preventive maintenance is being made easier by the vehicle manufacturers. More than 50% of VMs (Vehicle Manufacturers) now either have A/C-related scheduled maintenance or run promotions via their dealer networks to generate business and build awareness.

Opportunities for A/C service/repair include the following.

Refrigerant loss

Systems lose refrigerant naturally at around 10–15% per annum causing a range of secondary problems including:

1. Poor cooling – most noticeable in hot weather.
2. Impaired lubrication – often eventually causing compressor failure.
3. Air in system – can cause reduction in cooling and speed up internal system corrosion.
4. Moisture in system – can cause blockage and contribute to corrosion.

Servicing A/C systems regularly will not only improve performance and fuel economy it will also minimise the risk of damage.

Odour problems – anti-bacterial treatment opportunity

Stale or unpleasant smells are a common and unfortunate side-effect of A/C. These smells and even 'sick car syndrome' (where vehicle occupants experience flu-like symptoms) are the result of bacteria build-up in the evaporator. This problem can be easily treated.

General system diagnostics and repairs

Unfortunately, as preventive maintenance has not always been standard practice, A/C systems have suffered years of neglect. The result is often component or system failure. The positioning of the condenser at the front of the vehicle also makes it vulnerable to salt corrosion, stone damage etc.

Technological development

The development of new technology requires new methods and procedures as well as possible certification. Training agencies can provide educational support to promote working practices and equipment suppliers will have opportunities to provide new machines that work with new A/C systems or refrigerant. If CO_2-based systems prevail (2008) then new practices will need to be introduced. This is an opportunity for experts in the field to provide training and resources to help with this learning transition.

Training opportunities

The UK has a number of industrial and educational training providers. Most receive accreditation as training centres under the guidance of awarding bodies. The automotive sector uses a number of awarding bodies – City & Guilds, and the Institute of the Motor Industry (IMI) are among a few of them. The criteria set by these bodies are generated through the collaboration between education and industry. Currently there is no legislation requiring the motor vehicle sector to attend training courses enabling them to work on vehicles with air-conditioning. The only requirements under the EPA Act are the safe handling of refrigerants. This is also required in the US.

Compulsory training – Safe handling of refrigerants – CITB (Construction Industry Training Board)

The Environment Protection Act forbids the deliberate discharge of refrigerants into the atmosphere. Where refrigerant has to be removed from a system and cannot be immediately reused, it is recovered and sent for recycling or disposal by suitably regulated companies who employ competent engineers to undertake this task.

These engineers need to be up to date with the latest legislation and practice if they work on the installation, commissioning, servicing or repairing of refrigerant equipment.

The BES refrigerants scheme covers safe handling of refrigerants.

Passing the appropriate assessments awards a CITB-Construction Skills certificate and card, which is used to prove competence to work with refrigerants in accordance with the Environmental Protection Act.

Voluntary education – air-conditioning

City & Guilds 6048 Motor Vehicle Air Conditioning (Service and Repair)

The City & Guilds 6048 Motor Vehicle Air Conditioning certificate is a recognised qualification and the course covers the underpinning knowledge of vehicle refrigeration and air-conditioning. This course is accredited by City & Guilds and monitored and assessed by Air Parts Europe Ltd. The course length is approximately 20 hours. The course is ideal for part-time attendance as a short professional course allowing students to apply their knowledge in their working context. Most candidates work in the motor vehicle service industry and require knowledge to expand their business into the A/C service and repair industry. The underpinning knowledge required to complete the course is covered in this book.

City & Guilds Progression Award Level 3

City & Guilds offer air-conditioning training at level 3 of their Progression Award, Unit 7 – Diagnose faults and repair systems and components – air-conditioning and climate control.

The unit is currently optional although it is hoped that eventually it will become compulsory due to the possible future standard fitment of A/C systems. The underpinning knowledge required to complete the unit is covered in this book. This unit is generally delivered with core and other optional units enabling the candidate to gain full certification.

Institute of the Motor Industry (IMI)

The IMI is the leading awarding body for the retail motor industry. They offer a range of qualifications from level 1 to level 5:

- *Level 1* – vocational qualifications are pre-apprenticeship programmes for students from 14 years old, in school or further education. These qualifications teach basic knowledge and routine tasks.
- *Level 2* – suitable for those who have a level 1 qualification, or are likely to achieve GCSE grades D–F in English, mathematics and a science-based subject. These qualifications cover routine tasks and require previous knowledge or work experience.
- *Level 3* – suitable for those who have achieved a level 2 qualification, or are likely to achieve GCSE grades A–C in English, mathematics and a science-based subject. These supervisor level qualifications cover non-routine, more complex tasks and require previous knowledge or work experience.
- *Level 4* and above – management and master technician level qualifications, for those who have already achieved a level 3 qualification. These enable progression to higher education, management and level 5 qualifications and give a good grounding in the skills required to run a business.

The IMI offer QCA (Qualifications and Curriculum Authority) courses and qualifications in A/C as well as A/C within other units of study. The IMI also offers Automative Technician Accreditation.

Automotive Technician Accreditation (ATA)

ATA is a voluntary assessment programme for technicians working in the retail motor industry. It has the backing of major vehicle manufacturers, independent service and repair organisations and Automotive Skills, which is the Sector Skills Council for the retail motor industry.

Automotive Technician Accreditation is governed by the Institute of the Motor Industry (IMI), and ATA registered technicians sign and are bound by a special Code of Conduct. They are issued with a photo identity card and their details are included on the ATA website.

ATA brings major benefits for consumers, technicians and employers including:

- Consumer confidence in ATA registered technicians and the organisations employing them.
- Proof of current technical competence and professional responsibility.
- A benchmark for technician recruitment and training.

United states

Compulsory education – safe handling of refrigerants

All technicians opening the refrigeration circuit in automotive air-conditioning systems must now be certified in refrigerant recovery and recycling procedures and be in compliance with Section 609 of the Clean Air Act Amendments of 1990.

MACS (Mobile Air Conditioning Society) offer a booklet and test which can be downloaded from the internet or obtained by post. Upon completion the candidate is awarded a certificate in the Safe Handling of Refrigerants in line with the requirements set out by the EPA.

ASE certification

ASE's mission is to

> *improve the quality of automotive repair and service through the voluntary testing and certification of service professionals.*

Approximately 400 000 professionals hold current ASE credentials.

Becoming an accredited automotive HVAC training provider can offer opportunities to generate income from educating and develop links with the service industry for skill and knowledge updating. This could also lead to aftermarket supply of tools and consumables. Educational training could certainly be an area for growth with the introduction of CO_2-based A/C systems especially if the technicians are required to be licensed due to the hazards of working with such systems.

1 Air-conditioning fundamentals

The aim of this chapter is to:

- Give an overview of the historical development of the heating and ventilation system and introduction of the air-conditioning (A/C) system.
- Provide the reader with a case study on the design and optimisation of an air-conditioning (A/C) system.
- Enable the reader to understand the fundamental principles and operation of the heating, cooling, ventilation and air-conditioning system.
- Introduce the possible replacement refrigerant/system to R134a.

1.1 History of automotive air-conditioning systems

The early history of transportation systems starts mainly with the horse drawn carriage. This was eventually surpassed by the invention of the automobile. Early automobiles had cabin spaces that were open to the outside environment. This means that the occupants had to adjust there clothing to allow for different weather conditions. Closed cabin spaces were eventually introduced which required heating, cooling and ventilating to meet customer expectations. Early heating systems included heating clay bricks and placing them inside the vehicle or using simple fuel burners to add heat to the vehicle's interior. Ventilation inside the vehicle was achieved through opening or tilting windows or the windscreen; vents were added to doors and bulkhead to improve air circulation and louvred panels were the equivalent to our modern air ducts. Air flow was difficult to control because it was dependent upon the vehicle speed and sometimes would allow dirty, humid air which contained fumes to enter the interior from the engine compartment. Cooling could be as simple as having a block of ice inside the vehicle and allowing it to melt! Eventually a number of design problems were overcome, these included air vents at the base of the windscreen for natural flow ventilation and electric motors to increase the flow at low speeds. Eventually heat exchangers were introduced which used either the heat from the exhaust system or water from the cooling system as a source, to heat the inside of the vehicle cabin. Early cabin cooling systems were aftermarket sourced and worked on evaporative cooling. They consisted of a box or cylinder fitted to the window of the vehicle. The intake of the unit would allow air to enter from outside and travel through a water soaked wire mesh grille and excelsior cone inside the unit. The water would evaporate due to absorbing the heat in the air and travel through the outlet of the unit which acted as a feed to the inside of the vehicle. The water was held in a reservoir inside the unit and had to be topped up to keep the cone wet otherwise the unit would not operate. The air entering the vehicle would be cool if the relative humidity of the air entering the unit was low. If the relative humidity of the air was high then the water could not evaporate. When the unit was working effectively it would deliver cool saturated water vapour to the inside of the vehicle which raised the humidity levels. These units were only really effective in countries with very low humidity.

In 1939 Packard marketed the first mechanical automotive A/C system which worked on a closed cycle. The system used a compressor, condenser, receiver drier and evaporator (fitted inside the boot/trunk) to operate the system. The only system control was a blower switch. Packard marketing campaign included: 'Forget the heat this summer in the only air-conditioned car in the world.' The major problem with the system was that the compressor operated continuously (had no clutch) and had to have the belt removed to disengage the system which was generally during the winter months. Over the period 1940–41 a number of manufacturers made vehicles with A/C systems but these were in small volume and not designed for the masses. It wasn't until after World War II that Cadillac advertised a new feature for the A/C system that located the A/C controls on the rear parcel shelf, which meant that the driver had to climb into the back seat to switch the system off. This was still better than reaching under the bonnet/hood to remove the drive belt. In 1954–55 Nash-Kelvinator introduced air-conditioning for the mass market. It was an A/C unit that was compact and affordable with controls on the dash and an electric clutch.

The design and optimisation of an air-conditioning system

Case study – the air handling system

Experimental approach

In the past, the only way to evaluate a proposed air handling system design was to build a prototype and test it in the laboratory. The air handling components were placed on a test stand, conditioned air was supplied at the inlet and the airflow and temperature distribution at critical locations were measured. This approach takes a considerable amount of time and requires the construction of expensive prototypes. In addition, it provides little or no understanding of why a design performed the way it did. In particular, testing is unable to detect details of recirculating areas, turbulence, temperature stratification and constrictions that adversely impact performance and pressure loss. In addition, the performance of the system usually needs to be evaluated in many different configurations. For example, it sometimes is necessary to evaluate the air handling system in different modes of operation – vent, floor, defrost and mixed – at each of eight different temperature controls.

Modern methods of design

The design process of modern vehicle systems improved with the introduction of Computer Aided Design (CAD), Computer Aided Engineering (CAE) and Computer Aided Manufacturing (CAM). CAD allows designs to be generated and visually appreciated on a computer. Standard components can be shared among manufacturers and suppliers to ensure that components assemble correctly. Designs can be sent to clients for verification and feedback. Designs can be modified and rechecked within short periods of time in a number of different formats, e.g. an STL file (stereolithography). Complex parts and assemblies can often be manufactured very quickly using rapid prototyping facilities (CAM). CAD also includes the facility to provide virtual testing. This is generally provided using additional modules or add-ins converting CAD to CAE. The software is even now used among a number of secondary schools in the UK who have the use of Solidworks as a CAD package for their technology departments which include add-in modules like Cosmos Works for Finite Element Analysis and Computational Fluid Dynamics. Finite Element Analysis (FEA) is basically mechanical stress analysis and Computational Fluid Dynamics (CFD) analyses the flow of a fluid like air through or over complex geometry. These additional features are all computer-based and use mathematical equations built into the software to predict variables like the stress distribution of a component or assembly (FEA) or the flow of air through an air vent (CFD). All these tests would have originally been carried out manually with continual adjustments being made to a model to optimise it.

Figure 1.1 Computer generated model designed using CAD (without ducting and vents)

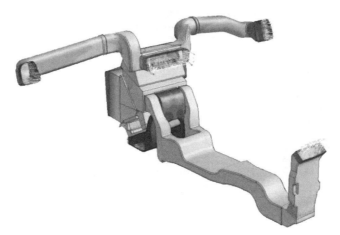

Figure 1.2 Air pressure loss predicted by CFD

The process

The A/C system begins life as an idea driven by consumer needs and government legislation. This leads to a specification. The specification will include minimum performance requirements, temperatures, control zones, flow rates etc. This will lead to a number of concept designs. The designs will have a number of computer generated models which will be presented as possible solutions to the original specification. These need to be tested for their performance.

Performance testing using CFD may include fluid velocity (air flow), pressure values and temperature distribution. Using CFD enables the analysis of fluid through very complex geometry and boundary conditions. The geometry typically includes ducts that expand and contract, change from round to square cross-sections, go through complex curves throughout their length, and have many branches and internal walls.

As part of the analysis, a designer may change the geometry of the system or the boundary conditions such as the inlet velocity, flow rate etc. and view the effect on fluid flow patterns.

CFD is an efficient tool for generating parametric studies with the potential of significantly reducing the amount of physical experimentation required to optimise the performance of a design.

A fan performance curve can be inputted into a model. Without this feature, the user has to guess the flow within the fan enclosure, calculate the pressure using CFD and see if it matches

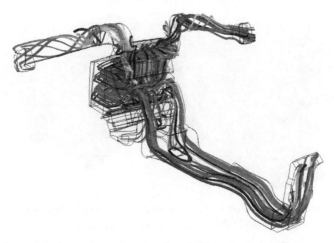

Figure 1.3 Streamlines showing flow field in an air handling system

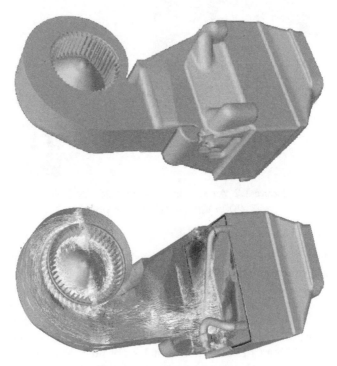

Figure 1.4 Fan flow optimisation

the fan's characteristics. If the pressure doesn't match, then another guess has to be made. Normally, at least three iterations (test runs) are required to make a match.

The software has the facility to enter a fan performance curve directly into the model. Each analysis run then interacts with the fan curve to determine the precise operating conditions of the fan as part of the regular analysis. Using this technique, engineers can easily determine what

Figure 1.5 Improved fan design

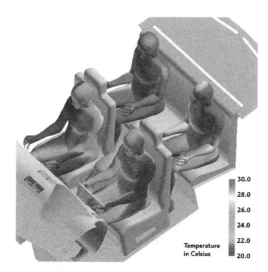

Temperature
in Celsius

30.0
28.0
26.0
24.0
22.0
20.0

Figure 1.6 Human modelling for temperature distribution

type of fan is required to meet air flow requirements within the vehicle, normally 158 cubic feet per minute (75 litres/second) for heating and 300 cubic feet per minute (141.6 litres/second) for cooling.

As a typical example of improvements consider the typical design specification of the HVAC system with respect to the temperature dial on the instrument panel. In other words, moving the dial from position one to position two should have the same impact on temperature as moving from position two to position three. In the past, the linearity of the temperature dial could not be estimated until full vehicle prototypes were constructed. At that point, changes were costly and the testing data provided little or no input on what type of changes were required.

Figure 1.7 Prototype HVAC unit for testing

Now, engineers can determine the linearity of a proposed design as soon as the solid model in CAD has been created in a matter of days. They typically set up a series of analysis runs that evaluate eight different temperature settings at each of the three HVAC system modes. In less than a week, they can determine outlet air temperature at each setting.

Once all CFD modelling is complete the prototypes are made to ensure the physical models operate as predicted by the computer models. The accuracy of simulated and actual system performance can vary up to 10–15%. Generally, lead times are reduced and designs can be evaluated much quicker allowing more time to optimise their working performance.

1.2 Introduction to heating and ventilation

Heating and ventilation in automotive transport is not just a function of temperature control. The safety of occupants to reduce driver fatigue, ensure good visibility and maintain comfort is key to the successful design of such systems. A continual flow of air through the vehicle's interior reduces carbon dioxide levels, acts as a demister and prevents the build-up of odours. Carbon dioxide in high concentration can cause a driver to be less responsive. There are recommended ventilation rates which specify the number of times the internal cubic capacity (air space) of the vehicle must be replaced per hour. Included in this calculation are the number of possible occupants and the internal volume of the vehicle. In some countries the performance of a heating and ventilation system is governed by legislation. The heating and ventilation system combined with an air-conditioning system provide a temperature range for occupants to select from. This can be a real challenge due to some extreme weather conditions experienced across the globe. Often auxiliary booster devices are required to provide additional 'heating' or 'cooling' of the interior.

The car heating system

Heat is a form of energy which means it cannot be destroyed. The principle of the heating and ventilation system is to transfer enough heat from one point to another. The heater is a device which heats the air entering or already inside the vehicle (recycled air). The heated air is then directed to a combination of different places via a distribution of air ducts within the vehicle.

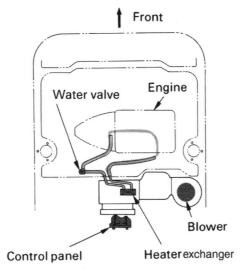

Figure 1.8 Water cooling system
(with the agreement of Toyota (GB) PLC)

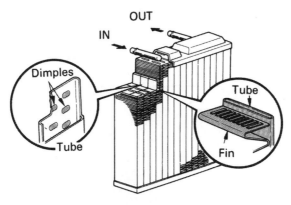

Figure 1.9 Heat exchanger

There are a number of different methods available to heat the air – exhaust heater, heat as a by-product of combustion, electric heater etc. Generally motor vehicles use heat from combustion which is transferred through water or air depending on whether the engine is water or air cooled. If the vehicle is air cooled then a system of shrouds is used to direct the heat from the external surface of the engine, exhaust or in some cases from the lubrication system towards the inside of the vehicle.

Water cooled engines
The engine has a water cooling system which is used to maintain engine temperature by transferring combustion heat (as a by-product of the combustion process) away from the combustion chamber. The heated coolant is then carried from the combustion chamber through pipes to a heat exchanger.

The heater exchanger (heater matrix)
The heat exchanger (Fig. 1.9), often called the matrix, is situated inside the vehicle housed within the heater assembly (Fig. 1.8). It is designed to have a large surface area enabling air to

pass over the surface of its fins. Fresh or recycled air (air from inside the vehicle) is directed under force over the surface of the heater core and then distributed via air and panel vents into the vehicle's interior. The heater core is made up of tubes and fins which are made from aluminium alloy and have aluminium or plastic tanks attached to the core with inlet and outlet ports.

Heat control

Heat control is determined by the occupants of the vehicle. This is done by selecting the required interior temperature which the occupants require via a control panel. The control panel will either control components to allow more or less water to enter the heat exchanger or allow more or less air to flow over its external surface. These two systems used to control the heater's thermal output are referred to as:

1. Water flow type.
2. Air mix type.

Water flow type

This system controls the amount of coolant flowing from the engine cooling system to the heat exchanger using a control valve (Fig. 1.10). The control valve (Figure 1.11) varies the flow of coolant going inside the heat exchanger which in turn varies the temperature of the heater core. Regulation in such a system can be difficult especially with the coolant flow and temperature being dependent upon engine speed and load. The system does not respond immediately to

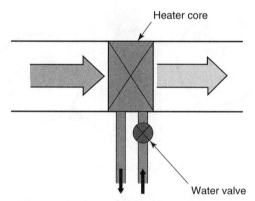

Figure 1.10 Heat controlled by a water valve
(with the agreement of Toyota (GB) PLC)

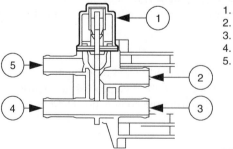

1. Control solenoid
2. Flow from engine
3. Flow back to engine
4. Flow from the heater core
5. Flow to the heater core

Figure 1.11 A solenoid operated water control valve
(reproduced with the kind permission of Ford Motor Company Limited)

change, for example if a lower temperature is required to the interior then the control valve will restrict the flow of coolant to the heater core. To achieve the reduced temperature the heater core must lose the heat required to cool to the new selected temperature, thus giving off heat; this takes time. A benefit of regulating the heater core using a water control valve is that it allows the whole volume of air to flow through the heater core itself, improving heating performance.

Air mix type

This system controls the volume of air allowed to flow over the surface of the heat exchanger using an air mix/blend control door fitted inside the heater assembly. The internal door directs air over or bypassing the heater core depending on its position. The position is determined by the occupants who select a temperature range from hot to cold. If a mid-range temperature is selected then the quantity of air will flow over the heater core (Fig. 1.12) and a quantity will bypass the heater core (Fig. 1.12). This air will then mix later in a mixing chamber to reach the final required temperature before leaving the heater assembly. The air mix control doors are generally operated by Bowden cable, vacuum or electronic servo. The negative aspect of such a design is that the use of a mixing chamber means that the heater assembly tends to be larger for the air mix type than the coolant controlled type. When not in use heat can radiate from the heater core warming natural air flow which is transferred to the cabin space although this can be overcome by using a shut-off solenoid to stop the flow of coolant when maximum cooling effect is required. Positive aspects include quick response to changes in temperature and more accurate control of temperature variations.

Air distribution through the interior of the vehicle

A ventilator is a device used to direct air through the inside of a vehicle. There are generally two types of ventilator used on a vehicle:

1. Natural flow ventilator.
2. Forced flow ventilator (blower).

Natural flow ventilator

This is created by air pressure outside of the vehicle caused by the forward motion of the vehicle. As the vehicle moves in a forward direction positive and negative pressure is created on the vehicle's surface due to its aerodynamic shape. Areas where positive pressure is created are ideal places for air vents which allow air to enter the vehicle, travel through the interior and then exit the vehicle via a vent often positioned in a negative pressure region. The air intake is positioned

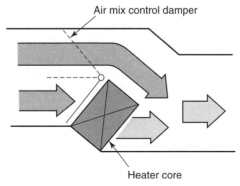

Figure 1.12 Heat controlled by an air mix control damper
(with the agreement of Toyota (GB) PLC)

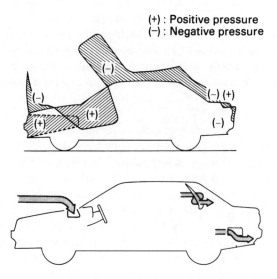

Figure 1.13 Positive and negative pressure across the surface of the vehicle
(with the agreement of Toyota (GB) PLC)

at the bottom of the windscreen where static pressure is high so air under pressure can flow into the vehicle. There are a number of downsides to this position:

1. The engine compartment must be adequately sealed so no dissatisfying smells find their way into the interior of the vehicle.
2. Air flow is proportionate to vehicle speed causing lack of flow at low speeds and possible excessive noise and draughts at high speed.

This is generally reduced by only allowing a small pressure differential across the intake and exhaust points and adequate fan assistance. Air exhausts through outlets, which are generally located in a rear pillar hidden behind a trim panel or behind the rear bumper (Fig. 1.13).

Air inlet and outlets
Figure 1.14a illustrates the position of the inlet housing which is used to separate dirt particles and water from the air taken into the passenger compartment, the fresh air is fed through a cowl panel grille and pollen filter housing (2). This often houses air filtration systems like pollen filters (1). The pollen filter is able to clean the fresh air trapping smaller particles, such as dust and pollen which are able to get through the cowl panel grille.

The air outlets are arrangements of rubber flaps, mostly hidden behind the rear bumper (Fig. 1.15) or behind a trim panel on the rear pillar. Because the pressure in the passenger compartment is created by the blower motor and natural air flow, the air outlets open enabling air exchange to take place. If there is no air flow through the vehicle interior, the flaps close to prevent exhaust fumes from entering. If vehicle ventilation is unsatisfactory, check the flaps for freedom of movement.

In case of exhaust fumes reaching the interior compartment, check the closing function of the flaps and their restriction of air flow (Fig. 1.15).

Forced air flow

In forced air ventilation systems an electric fan is fitted inside the vehicle. Fans are generally used when vehicle speeds are low or comfort demands are high (demisting, heating and cooling).

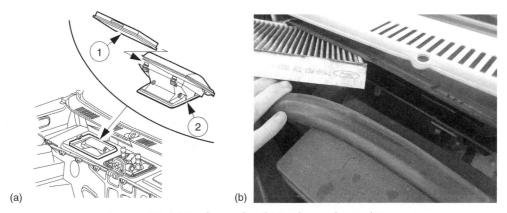

Figure 1.14 (a) Fresh air inlet; (b) Fresh air inlet Ford Fiesta
(reproduced with the kind permission of Ford Motor Company Limited)

The blower fan can force air over the heater core allowing the heat to be transferred to the air which is distributed around the vehicle. Intake and outlet vents to the interior are generally located in the same position as the natural flow ventilator. Figure 1.8 shows the position of the fan (blower) within the system.

Fan design

All fans are driven by an electric motor. The motor can rotate at varying speeds depending on the current supplied to it (see also section 3.3). The fan allows the air flow to be adjusted according to the requirements of the occupants. Fans are generally divided into axial and centrifugal flow types. In axial type fans the air is drawn in and forced out parallel to the rotating axis. In the centrifugal type the air is drawn in on the rotating axis and forced out perpendicular to the rotating axis (the direction of centrifugal force) (Fig. 1.17). The shape of the vanes of the fans is often profiled to maximise flow and volume and minimise size and noise. The blower is a dominant low-frequency source of noise, while at higher frequencies, additional air flow noise sources exist. These include high shear regions within the ducting, separation of flow due to flow obstructions, and the exit flow from air vents. Flow optimisation can be achieved using CFD (Computational Fluid Dynamics), which allows an engineer to analyse flow patterns and pressure regions within the system and make adjustments on the size, shape and position of components in efforts to make the system as aerodynamically efficient and quiet as possible. Other efforts included noise isolation through the use of padding or positioning sources outside of the interior.

Air filtration

Pollen filter

The filter is located in the heater assembly housing before the heat exchanger (Figure 1.14a and b). Fibres in the filter prevent large particulates from entering the system and trap the really small ones, the filter is electrostatically charged. Due to this electrostatic charge, the filter attracts particles like a magnet attracts iron filings. Besides removing visible particles, the filter also removes pollen, spores and different types of dust etc. from the cabin air.

Carbon filter and germicidal lamp

An active carbon combination filter and/or a germicidal lamp are generally fitted as an option in place of the pollen filter. The active carbon combination filter has the same advantages as the

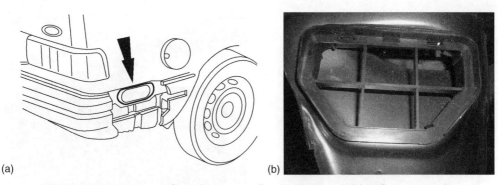

(a)

(b)

Figure 1.15 (a) Rear air flap. (b) Rear air flap Ford Fiesta (rubber flap removed) (reproduced with the kind permission of Ford Motor Company Limited

Figure 1.16 Centrifugal type fan assembly

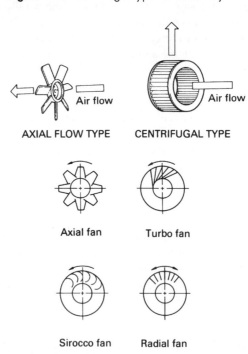

AXIAL FLOW TYPE   CENTRIFUGAL TYPE

Axial fan   Turbo fan

Sirocco fan   Radial fan

Figure 1.17 Axial and centrifugal type fan assembly

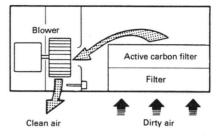

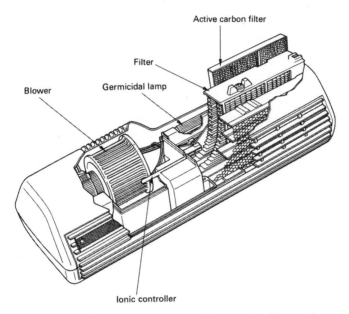

Figure 1.18 Air filtration system with pollen, carbon filter and germicidal lamp
(with the agreement of Toyota (GB) PLC)

pollen filter plus an effective active carbon layer. The active carbon layer neutralises unpleasant odours and keeps the air free of ozone. It also reduces diesel exhaust fumes from entering the interior of the vehicle. A germicidal lamp is used in air-conditioning systems to kill any bacteria which enters or forms within the air filtration system. This also stops odours in the system through the build-up of bacteria (see also section 5.5).

Photo catalytic filter
Photo catalytic filters destroy pollutant gases and micro-organisms entering the vehicle. In less than five minutes the entire cabin air can be purified.
   Benefits:

1. Continuous protection against potentially harmful external/internal pollutants and from discomforting odours.
2. Alleviation for allergy sufferers – micro-organisms causing allergies are destroyed.
3. Extended service life of filters – 2000 hours, equivalent to approximately five years' average vehicle use.
4. Complete destruction of pollutants compared to today's carbon filters.

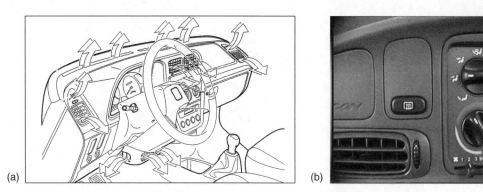

(a)   (b)

Figure 1.19 (a) Air distribution showing panel, face and floor vents (b) Manual control panel (reproduced with permission of Peugeot)

Photo catalytic oxidation converts toxic compounds like carbon monoxide and nitrous oxide into benign constituents such as carbon dioxide and water without wearing out or losing its effectiveness. When light strikes titanium oxide, hydrogen peroxide (H_2O_2) and hydroxyl radicals (OH) are formed. These two substances possess powerful oxidising properties and through mutual interaction are able to decompose odorous substances into odourless carbon dioxide and water. A powerful oxidative also removes bacteria and deactivates viruses.

Air quality sensing

An Air Quality Sensor (AQS) can be located in the main air inlet duct of the HVAC system. When a threshold for carbon monoxide or nitrogen dioxide is reached, the AQS communicates to the HVAC system to initiate the air recirculation mode. For more information see section 3.2.

Directing air flow and controlling temperature range can be manually selected by the occupants or electronically controlled via a control module. The heating system is designed to offer a temperature range. Research into a comfort zone for passengers exists but is subjective due to different nationalities that are acclimatised by the weather on their continents. The basic heating and ventilation system control panel contains a temperature control knob and a number of air distribution options.

Air distribution unit

The air distribution unit is generally located under the instrument panel of the vehicle. Inside the air distribution unit there is a system of ducts and mixing/directing doors. In addition the unit houses the blower motor, the heater core and for vehicles with an air-conditioning system, the evaporator. The filtered incoming air from the intake panel grille is induced by the blower motor and is forced through the air distribution unit. The air coming from the blower is directed to the different air ducts via the moving doors in the air distribution unit. The temperature is regulated by mixing warm and cold air. The air is then directed to different air outlets/air nozzles and panel vents. There are basically two ways for the ventilation system to take in air: fresh air from the outside and recirculated air from the interior. Therefore the air distribution unit has two air inlets which are alternately closed by a door. Operating in recirculation mode allows it to keep away unpleasant outside smells from the inside and it also improves the cooling output of the air-conditioning system. When recirculation mode is switched on for a longer period of time, the humidity level inside the vehicle will increase because of the moisture content of the breath of the passengers. This can lead to fogging on the windows. Switching to fresh air mode with the air-conditioning system reduces the humidity of the inside of the vehicle.

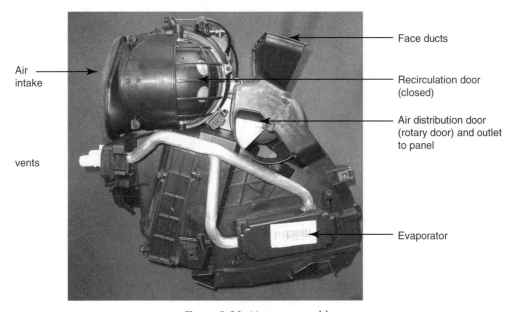

Face ducts

Air intake

Recirculation door (closed)

Air distribution door (rotary door) and outlet to panel

vents

Evaporator

Figure 1.20 Heater assembly

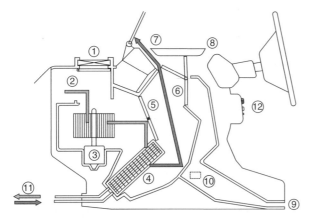

1. Air filtration
2. Air recirculation door
3. Blower motor and centrifugal fan
4. Heat exchanger
5. Temperature blend door
6. Air distribution door
7. Panel vent (not adjustable)
8. Face/head vent (adjustable)
9. Panel vent – rear passengers' foot well
10. Ducting to passenger foot well
11. Flow of coolant
12. Control panel

Figure 1.21 Air recirculation

Simplified view of system components

Figure 1.21 shows the air intake door (recirculation door) is closed so no external air will enter the vehicle except through a port which feeds the blower motor from the interior. The blower is operating so air inside the vehicle will be recirculating around the vehicle's interior. The heat exchanger will still heat the air as required as long as there is a difference in temperature (between air and heater) and the occupants have selected so via the control panel, which will vary the position of the temperature blend door (2). While the air is recirculating there is a danger that water vapour will condense on the inside of the vehicle's windscreen. This is affected by the following:

- external air temperature;
- interior temperature;
- number of occupants;
- relative humidity of the air inside the vehicle.

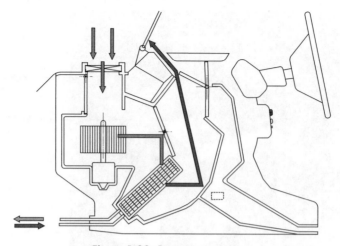

Figure 1.22 Demisting position

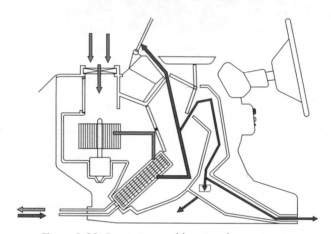

Figure 1.23 Demisting and heating the occupants

If this occurs then air recirculation must stop and external air must enter the vehicle through the air intake door. Recirculation of air is often selected when driving in polluted areas, e.g. heavy traffic.

In the demisting position (Fig. 1.22) the air from outside is moved under force from the blower motor to the temperature blend door which is fully closed. This forces the total volume of air to flow through the heater core where it will be heated and then directed by the top distribution door towards the windscreen and side windows. Note that no air is directed towards the occupants. This allows the maximum volume of air to flow to the windscreen to aid the demisting process. This is done through the evaporation of the condensed water droplets on the screen (see section 1.3 for more information).

In the demisting and heating position (Fig. 1.23) the air intake door is fully open allowing external air to flow through to the blower. The blower forces air towards the temperature blend door which is fully closed forcing all the air to flow through the heater core. All the air flows through the heater core and is then directed to the top distribution door where a portion is directed towards the windscreen and side windows and the rest is directed to the foot vents which includes passengers in the rear of the vehicle.

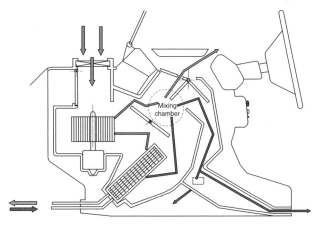

Figure 1.24 Heat directed to occupants' feet and face vents

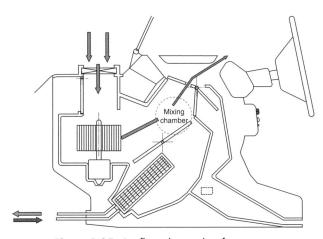

Figure 1.25 Air flow directed to face vents

Figure 1.24 shows the air intake door is fully open allowing air to flow through to the blower. The blower forces air towards the temperature blend door. The blend door directs a volume of air towards the heater core and the rest towards the distribution door allowing air to flow to the face vents. The air going through the heater core is then directed towards the back of the blend door and then the distribution door, where it is distributed by the feet vents.

There will be a temperature difference between the face vent and feet vent of approximately 7°C. This is due to humans feeling comfortable with their feet being warmer than their head in cold conditions.

The air intake is fully open to allow air to flow through the blower. The blower forces air towards the temperature blend door and depending on its position it will direct air straight to the top distribution door and the face vents or it will direct a portion of the air towards the heater core to raise the temperature of the interior and improve the comfort levels of the occupants (Figure 1.25). The interior temperature is generally controlled by the occupants via the control panel. This selection offers the occupants fresh outside air straight to the head which is beneficial in hot weather conditions removing heat from the occupants by convection. This increases the occupants' comfort, especially if perspiring, allowing the latent heat of evaporation to remove sweat producing rapid cooling, relative humidity permitting (see section 1.3 for more information).

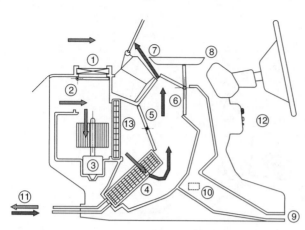

1. Air filtration
2. Air recirculation door
3. Blower motor and centrifugal fan
4. Heat exchanger
5. Temperature blend door
6. Air distribution door
7. Panel vent (not adjustable)
8. Face/head vent (adjustable)
9. Panel vent – rear passengers' foot well
10. Ducting to passenger foot well
11. Flow of coolant
12. Control panel
13. Evaporator (models with A/C)

Figure 1.26 System with evaporator fitted

System components with A/C (including evaporator)

Figure 1.26 illustrates the position of the evaporator in the heating and ventilation system. All air passes through the evaporator irrespective of whether the system is operating. When the A/C system is running the evaporator temperature is approximately 2–6°C (35–42°F). This causes the temperature of the air to reduce and moisture in the air to condense producing water droplets on the evaporator's surface. This reduces the moisture content (dehumidifying) of the air and also helps to remove dirt particles (purifying) suspended in the air stream. The water covers the surface of the evaporator trapping dirt particles and eventually dripping off the surface on to a drain tray which directs the water to the outside of the vehicle.

Note – if the drain pipe becomes blocked then water will enter the inside of the vehicle.

Air vents

Air vents must be ergonomically designed to avoid draughts. Directional air vents generally have three adjustments, up and down, left and right and open and closed. Circular vents are also used which are free to rotate within a given circumference. The vents are generally used for face/head heating. Panel air vents are fixed and cannot be adjusted. These are generally used for windscreen and side screen demisting and floor heating.

Air diffuser system

The soft air diffusion system has been specifically designed to produce an evenly distributed blanket of air that provides all vehicle occupants with the same high level of climatic comfort.

A range of diffuser systems are used within commercial heating and ventilation systems. They are now being implemented within automotive climate controlled systems. There are a range of diffuser types depending on the required air flow characteristics:

● Linear diffusers provide continuous air flow across the length of an outlet. Air flow is quiet and comfort increases while reducing draughts.
● Mini-flow is used when quiet delivery and a low velocity of air are required.
● Gentle-flow air jets with diameters ranging from 1/4″ to 1/2″ diffuse more quickly.
● Super-flow air jets with diameters ranging from 1″ to 6″ can be used for a wide variety of applications. The long throw of the air jets effectively propel air to greater distances.

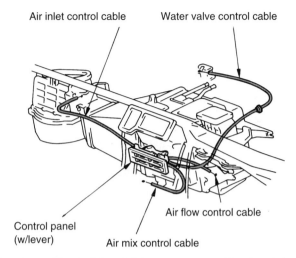

Figure 1.27 Bowden cable used for air inlet, temperature blend and water flow control (with the agreement of Toyota (GB) PLC)

Benefits of using diffusers

- Alleviates discomfort – eliminates draughts from A/C system experienced by occupants of front seats.
- Improves climatic comfort for all occupants – air is distributed evenly throughout the cabin and is particularly beneficial to back seat passengers.

Air door actuators

The internal heating and ventilation doors are opened and closed by Bowden cables (Fig. 1.27), pneumatic control motors or electrical control motors. Manually regulated systems use Bowden cables in most cases. Automatic and semi-automatic systems require control motors. These can be operated electrically or pneumatically. The pneumatic control motors are also controlled electrically by the solenoid valve in the vacuum lines.

Bowden cable

Doors can be operated mechanically by Bowden cables. The rotation or sliding of a control switch provides movement which is transmitted by cable to the doors.

Pneumatic control

Pneumatic control actuators (Fig. 1.28) consist of a diaphragm unit attached to an actuator rod. The diaphragm has a spring acting on its surface holding it in position. To move the diaphragm the spring pressure must be overcome. This is achieved by applying a vacuum via the vacuum connection (1). The vacuum is supplied by the engine inlet manifold often via the brake vacuum servo connection (petrol engine) or brake vacuum pump (diesel engine). The vacuum creates a pressure above the diaphragm which is lower than atmospheric pressure. The diaphragm housing has a hole in its base allowing atmospheric pressure to act on the bottom surface of the diaphragm thus the force of atmospheric pressure is used to overcome the spring tension. The rate of movement is dependent on the spring tension and the difference in pressure between the upper (vacuum) and lower (atmospheric) sections of the diaphragm. As the diaphragm moves the actuator rod

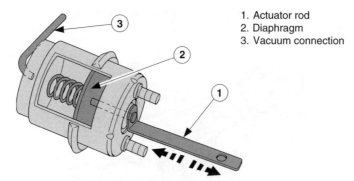

1. Actuator rod
2. Diaphragm
3. Vacuum connection

Figure 1.28 Recirculation air actuator vacuum operated
(reproduced with the kind permission of Ford Motor Company Limited)

moves as well. The actuator rod is attached to a door inside the HVAC unit. The door opens or closes airways for recirculation (Fig. 1.26). Pneumatic control actuators generally have two positions – open and closed. They can be controlled by varying the vacuum applied using a variable orifice or by applying two connections of different diameter to apply different pressures. If variable control is required it is easier to employ an electric motor.

A vacuum accumulator is generally employed to control the fluctuating pressure applied to the diaphragm unit through the use of a non-return valve. A solenoid is also fitted to the system to control fluctuating pressure when vacuum is applied. The valve can be operated manually by means of switches or automatically by a control module.

This system has a number of drawbacks:

1. The vacuum is taken from the inlet manifold and uses critical energy otherwise used to aid combustion.
2. If a leak is present in the system then combustion efficiency will be greatly reduced and pollutants from exhaust emissions will increase.
3. The unit can only fully open a flap or fully close a flap if there is no position between these points.

These units have generally been replaced with electric motors giving greater control.

Electronic control

Electrical control motors are used for fine adjustment of blend/distribution doors (Fig. 1.29). These motors are usually used for operating the temperature blend door and air distribution door, as this door has to be moved proportionally. Systems with electronic temperature control often have an integrated potentiometer in the control motor, which gives feedback to the control module about the position of the door (for a full explanation see Chapter 3).

Classification of heating and ventilation systems by zone

A zone is an area of the internal space of the vehicle that can be cooled or heated to a specific temperature. For example, a driver may feel hot due to their clothing and require cooling and a passenger may feel cold and require heating. Both occupants have a set of controls to adjust the temperature and ventilation rate of their personal space. Systems that have more than one zone are generally electronically controlled. Heating and ventilation alone can be split for

Figure 1.29 Electrical control motor and reduction gear

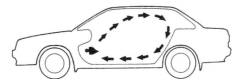

Figure 1.30 Dashboard installed HVAC system
(with the agreement of Toyota (GB) PLC)

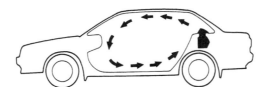

Figure 1.31 Boot installed single zone HVAC system
(with the agreement of Toyota (GB) PLC)

zone control but generally if systems have this facility they include heating, ventilation and air-conditioning (HVAC).

Dash HVAC

Installed under the dashboard with one single zone which is the interior space. The dashboard type has the benefit of forcing cold air directly to the occupants enabling the cooling and heating effect to be felt to a much greater degree than the system's capacity to cool or heat the entire space. Example – the output at the air vent on an HVAC system might be 2°C which can be blown directly on to the occupant's face for immediate cooling. The interior space will generally cool to approximately 22°C (depending on load).

Boot HVAC

Installed in the boot which has a large space available for the heating and cooling units. The outlets are positioned at the back of the rear seat. Negative aspects of this design include loss of boot space and cool air streams flowing from the rear of the vehicle.

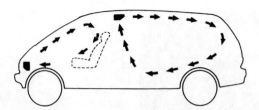

Figure 1.32 Dual zone dashboard installed HVAC system
(with the agreement of Toyota (GB) PLC)

Dual HVAC

Generally installed at the front of the vehicle under the dashboard and extended to the rear. Dual systems can include up to three zones, driver, front passenger and the rear passengers. All zones have a set of HVAC controls to select the desired level of comfort. This system is common on high specification vehicles and MPVs (Multi Passenger Vehicles) – vehicles with a high capacity.

Booster heater systems

Booster heating systems are generally used for the following reasons:

1. Large interior cabin space.
2. Efficient engine combustion with low heat output, additional heat input required.
3. Large interior space (MPV – Multiple Passenger Vehicle).
4. Vehicles operating in extreme weather conditions.

The benefits of such a system are as follows:

1. Improved visibility due to rapid demist.
2. Shorter cold start period improving catalytic converter efficiency and less engine wear.
3. Improved passenger comfort.

PTC heaters

Booster heaters can be as simple as an additional water pump fitted to accurately control/boost coolant through the heat exchanger fitted inside the vehicle to improve heating capacity, or a separate unit which can provide additional heat input to the coolant or air distribution by burning fuel (fuel heaters) or electricity (PTC – Positive Temperature Coefficient – heaters). Booster heaters should not be confused with dual heating and air-conditioning systems. Dual systems are extensions of the same system and provide heating and cooling control within designated zones inside a vehicle.

The PTC heater unit has an element mounted on a ceramic base and installed in the heater casing. It directly heats up the air flow entering the passenger compartment.

The key characteristics of a PTC additional heater are:

- rapid heating after starting the vehicle;
- light and compact design;
- the unit cannot overheat;
- maintenance free.

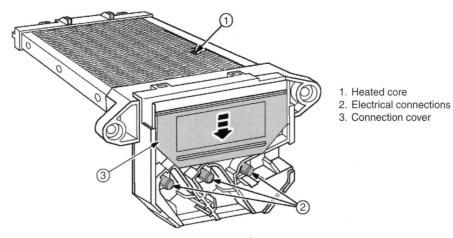

1. Heated core
2. Electrical connections
3. Connection cover

Figure 1.33 PTC heater unit
(reproduced with the kind permission of Ford Motor Company Limited)

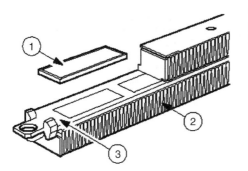

1. Metallised ceramic plate
2. Aluminium elements
3. Frame

Figure 1.34 Heating element
(reproduced with the kind permission of Ford Motor Company Limited)

The PTC heater element consists of small metallised ceramic plates, which are layered alternately along the unit core (1), see Figures 1.33 and 1.34 with aluminium radiator elements. These layers are held together by spring elements in a frame. The aluminium elements provide the electrical contacts. They also transfer heat to the passing heater air flow. In order to prevent electrical short circuits due to metal foreign bodies, a heat-resistant plastic mesh with an aperture of 0.8 mm is located on the heater element. The heater element is divided into separate heating circuits with a ratio of one third to two thirds so that the heating power can be adapted to suit different requirements.

PTC heater elements act as a positive temperature coefficient resistor. This means that its resistance value is relatively small at low temperatures and increases with higher temperatures. A high current flows initially when a voltage is applied to the cold PTC element, as a result of which it heats up. As the temperature rises, so does the resistance. This results in a reduced current draw. The time taken to stabilise the current is approximately 20 seconds. The temperature of the additional heater depends on the rate of heat transfer to the surrounding area. If the rate of heat transfer is good, for example, a cold low humidity condition will have a greater temperature difference and allow for a greater rate of heat transfer, therefore the resistance will remain low. Once the air has warmed and the heat transfer rate reduces the PTC unit will start to increase in temperature due to the inability to give up heat. This will cause the

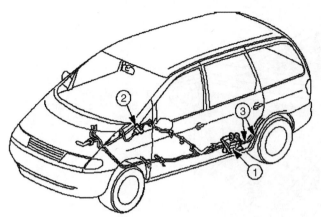

1. Heater unit
2. Fuel metering unit
3. Coolant hoses to rear heat exchanger

Figure 1.35 Diesel heating booster system location
(reproduced with the kind permission of Ford Motor Company Limited)

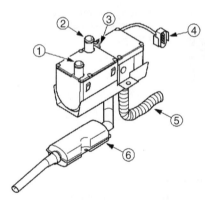

1. Coolant inlet from the engine cooling system
2. Coolant outlet to rear heat exchanger
3. Fuel feed from metering unit
4. Electrical connection
5. Air intake hose
6. Exhaust with built-in damper

Figure 1.36 Booster unit
(reproduced with the kind permission of Ford Motor Company Limited)

resistance on the unit to increase and thus reduce the current flow. The reduced current flow maintains or reduces the temperature. If the opposite occurs and the heater manages to give up heat to the surrounding air then the unit will cool and the resistance will reduce. This will increase current flow through the unit and increase the unit's temperature. As a result of these specific resistance characteristics, it is not possible for the PTC element to overheat. The maximum surface temperature is around 165°C.

The unit is only operated at low ambient temperatures (<15°C, supplied by air temperature sensor) and when insufficient heat can be supplied via the coolant-based heating system (<73°C, information supplied by coolant temperature sensor) and low alternator (generator) loadings. The PTC heater consumes a great deal of current so it can only be operated when the engine is running and its operation is phased. Phased operation means that the unit is treated like a three bar electric heater. One bar at a time is switched on so the load on the alternator (generator) is progressive and not sudden. The load will also affect the idle speed and emissions.

The heating power is around 1 kW when fully operating. This places additional loading on the vehicle electrical system. Vehicles with PTC additional heaters are fitted with a more powerful generator.

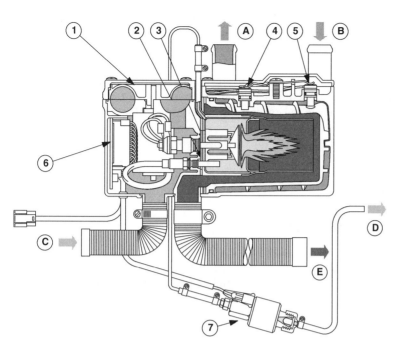

1. Combustion blower
2. Flame sensor
3. Glow plug
4. Temperature sensor
5. Overheating sensor
6. Control module
7. Metering pump
A. Coolant outlet
B. Coolant inlet
C. Fresh air supply
D. Fuel return line
E. Exhaust

Figure 1.37 Booster heater operation
(reproduced with the kind permission of Ford Motor Company Limited)

Diesel fuel booster system

If a vehicle cannot provide adequate heated coolant to a heat exchanger then it may use a fuel booster to provide the additional heat input. There is a number of manufacturers using such a system due to ultra efficient diesel combustion systems and the large interior space that requires heating (e.g. Multi Passenger Vehicles (MPVs); Fig. 1.35 and 1.36). This is often between 1 and 3 kW. At maximum output the fuel penalty is 0.38 litres/hour and the operating temperature is between −40 and +80°C. Once above 80°C the unit shuts down. Such units can provide a variable output often dependent upon the measured temperature of the coolant flowing to the heat exchanger and the exterior air temperature. Such a unit may use air or coolant to transfer heat from the unit to the interior of the vehicle.

All the starting, regulating and run-on functions are controlled fully automatically. The main unit is divided into four main sections – heat exchanger, combustion chamber, fan assembly and control module. Figure 1.37 shows the unit with the coolant inlet (B) and outlet (A) on top of the unit. This feeds the heat exchanger section which allows coolant to flow around the outside of the core. The inside of the core is the combustion chamber which contains a glow plug (3), flame sensor (2) and fuel and air inlet. The combustion is initiated by a glow plug fed with fuel which is quantified by the metering unit (7). The glow plug is switched off after a certain time when the flame has stabilised. The fan assembly/combustion blower (1) provides a fresh supply of air to the combustion chamber. After combustion the heat flows from the inside of the heat exchanger which is then conducted through the exchanger's walls by the coolant. Because the heating power output depends on coolant temperature, sensors are fitted to the heat exchanger housing monitoring coolant temperature to and from the unit. This information is used to govern the unit's heat output to the vehicle's heat exchanger. The output temperature sensor and combustion sensor are used for failsafe purposes in case the unit fails, e.g. overheats. All information is monitored by the ECU (6) which has the ability to communicate with diagnostic equipment. Because the system is electronically controlled, integration into an existing HVAC (Heating, Ventilation and Air-Conditioning) system can be easily achieved.

The booster heater is switched off due to a malfunction if:

1. no ignition takes place after a second attempt at starting (after 90 seconds);
2. the flame is extinguished during operation and a fresh attempt at starting fails;
3. the overheating sensor responds in the event of overheating (125°C);
4. the voltage is over- or undershot;
5. the glow plug, metering pump or temperature sensors or fresh air blower are faulty.

The booster heater can be tested as soon as the ignition is switched on. A fault code can be read out during a diagnostic check.

Use of an A/C system as a heat pump

With the use of highly efficient diesel engines, hybrid vehicles, and electric fuel cell technology, water-based cooling systems often do not offer the heating capacity for passenger compartments. The vapour compression cycle A/C system can be used as a heat pump. This process is explained in section 1.5.

1.3 The basic theory of cooling

An air-conditioner is a generic term for a unit which maintains air within a given space at a comfortable temperature and humidity. To achieve this, an air-conditioning unit must have a heater, cooler, moisture controller and a ventilator.

The principle of an HVAC system:

● heater – adding heat by transferring it;
● cooler – removing heat by transferring it;
● humidity – removing or adding moisture;
● purification – by filtration;
● ventilation – air movement through the vehicle.

The HVAC system creates a comfort zone for the occupants which can be adjusted within a range. Not all humans desire exactly the same environment, variations occur due to different continents, countries, cultures, gender, age or simply due to the type/amount of clothing being worn when inside the vehicle. The system must provide a way of controlling the climate inside the vehicle which is generally referred to as 'climate control'.

Heat

Heat is a basic form of kinetic energy which cannot be destroyed; it can only be converted to or from other forms of energy. In accordance with scientific laws all heat will travel from a hot to a cold surface until the temperatures have equalised. The rate at which heat will travel is dependent upon the difference in temperature between a hot (more energetic molecular movement) and cold (less energetic molecular movement) area. The SI unit for heat energy is the joule. Other units include calories and BTU (British Thermal Units). The effects of heat energy are measured using temperature.

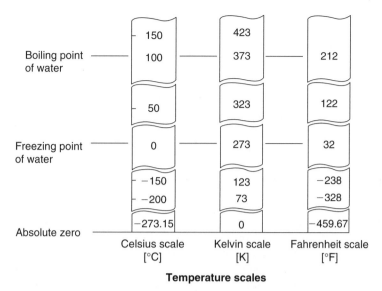

Figure 1.38 Temperature comparison
(with the agreement of Toyota (GB) PLC)

Heat intensity

The SI unit for heat intensity is the kelvin, the derived unit is Celsius or Fahrenheit. The kelvin scale is a theoretical scale based on the laws of thermodynamics using absolute zero as the start of its scale instead of zero as used with °C. Scientists state that a temperature called 'absolute zero' is the point at which all heat is removed from an object or substance (a complete absence of molecular movement). The intensity of heat can be measured using a thermometer. This only gives the heat intensity of a substance and not the heat quantity.

The scales on the chart in Figure 1.38 show the boiling point and freezing point of water. It must be noted that this is only true when it occurs at sea level – 101.3 kPa (1.013 bar). Heat intensity within a vehicle for comfort should be between 21 and 27°C (65 and 80°F).

Sensible heat

Heat which causes a change in temperature is called *sensible heat*. As previously stated it can be 'sensed' using a thermometer or pyrometer. Theory tells us that by adding heat to a liquid such as water there will be a proportional increase in its temperature which can be measured on a thermometer scale, e.g.:

The amount of heat required to raise the temperature of 1 kg of water by 1°C is 4.2 kJ.

For example:

420 kJ of heat must be applied to 1 kg of water at 0°C to bring it up to the boiling point of 100°C.

Conversely, the same amount of heat must be removed from the boiling water to cool it down to freezing point.

Alternatively:

1 BTU heat quantity changes the temperature of 1 lb water by 1°F.

Specific heat capacity

Different substances absorb different amounts of heat to cause the same increase in temperature. Specific heat capacity is used to measure the amount of heat required to cause a change in the temperature. The basic unit for specific heat capacity is the *joule per kilogram kelvin (J/kg K)*. Different materials have different specific heat capacity values.

Latent heat and a change of state

Latent heat (hidden heat) is the heat energy required to change the state of a substance without changing its temperature. Heat can have a direct effect on substances when they change state. Evaporation is the term used when enough heat is absorbed by a substance to cause it to change into a vapour. Condensation is when enough heat is removed from a vapour causing it to change to a liquid. When a change of state occurs a great deal of energy is either absorbed or released. This is called the latent heat of vaporisation and the latent heat of condensation.

Example:

2260 kg of latent heat is absorbed when 2 kg of water at 100°C changes state from a liquid to a vapour.

970 BTU of latent heat is absorbed when 1 lb of water at 212°F changes state from a liquid to a vapour.

This will occur without any change in the thermometer reading.

If ice is heated it will reach 0°C (32°F) then start to melt. This means you will have liquid and a solid existing together. The more heat you add the more the ice will melt with no increase in temperature until no ice exists, then the water will start to increase in temperature (sensible heat). When water reaches 100°C (212°F), any more heat energy added will result in some or all the liquid changing into a vapour. The amount of vapour produced depends on the heat energy available and the pressure above it.

When a substance changes state it can absorb hundreds of times more energy than when it is just increasing in heat intensity. This latent heat is the common process used to transfer heat from the interior of the vehicle to the exterior. The key to successful cooling is to get the liquid to the point where it wants to change state (evaporate) at the right point within the A/C system. To do this we manipulate the system pressure.

The function of the heating system is to increase or reduce heat input to the inside of the vehicle. A typical automotive combustion engine is only about 30% efficient, which means only 30% of the fuel delivered to the combustion engine is converted into usable energy. Some of this energy is transferred to the cooling system which is where the heating system will obtain its heat source for the inside of the vehicle. The rest is lost through exhaust gases and radiation.

Heat can be transferred using one or a combination of three processes – conduction, convection or radiation.

Conduction, convection and radiation

Conduction

Figure 1.40 illustrates the direct transmission of heat by conduction within a substance, for example, if heat is applied to one end of a steel bar, the other end will eventually increase in

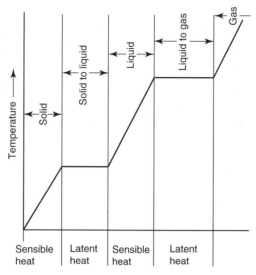

Figure 1.39 Sensible heat and latent heat
(with the agreement of Toyota (GB) PLC)

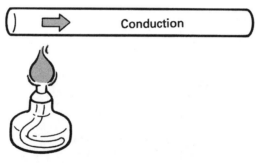

Figure 1.40 Conduction
(with the agreement of Toyota (GB) PLC)

temprature. Some materials are excellent conductors of heat – aluminium, copper – and other materials act more like insulators – polymers (plastics).

Convection

Convection is heat movement *through a medium* like a liquid in a saucepan (Fig. 1.41). Convection is a continuous movement of the medium and heat. The medium, liquid or gas, moves and releases heat to the surrounding areas. When heating occurs in a liquid or gas, expansion occurs and parts of the substance become lighter than other parts which contain less heat. Natural convection currents occur in any substance which is not heated evenly.

Radiation

Heat can *travel through heat rays* and pass from one location to another without warming the air they travel through (Fig. 1.42). An example is ultraviolet radiation which travels from the sun. Radiant heat can travel from a warmer object like the sun to a cooler object like the earth's surface. The surface colour and texture affect the heat emitted and absorbed. Colour is not as important as texture; dark *rough* surfaces make better heat collectors than light *smooth* surfaces.

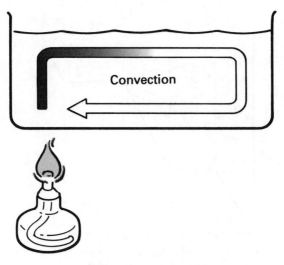

Figure 1.41 Convection
(with the agreement of Toyota (GB) PLC)

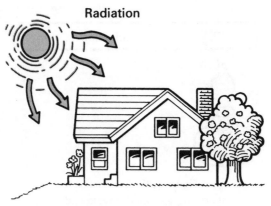

Figure 1.42 Radiation
(with the agreement of Toyota (GB) PLC)

From the engine to the interior

Convection occurs when material, such as an engine, passes heat to the cooling system of the vehicle. As the potential energy of the fuel is converted to mechanical and heat energy by the engine combustion process, the heat of the engine must be removed. The liquid in the cooling system is pumped through the engine, and the convection process transfers engine heat to the liquid. The cooling system liquid then takes this heated coolant to the radiator. The metal heat exchanger uses the conduction process to remove the heat from the liquid coolant and to the exchanger fins. The radiator fins then pass the heat of the radiator to the passing air flow through the heat exchanger.

Enthalpy

Enthalpy is the measure of the usable energy content of a substance. When a liquid increases in temperature it also increases in enthalpy. When the liquid changes into a vapour through

Table 1.1 Relationship between vacuum and temperature at which water boils

Boiling point of water °C	Boiling point of water °F	kPa	bar
48.9	120.02	−89.60	−0.896
43.3	109.94	−92.50	−0.925
37.8	100.04	−94.80	−0.948
32.2	89.96	−96.50	−0.965
26.7	80.06	−97.80	−0.978
21.1	69.98	−98.80	−0.988
15.6	60.08	−99.60	−0.996
10	50	−100.40	−1.004
4.4	39.92	−100.60	−1.006
−1.1	30.02	−100.80	−1.008
−6.7	19.94	−101.00	−1.01
−12.2	10.04	−101.10	−1.011
−15	5	−101.20	−1.012

latent heat of evaporation it does not increase in temperature but does increase in enthalpy because the energy within the substance increases.

Pressure

Pressure is defined as the force exerted on a unit area by a solid, liquid or gas. The SI unit used to indicate pressure is the pascal (Pa). Other units of pressure are pounds per square inch (psi), kilograms of force per centimetre squared (kgf/cm^2), atmospheres (atm) and millimetres of mercury (mmHg). Atmospheric pressure at sea level is 101.325 kPa (14.6 psi). This is generally shown on a gauge as zero (gauge pressure) unless it is a gauge which measures atmospheric pressure. This means that absolute pressure is atmospheric pressure plus gauge pressure.

Previously it was discussed that water changes state at 100°C. This is if the liquid is at sea level under atmospheric pressure. If you reduce the pressure on the liquid by moving it above sea level or apply a vacuum to it then the boiling point is lowered. If you increase the pressure applied to the liquid by moving it below sea level or specifically applying a pressure on it, then the boiling point is raised.

Table 1.1 shows that if a deep vacuum (1 bar vacuum) is created in a closed system like an A/C system water will boil at 10°C (50°F). This enables technicians to remove moisture from the A/C system using a vacuum pump (refer to Chapter 5). The chart also presents the importance of creating a deep enough vacuum under cold ambient conditions.

A pressurised liquid

Consider Figure 1.43. If we place a liquid that will readily evaporate at atmospheric pressure inside the box but close the tap, the pressure will build up and increase the boiling point of the liquid (e.g. at a pressure of 5 bar water only boils at 152°C). When the tap is open the pressure will suddenly decrease and the liquid will readily evaporate lowering the temperature (by absorbing the heat inside the box).

Critical temperature and pressure

There is a critical temperature which is the maximum point at which a gas can be condensed and a liquid can be vaporised by raising the pressure. Refrigerants R134a and R12 have critical

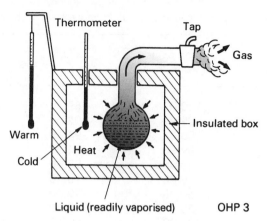

Figure 1.43 Liquid removing heat by changing state
(with the agreement of Toyota (GB) PLC)

pressures and temperatures that dictate the maximum pressure/temperature they can be subjected to. If an air-conditioning system operates below the critical points of its refrigerant then they are termed subcritical systems. If they exceed their critical pressure/temperatures then they are termed transcritical systems.

Refrigerants

Refrigerants are the working fluids of the A/C system. An ideal refrigerant would have the following properties:

1. Zero ozone depleting potential and zero global warming potential.
2. Low boiling point.
3. High critical pressure and temperature point.
4. Miscible with oil and remain chemically stable.
5. Non-toxic, non-flammable.
6. Non-corrosive to metal, rubber, plastics.
7. Cheap to produce, use and dispose.

Refrigerant CFC12 – dichlorodifluoromethane

R12 is a CFC (Chloro Fluoro Carbon). The refrigerant consists of chlorine, fluorine and carbon, and has the chemical symbol CCL_2F_2. It was used for many years from the early development of A/C systems up to the mid-1990s when it was progressively phased out leading to a total ban on 1 January 2001 due to its properties which deplete the ozone and contribute to global warming (see Chapter 6, section 6.1). A benefit of R12, when it was originally designed, was its ability to withstand high pressures and temperatures (critical temperature and pressure point) without deteriorating compared to other refrigerants that were around at that time. R12 mixes well with mineral oil which circulates around an A/C system. It is non-toxic in small quantities although it does displace oxygen and is odourless in concentrations of less than 20%. R12 can also be clean/recycled. You must not burn/heat R12 to a high temperature ($>300°C$) with a naked flame because a chemical reaction takes place and phosgene gas is produced. A lethal concentration of phosgene is 0.004% per volume.

R12 properties:

1. It is miscible with mineral oils.
2. It does not attack metals or rubber.
3. It is not explosive.
4. It is odourless (in concentrations of less than 20%).
5. It is not toxic (except in contact with naked flames or hot surfaces).
6. It readily absorbs moisture.
7. It is an environmentally harmful CFC gas (containing chlorine which destroys the atmospheric ozone layer).
8. It is heavier than air when gaseous, hence the danger of suffocation.

Refrigerant HFC134a – tetrafluoroethane

R134a is a known substitute for R12. R134a is an HFC (Hydro Fluoro Carbon). The refrigerant consists of hydrogen, fluorine and carbon, and its chemical symbol is CH_2FCF_3. Because the refrigerant has no chlorine it does not deplete the ozone. R134a is non-toxic, non-corrosive and does contribute to global warming; it is not miscible with mineral oil so synthetic oil, called PAG (Poly Alkaline Glycol), was developed. PAG oil is hygroscopic and absorbs moisture rapidly which means when in use you must ensure that the container is resealed as quickly as possible. R134a cannot be mixed with R12 and is not quite as efficient at high pressures and temperatures. R134a can be cleaned and recycled.

R12 and R134a have different size molecules, the R12 molecule is larger. This means the quantity of refrigerant required for an R134a system is higher than an R12. This also requires the flexible hoses and seals including oil to be replaced with compatible R134a components if a conversion from R12 to R134a is required (see Chapter 5). R134a contributes to global warming and will eventually be replaced, certainly within Europe, with another cooling medium which is reported to be less harmful to the environment. R134a is not a drop-in replacement for R12 A/C systems. A number of modifications are required to the system components to allow an R12 system to use R134a as a cooling medium (see Chapter 5).

Properties of R134a:

1. It is only miscible with synthetic polyalkylglycol (PAG) lubricants, not with
2. mineral oils.
3. It does not attack metals.
4. It attacks certain plastics, so only use special seals suitable for R134a.
5. It is explosive.
6. It is odourless.
7. It is not toxic in low concentrations.
8. It readily absorbs moisture.
9. It is inflammable.
10. It is heavier than air when gaseous, hence the danger of suffocation near the ground.

Refrigerant blends

The use of alternative refrigerants (Table 1.2) to R134a and R12 are not accepted by Original Equipment Manufacturer's (OEM's) standards. Manufacturers only use the approved R134a refrigerant in vehicle A/C systems. The US has a range of Snap Approved Refrigerants which can be used as 'alternatives' to R12. If blends are used then separate approved service units and accessories must be used to avoid contamination. Strict procedures must be adhered to with regard to record keeping, barrier hose fitment, high pressure release device replacement,

Table 1.2 Substitutes acceptable subject to use conditions for CFC-12 in MVACs

Substitute	Trade name	Retrofit/new
HCFC22		R, N (buses only)
HFC134a		R, N
R406A	GHG, GHG-X3, GHG-12, McCool, Autofrost X3	R, N
GHG-X4, R414A (HCFC Blend Xi)	GHG-X4, Autofrost, Chill-it, Autofrost X4	R, N
Hot Shot, R414B (HCFC Blend Omicron)	Hot Shot, Kar Kool	R, N
FRIGC FR-12, (HCFC Blend Beta), R416A	FRIGC FR-12	R, N
Free Zone (HCFC Blend Delta)	Free Zone/RB-276	R, N
Freeze 12	Freeze 12	R, N
GHG-X5	GHG-X5	R, N
GHG-HP (HCFC Blend Lambda)	GHG-HP	R, N
Ikon 12, Ikon A (Blend Zeta)	Ikon 12	R, N
SP34E	SP34E	R, N
Stirling Cycle		N
CO_2		N
RS-24	RS-24	R, N
Evaporative Cooling		N

Key: R = Retrofit uses, N = New uses

service connectors fitment, strict labelling requirements, oil replacement and possible seal replacement. Blends currently cannot be recycled using an approved service machine. This means refrigerant has to be sent back to the supplier to be recycled. Blends are compounds made of other refrigerants, R22, R134a, etc. They are either *azeotrope* or *zeotrope*. Azeotrope has a single boiling point while zeotrope are a boiling point range. Zeotrope blends' boiling point range starts when the lighter elements start to boil and ends when the heavier elements boil off. If a leak occurs the blends' lighter elements will vaporise and escape leaving the heavier elements. This is called *fractionising*, which changes the characteristics of the blend.

Note – R134a is the only refrigerant that should be used as a replacement for R12. Retrofitting must be carried out in accordance with SAE requirements (see legislation).

Substitutes
Substitutes are reviewed on the basis of ozone depletion potential, global warming potential, toxicity, flammability, and exposure potential. Lists of acceptable and unacceptable substitutes are updated by the EPA (Environmental Protection Agency) several times each year.

Refrigerant service connectors
To prevent accidental mixing of the refrigerants the SAE (Society of Automotive Engineers) developed guidelines for different service valve connectors for R12 and R134a (Figs 1.44 and 1.45). If refrigerants are mixed severe damage will occur to the system. R12 uses threaded connections and R134a quick release couplings.

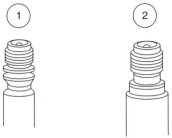

1. R12 high side connector (3/16" flare 3/8-24 threads)
2. R12 low side connector (1/4" flare 7/16-20 threads)

Figure 1.44 R12 threaded service connectors
(reproduced with the kind permission of Ford Motor Company Limited)

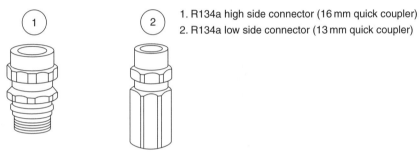

1. R134a high side connector (16 mm quick coupler)
2. R134a low side connector (13 mm quick coupler)

Figure 1.45 R134a quick release type service connectors
(reproduced with the kind permission of Ford Motor Company Limited)

Figure 1.46 R134a blue low pressure connector quick release coupling
(courtesy of Autoclimate)

The service connector forms a valve with the valve core screwed inside it. This valve allows the disconnection of a pressure gauge, A/C machine or control switches/sensors to be removed without draining the system. These are called Schrader type spindle valves (Fig. 1.48). They are similar in design to tyre valves. The needle is held in the closed position by spring force.

Before the coupling is fitted the shut-off valve (blue) must be closed ensuring the valve is not open. Once connected the valve is opened to obtain a reading on the service unit. Blue is allocated to the low pressure side or suction side and red is allocated to the high pressure side or discharge side.

Figure 1.47 shows a threaded low pressure R12 connector with ball valve preventing refrigerant loss.

Figure 1.47 R12 low pressure connector 1/4" SAE Schrader valve 45°
end fitting with Goodyear GY5 barrier hose
(courtesy of Autoclimate)

Figure 1.48 Schrader valve
(courtesy of Autoclimate)

Figure 1.49 Valve core remover and installer with shut-off valves
(courtesy of Autoclimate)

The protective cap prevents the valve getting dirty and also provides an additional seal when the system is working. The protective caps must be screwed on again after the system is filled. The service connector valves must seal completely. To check, apply a few drops of compressor oil to the needle. If bubbles are formed, the valve is leaking and the valve core must be renewed.

Tools are available to remove Schrader valves without draining the system, providing there is sufficient space.

Hose material

A/C systems are designed to use as little flexible hoses as possible due to leakage. Aluminium extruded tubing is generally used with the exception of the compressor which uses flexible hoses because it is connected to the engine. Modern hoses for R134a use an inner lining of nylon due to the size of the R134a molecule and to reduce moisture ingression. Covering the nylon is an external tubing of Neoprene. Polyester braid is used as reinforcement with a final covering of PVC (Polyvinyl Chloride). R12 hose is constructed in the same way but without the nylon lining due to the larger size molecule. Electric compressors may remove the need for

Table 1.3 Comparative data

	Unit	R12	R134a
Chemical name		Dichlorodifluoromethane	Tetrafluoroethane
Colour of storage bottle		White	Light blue
Thread on container		7/16″	8/16″
Connector size/type		1/4″ flare 7/16-20 threads	Low quick couple 13 mm
		3/16″ flare 3/8-24 threads	High quick couple 16 mm
Lubricant		Mineral oil	PAG
Desiccant		XH-5	XH-7 or XH-9
Boiling point at 1 bar	°C	−29.8	−26.5
Critical temperature	°C	+112	+101.15
Critical pressure	bar	41.48	40.68
Molar mass	g/mol	120.92	102.03
Liquid density +20°C	kg/l	1.33	1.23
Ozone reduction factor		0.96	0.0
Hot house factor		3.1	0.3
Coldness gain	kj/m³	1495	1455
Cooling efficiency		4.3	4.2

any flexible hoses. The government and environmental groups are placing increased pressure on zero leak rates.

OEM will always advise you to use the correct refrigerant for the correct system. Deviation from the original set-up should only occur if you are converting an R12 system to an R134a. If converting from R12 to R134a then hoses must be replaced with the curent type.

The pressure/temperature relationship of R134a

The graph in Figure 1.50 shows the pressure/temperature curve for refrigerant R134a.

The graph shows the refrigerant to be in a gaseous/vapour state above the curve and a liquid below the curve. The curve represents the boiling point of the refrigerant under varying pressure and temperature relationships.

1. The refrigerant is in a gaseous/vapour state and if the temperature is kept constant and the pressure is increased then the refrigerant will condense into a liquid.
2. If the pressure is kept constant and the temperature is reduced then the refrigerant can be condensed into a liquid.
3. If the temperature is kept constant and the pressure is reduced then the refrigerant will evaporate into liquid/vapour.
4. If the pressure is kept constant and the temperature is increased then the refrigerant will evaporate into liquid/vapour.

The A/C system is designed to manipulate these relationships to enable the refrigerant to transfer heat from the cabin space.

Comfort – humidity

Humidity is the term used to describe the wetness or dryness of air. The air around us contains a percentage of water vapour. Humans perspire (sweat) which evaporates and cools the surface of the skin by convection. If humans are in hot humid conditions then this makes us feel sweaty,

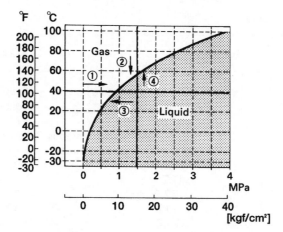

Figure 1.50 Pressure/temperature curve R134a
(with the agreement of Toyota (GB) PLC)

uncomfortable, anxious and can induce stress. A fan to force air over the occupant of a vehicle can improve the evaporation rate and improve comfort.

There are generally two ways to measure humidity, relative humidity and absolute humidity (Fig. 1.51). Relative humidity is the most common measurement and tells us how much water vapour by weight the air actually contains compared to how much it could contain at that given temperature. As an example, if the relative humidity is 50%, the air could hold twice as much vapour as it does at that given temperature. The amount of vapour that the air can hold changes with its temperature. If the air warms up then it can hold more vapour which would reduce the relative humidity because it could hold more vapour than what it was actually carrying. If the air cools then its relative humidity reduces because it can now hold less.

Absolute humidity is the amount (by weight) of vapour that the air contains compared with the amount of dry air. When air becomes saturated with water and then cools the relative humidity will eventually (depending on the rate of cooling) become 100%. The temperature is called the 'dew point' of air for the absolute humidity. If the air cools any further then the vapour it contains will condense.

The significance of this information is the control of the humidity within the A/C system. Humidity is controlled by the surface area of the evaporator and the volume and flow of air travelling through it. The cold surface of the evaporator causes the moisture in the air to condense and cover the surface in water droplets. This reduces the moisture content thus drying the air and improving comfort. Relative humidity for comfort levels is generally about 60%.

Dry bulb temperature

This is the temperature indicated by an ordinary thermometer used to measure air temperature.

Wet bulb temperature

In a wet bulb thermometer the heat sensitive bulb of a glass tube thermometer is wrapped in a gauze, one end is suspended in a water container to allow the water to be drawn upwards by capillary action and moisten the bulb. The moisture robs a percentage of the heat surrounding the bulb which is dependent upon how easily the water can evaporate. The temperature that is registered is referred to as the 'wet bulb temperature'. Some equipment suppliers sell as an accessory a probe with a wet sock which attaches to a multi-meter allowing the electronic

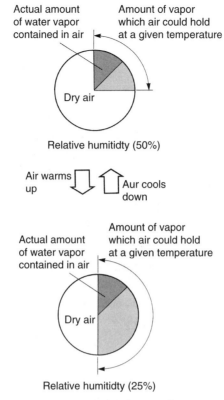

Figure 1.51 Humidity graph
(with the agreement of Toyota (GB) PLC)

measurement of temperature. The sock can be removed to obtain the dry bulb temperature. As discussed, relative humidity is measured by comparing the wet bulb temperature against the dry bulb temperature. Instruments that contain both measuring devices are called psychrometers.

As an example – in Figure 1.52 the graph shows the wet bulb temperature 19.5°C (follow diagonal line), dry bulb temperature 25°C (horizontal line) and relative humidity = 60% (point where both intersect).

After measuring the dry and wet bulb temperatures of the air entering the evaporator and the air exiting the centre vent inside the vehicle, the graph can be used to calculate the relative humidity and compare the two results ensuring good evaporator performance. Example, relative humidity of the ambient air entering the HVAC unit – 70%, relative humidity of the air exiting the centre vents inside the vehicle – 50%.

Humidity sensor

See Chapter 3, section 3.2.

1.4 Vapour compression refrigeration

Currently the most common cycle used in automotive applications is the vapour compression cycle, e.g. R12 and R134a closed systems. To understand the vapour compression cycle and

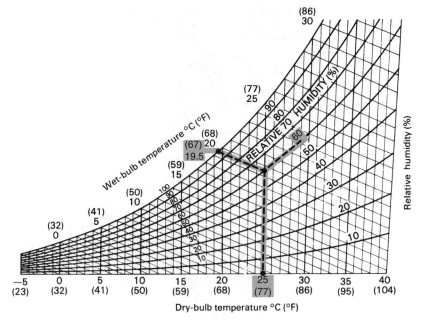

Figure 1.52 Psychometric chart
(with the agreement of Toyota (GB) PLC)

other cycles it is important to understand the types of changes that a refrigerant goes through when used in an A/C system. This is explained using a pressure/enthalpy diagram (Figure 1.53).

A refrigerant in the 'subcooled' liquid region point a is at a temperature which is below its boiling point. If heat is continually added while maintaining constant pressure, the refrigerant's temperature and enthalpy will increase. Its state will eventually approach 'saturated liquid' point b. This is where the liquid will start to vaporise. As the heat is continually added the liquid vaporises and continues to increase in enthalpy but not increase in temperature. The 'saturated liquid' vaporises until it becomes a 'saturated vapour' point b–c. The 'saturated vapour' at point c has no liquid because it has completely vaporised. The heat that is absorbed through the transition from 'saturated liquid' to 'saturated vapour' is called the latent heat of evaporation (heat is absorbed without any increase in temperature). With additional heat still available to the 'saturated vapour' the temperature increases causing the refrigerant to become 'superheated vapour' point d.

At this point the heat addition causes an increase in temperature. The saturated liquid and saturated vapour curves meet at a point called the 'critical point'. This point has a corresponding critical pressure and critical temperature. Above the critical pressure the refrigerant is in a state called the 'supercritical region'. The supercritical region is where heat addition or removal does not cause a distinctive liquid vapour phase transition.

The ideal vapour compression cycle

The following applies the vapour compression cycle as an ideal cycle within an automotive A/C system. The Figure 1.54 pressure/enthalpy diagram shows the beginning of the cycle as the refrigerant enters the compressor as a 'saturated vapour' at point 1. The refrigerant is compressed adiabatically (the compressor is 100% efficient and no heat is removed by the process) and becomes a 'superheated vapour' due to the increase in pressure, temperature and enthalpy

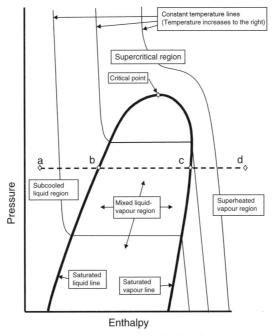

Figure 1.53 Pressure/enthalpy diagram

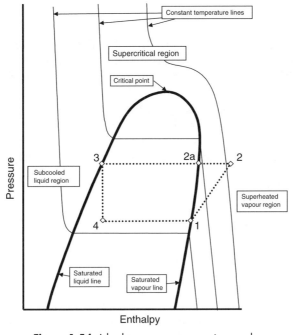

Figure 1.54 Ideal vapour compression cycle

as shown by point 2. The refrigerant by this point is above the temperature of the outside air. The refrigerant leaves the compressor and enters the condenser (heat exchanger). The condenser allows the heat to transfer to the outside air effectively removing enough heat to change into a saturated vapour from a superheated vapour point 2a. This has lowered the temperature of the refrigerant. Now the refrigerant is a 'saturated vapour' the pressure and temperature are kept constant but heat is still being removed and only the enthalpy continues to decrease. The vapour begins to condense to liquid (latent heat of condensation). Condensation continues until all the vapour is a 'saturated liquid' point 3. The refrigerant leaves the condenser as a saturated liquid and travels to an expansion valve or fixed orifice tube. The refrigerant now undergoes an 'isenthalpic' expansion process (constant enthalpy). The process significantly reduces the temperature and pressure of the refrigerant while the enthalpy remains the same. A small amount of liquid refrigerant (flash gas) vaporises during expansion but most of the refrigerant is liquid at a temperature lower than that of the outside air (air inside the vehicle or entering the vehicle) point 4. The refrigerant flows through the evaporator which acts as a heat exchanger that transfers heat from the air flowing through its fins to the refrigerant flowing through its coils. The refrigerant absorbs the heat increasing in enthalpy while the temperature and pressure remain the same. The liquid refrigerant vaporises until it becomes a 'saturated vapour'. The saturated vapour then travels to the compressor to start the cycle again.

The real world operation

The real world operation of the A/C system deviates from the ideal cycle. It is difficult for condensation and evaporation to end exactly on the liquid/vapour saturation lines. This is particularly difficult due to the fact that the system has to operate under so many varying conditions. To ensure an acceptable performance under all loads the condensers are designed to 'subcool' the refrigerant to a certain amount to ensure that only liquid refrigerant flows to the expansion device for optimum performance (also one of the jobs of the receiver drier). If vapour flows to the expansion device it reduces the flow of refrigerant significantly. Evaporators are generally designed to slightly 'superheat' (TXV systems) the refrigerant to ensure that only vapour flows to the compressor and no liquid (except for oil circulation 3%). There are also pressure drops across components like the condenser and evaporator which cause deviation from the ideal constant pressure process. Fixed Orifice Valve (FOV) systems use a fixed orifice diameter which is designed for optimum flow at high compressor and vehicle speeds. Poor performance at idle conditions can occur where there is a possibility of the evaporator being excessively flooded. Some flooding is advantageous with the FOV system. This is to increase cooling capacity and reduce 'hot spots', which are areas of reduced heat transfer caused by poor refrigerant distribution. Too much flooding reduces compressor performance.

1.5 Alternatives cycles

Pressure from the European Union for more environmentally friendly A/C systems has forced manufacturers to look for alternative refrigerants or technologies for HVAC units. A great deal of controversy exists on which technology the EU, US and the automotive industry wants to supersede R134a or in fact just improve the current R134a into a leak-free system. The EU will phase out R134a from 2011 with a complete ban on its use by 2014 to 2017 (dates to be finalised). Possible alternatives are within this section although it seems inevitable that CO_2-based A/C systems will replace the R134a system which is currently used.

A list of possible alternative are:

- CO_2-based system (R744);
- absorption refrigeration;
- secondary loop system (HFC152);
- gas refrigeration (R729);
- evaporative cooling;
- thermo electric cooling (Peltier effect).

The production of R744 (CO_2)-based HVAC systems will be included on vehicles being mass produced by 2008. The following information has been provided to aid the reader in predicting what refrigerant and accompanying technology will be implemented to supersede R134a.

The CO$_2$ (R744)-based refrigeration cycle (transcritical system)

Refrigeration and air-conditioning systems where the cycle incurs temperatures and pressures both above and below the refrigerant's critical point are often called transcritical systems. Transcritical systems are somewhat similar to the subcritical systems described above although they do have some different components.

Figure 1.55 illustrates the transcritical vapour compression process. It begins when the super-heated refrigerant enters the compressor at point 1. Its pressure, temperature and enthalpy are increased until it leaves the compressor at point 2 located in the supercritical region. Next the refrigerant enters the gas cooler whose function is to transfer heat from the fluid to the environment. Unlike the condensing process in the subcritical system, the refrigerant has not undergone a distinct phase change when it leaves the gas cooler at point 3. Note that this gas cooling

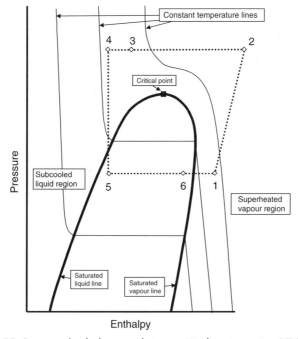

Figure 1.55 Pressure/enthalpy graph transcritical system using R744 (CO_2)

process does not occur at constant temperature. The cooled gas then enters an internal heat exchanger (sometimes called a 'suction line heat exchanger'), which transfers heat to that portion of the refrigerant that is just about to enter the compressor.

This results in additional cooling of the refrigerant to point 4 on the figure, improving performance at high ambient temperatures. From there, the flow undergoes a constant-enthalpy expansion process that decreases its temperature and pressure until it exits at point 5 in the mixed liquid/vapour region, at temperature and pressure well below the critical values. Next, the refrigerant enters an evaporator where it absorbs heat from the cooled space and its enthalpy and vapour fraction gradually increase until it exits at point 6. Finally, the flow enters the internal heat exchanger where it absorbs more heat, until it is ready to enter the compressor again at point 1 to repeat the cycle.

Note – the R744 cycle will also work in the subcritical region (i.e. some condensation in the gas cooler will take place) in case the ambient temperature is considerably lower than the critical temperature of R744, which is about 31°C.

System operation

Figure 1.56 shows an R744 closed loop A/C system capable of acting as a heat pump. The increased use of highly efficient diesel engines, particularly direct injection models as presently occurring

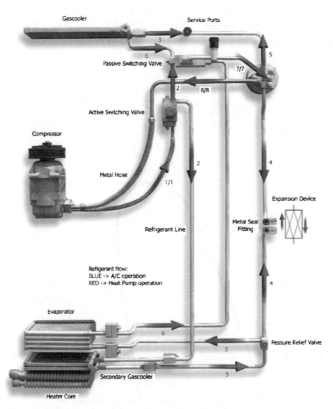

Figure 1.56 System layout for A/C and heat pump operation

in Europe, and the anticipated increase in the use of hybrid vehicles, means that engine coolant will no longer have the customary temperatures and capacities for acceptable passenger compartment heating and window defrosting/demisting operations. However, a heat pump can be used to heat the passenger compartment boosted to temperature levels to which vehicle occupants are accustomed.

The diagram (also reproduced in colour in the plate section) shows arrows in blue which represent refrigerant flow when in A/C mode and red arrows when in heat pump mode.

A/C operation (blue arrows)

1. Compressor
   Superheated refrigerant enters the compressor (temperature 30°C, pressure 35 bar). The refrigerant is compressed increasing its pressure, temperature and enthalpy (130 bar, 160°C). The given values represent a high load point.

2. The gas cooler (replaces the condenser)
   The refrigerant enters the gas cooler (via the active switching valve), upon entering the gas cooler the superheated gas allows heat to be transferred to the walls of the gas cooler and air travelling through it. The refrigerant does not go through a distinct phase change although its temperature (at the gas cooler outlet a few kelvin above inlet air temperature, for example 40°C for a 35°C ambient temperature) and enthalpy reduce. The refrigerant is still operating above its critical point.

3. The accumulator/internal heat exchanger
   The refrigerant flows to the high pressure side of the internal heat exchanger which removes heat by transferring it to the refrigerant that is about to enter the compressor. This again reduces the refrigerant's temperature (30°C).

4. Electronic expansion valve
   The refrigerant flows to the electronically controlled expansion device which creates a large pressure drop promoting the constant-enthalpy expansion process. The reduced pressure and temperature of the refrigerant now allows the device to operate below its critical point. The refrigerant is now a mixture of liquid and a flash vapour having the ability to change state with additional heat input.

5. Evaporator
   The refrigerant flows into the evaporator absorbing heat through evaporation until it exits the evaporator as a saturated vapour.

6/7. Accumulator/Internal heat exchanger
   The refrigerant flows (through the passive switching valve) to the accumulator/internal heat exchanger. This component combines the accumulator and internal heat exchanger functionalities into one part. The internal heat exchanger section enables the refrigerant to become slightly superheated. The accumulator portion separates the liquid and gaseous phase, stores the unused liquid refrigerant and allows the compressor oil together with the gaseous refrigerant to return to the compressor for lubrication.

8. Slightly superheated refrigerant flows back to the suction side of the compressor and the process repeats.

Note – temperatures and pressures are approximate and are dependent on system load.

Heating operation (red arrows)

Heating is achieved by directing the flow of heated refrigerant to a secondary gas cooler positioned inside the vehicle which heats the incoming/recycled air and is distributed through conventional ducting. The refrigerant then flows to the accumulator/heat exchanger and external gas cooler which is positioned at the front of the vehicle.

1. Compressor

   For an assumed ambient temperature of $-18°C$, the refrigerant enters the compressor at a temperature of about $-20°C$ and a pressure of 18 bar. The refrigerant will be compressed to about 90 bar at $90°C$.

2. The secondary gas cooler

   The active valve will divert the flow that – in an A/C cycle – would usually go to the gas cooler, to the secondary gas cooler. Upon entering the gas cooler the superheated/high temperature gas allows heat to be transferred to the walls of the gas cooler and the air travelling through it. The refrigerant is still operating above its critical point. Both temperature and enthalpy reduce.

3/4. The electronic expansion device

   The refrigerant flows to the electronically controlled expansion device which creates a large pressure drop promoting the constant-enthalpy expansion process. The reduced pressure and temperature of the refrigerant allows it to operate below its critical point. The refrigerant is now a liquid and flash vapour having the ability to change state with additional heat input. Note that for a heat pump cycle the expansion device needs to be a bi-directional type as flow enters from both sides.

5. The accumulator/internal heat exchanger

   The internal heat exchanger (IHX) section has no duties in the heat pump cycle (both the former high and low pressure sides of the IHX are now located on the low pressure side of the cycle) and is only a passage for the refrigerant.

6. The gas cooler

   The refrigerant flows from the accumulator/internal heat exchanger to the gas cooler, which in fact now acts as the evaporator for the heat pump cycle, where it will absorb more heat to change from a saturated vapour to become slightly superheated.

7. The accumulator/heat exchanger

   The passive valve collects the flow of the heat pump branch of the circuit and directs the flow to the accumulator/internal heat exchanger. Again, the IHX is pretty much without function, but the accumulator section acts exactly as in A/C operation.

8. The compressor

   The refrigerant flows from the accumulator/heat exchanger back to the compressor and the process repeats itself.

   Note – pressure and temperature figures depend on system load.

Heat pump operation will only be used at very cold ambient temperatures allowing for faster defrost and/interior warm-up. It does not replace the normal heater core and is just a supplement. Thus the mixed A/C/heater mode is still available, but there will be no combined A/C/heat pump operation (as both functions rely on the same refrigerant circuit). A/C will continue to be used for dehumidification in warmer ambient temperatures or during comfort drive, assisted by the normal heater core for reheating. Removing particles is the duty of the air filter that is located upstream of the evaporator.

R744 properties

R744 has a corrosive effect on polymers so metal pipes are used. R744 is non-flammable and relatively cheap compared to R134a. R744 is also easier to recycle than R134a.

Table 1.4 Properties of R744

	R744
Operational mode	Transcritical
High pressure	60–140
Low pressure	35–50
Max gas temperature	Up to 180°C
A/C line diameters	10–12
A/C line fittings type	Axial metal
Flexible hose material	Steel
Compressor displacement (cm³)	20–33
Compressor housing diameter (mm)	100–120
Expansion device	Electronic valve or mechanical orifice
Front end heat exchanger	Gas cooler

(Specifications and descriptions contained in this book were in effect at the time of publication. Visteon reserves the right to discontinue any equipment or change specifications without notice and without incurring obligation (07/05).)

Component information
Sanden and LuK have been assisting with the research and development of compressor designs for R744 systems.

- Research in 2000 was based on compressor Luk Variable Technology (LVT) (30–36 cm³), the parts are interchangeable with compressor Verband der Automobilindustrie (VDA) (160 cm³) – R134a.

Maximum pressures and temperatures to withstand:

- high side 16.0 MPa (2320 psi);
- low side 12.0 MPa (1740 psi);
- high side 180°C (356°F) (discharge temperature).

Possible lubricants for the system
POE or PAG are the most likely options. Surplus oil will be stored in the accumulator and sucked back to the compressor via an oil bleed hole in the accumulator through the suction line. Refrigerant filters will most likely be applied in an R744 system, one location could be within the accumulator, another one in front of the expansion device.

Accumulator/Internal heat exchanger
The accumulator is required to store lubricant for the compressor operation. The internal heat exchanger acts as a heat exchanger to increase cooling capacity which is required mainly at high ambient air temperatures. The accumulator also ensures that no liquid refrigerant enters the compressor during the system operation.

Figure 1.57 Mutli zone HVAC unit for CO_2 application

Gas cooler

The efficiency of the system's operation is highly dependent on the air flow through the gas cooler ensuring enough heat is removed or absorbed.

Expansion valve

A solenoid operated valve which can be operated by a high frequency pulse width modulated signal or an analogue DC voltage.

Visteon multi zone modular HVAC system

The modular multi zone HVAC system offers manufacturers the flexibility to personalise the HVAC system depending on model specification and provides up to four temperate controlled zones from one unit (this number of temperature controlled zones is usually provided by two or more units). Modular units allow for mass production of common components gaining efficiencies of scale while still providing flexibility to the customer.

System benefits:

- All metal sealing promotes no leak concept.
- Cross platform usage – modular multi zone design.
- Heat pump facility.
- Full electronic control.
- Low Global Warming Potential (GWP) (1) and zero Ozone Depleting Potential (ODP).
- Improvement in fuel economy.
- No additional fuel or electric heater required.
- Reduction in emissions.

Negative aspects:

- A higher level of technology is required and additional components are necessary.
- Additional cost to suppliers exists due to large investment in research and development (whether this cost will be passed on to the consumer or absorbed by the OEM is speculation).

- There will be an impact on the service and training industry requiring new knowledge, skills, equipment and possibly a certificate of competence to service and repair such systems.

Absorption refrigeration

The absorption refrigeration cycle is attractive when there is a source of inexpensive or waste heat readily available. This cycle uses a refrigerant that is readily soluble in a transport medium. In brief, the condensation, expansion and evaporation processes are identical to those of the vapour/compression cycle. But instead of the latter's compression process, the absorption cycle's liquid transport medium absorbs the refrigerant vapour upon leaving the evaporator, creating a liquid solution. This solution is then pumped to a higher pressure, and then heat is used to separate the refrigerant from the solution, whereupon the high pressure refrigerant flows to the condenser to continue the familiar cycle. The equipment used to accomplish the solution/dissolution processes is complex and heavy, but the advantage lies in the low work input requirement to raise the pressure of a liquid solution as compared to that required for compressing a gas. If the heat utilised is otherwise wasted heat, the low operating costs of absorption systems can be quite attractive. The two most common refrigerants used in absorption systems are ammonia, with water as the transport medium, and lithium bromide in water. However, toxicity issues with ammonia require safeguards, adding to system cost and complexity. Lithium bromide can be corrosive to most common materials, again adding to cost and complexity. Absorption systems are used mostly in large non-vehicular building applications, though occasionally there has been advocacy of their use as mobile air-conditioning systems.

Secondary loop system HFC152a

HFC152a (Fig. 1.58), which is flammable, must be used in a 'secondary loop' A/C system that uses a chiller to transfer cooling from the refrigerant in the engine compartment to coolant that is circulating into the passenger compartment. The secondary loop is required as opposed to a primary loop using a hydrocarbon-based refrigerant because if a leak occurs in the evaporator and hydrocarbons are released then an explosion could occur.

The primary loop (refrigerant circuit) operates in the same manner as the vapour compression cycle using a hydrocarbon-based refrigerant instead of R134a. It is positioned under the bonnet. The secondary system (coolant system) positioned inside the vehicle uses brine as a cooling medium which is under pressure by the circulating front and rear pump to transfer the heat from the front and rear coolers to the cooling medium. The reservoir is required to allow for the expansion of the coolant. The system is a dual A/C system with front and rear coolers. This system allows the easy addition of multiple cooling points with no additional expansion device. The refrigerant charge in the primary system is about half when compared to an R134a system with the same HVAC specification.

System benefits:

- Wide choice of refrigerants can be used.
- Enhances city traffic and idle cooling performance.
- Potential for targeted cooling (e.g. seats).
- Reduces refrigerant charge and leakage.
- Potential for elimination of heater core resulting in smaller HVAC case and cost savings (single heat exchanger for heating and cooling).
- Refrigerant NVH reduction – front and rear.
- Eliminates refrigerant maldistribution (coolant exhibits more uniform temperature distribution than refrigerant).

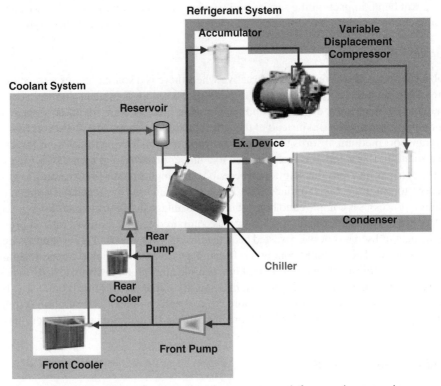

Figure 1.58 HFC152a secondary loop system with front and rear cooler (courtesy of Volvo)

System drawbacks:

- The use of flammable refrigerant.
- Cost of the system.
- Energy penalty.
- Weight of the system.

Gas refrigeration

The gas refrigeration cycle

In the past the gas refrigeration cycle, which is common on aircraft, has been considered for the automobile industry. Research on the use of an air refrigeration system exists and cannot be excluded as a future option due to this fact.

The gas refrigeration cycle is appropriately named due to the refrigerant remaining in a gaseous state throughout the entire cycle. R729, otherwise known as air, is used as the refrigerant medium.

The pressure enthalpy diagram for a closed gas refrigeration system operates outside of the phase transition of the refrigerant so that the saturation curves do not appear on the diagram unlike the vapour compression cycle.

To aid the explanation of the system operation a Jaguar aircraft has been used for illustration (see Figure 1.59).

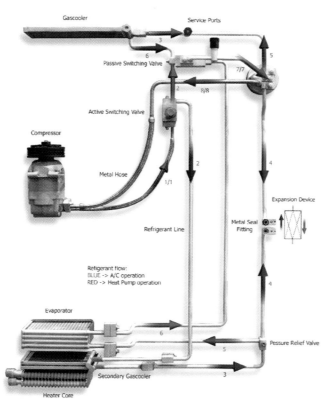

Plate 1 System layout for A/C and heat pump operation
(© 2005 Visteon All Rights Reserved).

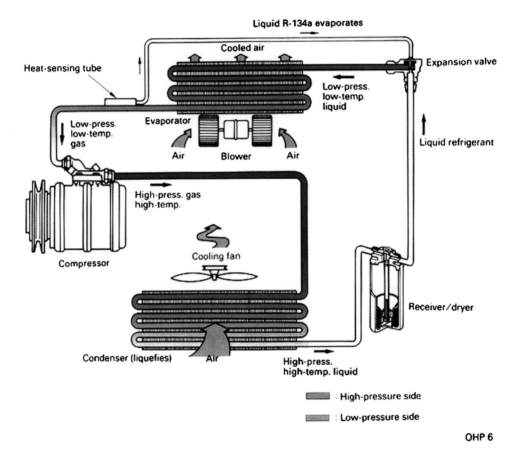

Liquid R-134a evaporates

Heat-sensing tube

Cooled air

Expansion valve

Low-press.
low-temp.
liquid

Low-press.
low-temp.
gas

Evaporator

Liquid refrigerant

Air   Blower   Air

High-press. gas
high-temp.

Compressor

Cooling fan

Receiver/dryer

Condenser (liquefies)   Air

High-press.
high-temp. liquid

: High-pressure side

: Low-pressure side

OHP 6

Plate 2 Expansion valve system – also reproduced in colour in the plate section
(with the agreement of Toyota (GB) PLC)

Figure 1.59 Jaguar aircraft
(courtesy of RAF Coltishall)

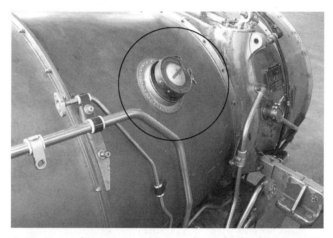

Figure 1.60 Gas turbine engine showing the outlet feed to the A/C system
(courtesy of RAF Coltishall)

The refrigerant (air) enters a rotary compressor to raise its pressure and enthalpy. Air charge is taken from both the compressors via an outlet as shown circled in Figure 1.60.

Only compressed gas at a temperature of about 190°C enters the A/C system because the feed is situated on the outlet of the compression stage and not ignition.

The air then travels to a primary heat exchanger where it gives up heat at constant pressure. The exchanger has ram air flowing through it to reduce the temperature of the air feed from the gas turbine engines.

In Figure 1.61 the primary heat exchanger is fitted along the spine of the aircraft (large circle) and is fed from the gas turbine engine (small circle).

The gas then flows to a turbine unit (cold air unit) where it expands reducing in enthalpy, temperature and pressure. Some units connect the turbine to a compressor to recover some of the energy given up during isenthalpic expansion. The air is at a temperature of about 100°C upon leaving the turbine (cool air unit).

Upon leaving the turbine the gas flows to a secondary heat exchanger (Figs 1.62 and 1.63), situated behind the cockpit. Cabin air circulates through the fins of the heat exchanger releasing heat and then a water extractor removes moisture in the air.

Figure 1.61 Exposed airframe
(courtesy of RAF Coltishall)

Figure 1.62 Secondary heat exchanger inlet
(courtesy of RAF Coltishall)

The air is then distributed around the cabin and auxiliary equipment situated on the aircraft. Because the cabin is pressurised, air is bled through two discharge valves in the aircraft fuselage. This means that the system is an open A/C system because the air does not flow back to the compressor to flow through the process again unlike a closed system.

The system can be operated manually or automatically to control the internal temperature of the cockpit. Heating is achieved by bypassing the heat exchangers thus transferring heat laden gas to the air distribution system. This is electronically sensed using thermistors and directed using control flaps.

Evaporative cooling

The latent heat of vaporisation of water can provide cooling to vehicle occupants. A crude approach is to spray one's face with a water mist, then place the head outside the window of a moving vehicle into the free air stream – the evaporating water carries away heat from the skin. There have been devices, mostly in the 1950s – 'The Weather Eye' – that worked on evaporative

Figure 1.63 Secondary heat exchanger
(courtesy of RAF Coltishall)

cooling. They consisted of a box or cylinder fitted to the window of the vehicle. The intake of the unit would allow air to enter from outside and travel through a water soaked wire mesh grille and excelsior cone inside the unit. The water would evaporate due to absorbing the heat in the air and travel through the outlet of the unit which acted as a feed to the inside of the vehicle. At one time, these devices had a certain attractiveness, particularly in hot, low humidity regions like the US southwest. However, their performance compared to modern vehicular air-conditioning systems is generally inadequate and they currently do not have significant popularity.

Thermo electric cooling (Peltier effect)

The basic concept behind thermoelectric (TE) technology is the *Peltier effect* – a phenomenon first discovered in the early 19th century. The Peltier effect occurs whenever electrical current flows through two dissimilar conductors; depending on the direction of current flow, the junction of the two conductors will either absorb or release heat.

Semiconductor material, usually bismuth and telluride, are generally used within the thermo-electric industry. This is due to the type of charge carrier employed within the conductor (see Chapter 3). Using this type of material, a *Peltier device* can be constructed. In its simplest form it consists of a single semiconductor 'pellet' which is soldered to electrically conductive mater-ial on each end (usually plated copper). In this 'stripped-down' configuration (Figure 1.65), the second dissimilar material required for the Peltier effect is the copper connection which also acts as a conductor for the power supply.

N type

Once impurities are added to a base material their conductive properties are radically affected. For example, if we have a crystal formed primarily of silicon (which has four valence electrons), but with arsenic impurities (having five valence electrons) added, we end up with 'free electrons' which do not fit into the crystalline structure. These electrons are loosely bound. When a volt-age is applied, they can be easily set in motion to allow electrical current to pass. The loosely bound electrons are considered the charge carriers in this 'negatively doped' material, which is referred to as 'N' type material. The electron flow in an N type material is from *negative to positive*. This is due to the electrons being repelled by the negative pole and attracted by the positive pole of the power supply.

P type

It is also possible to form a more conductive crystal by adding impurities which have one less valence electron. For example, if indium impurities (which have three valence electrons) are used in combination with silicon, this creates a crystalline structure which has 'holes' in it, that is, places within the crystal where an electron would normally be found if the material was pure. These so called 'holes' make it easier to allow electrons to flow through the material with the application of a voltage. In this case, 'holes' are considered to be the charge carriers in this 'positively doped' conductor, which is referred to as 'P' material. Positive charge carriers are repelled by the positive pole of the DC supply and attracted to the negative pole; thus 'hole' current flows in a direction *opposite to that of electron flow*.

Figure 1.64 shows two dissimilar materials, one is N type and the other P type. It is important to note that the heat will be moved (or 'pumped') in the direction of charge carrier movement throughout the circuit (it is the charge carriers that transfer the heat). Thus the electrons flow continuously from the negative pole of the voltage supply, through the N pellet, through the copper tab junction, through the P pellet, and back to the positive terminal of the supply.

The positive charge carriers (i.e. 'holes') in the P material are repelled by the positive voltage potential and attracted by the negative pole and flow through the positive pole through the P pellet and copper tab to the N pellet and the negative terminal. When a DC current is applied heat is moved from one side of the device to the other – where it must be removed with a heat-sink.

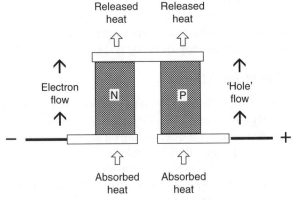

Figure 1.64 One simple semiconductor pellet
(courtesy of Tulleride)

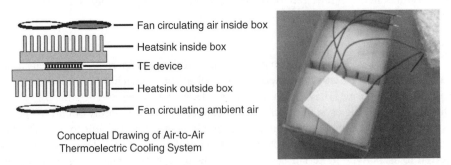

Fan circulating air inside box
Heatsink inside box
TE device
Heatsink outside box
Fan circulating ambient air

Conceptual Drawing of Air-to-Air
Thermoelectric Cooling System

Figure 1.65 Peltier module
(courtesy of Tullurex)

The 'cold' side is commonly used to cool. If the current is reversed the device makes an excellent heater.

Arranging N and P type pellets in a 'couple' and forming a junction between them with a plated copper tab, it is possible to configure a series circuit which can keep all of the heat moving in the same direction. Using these special properties of the TE 'couple', it is possible to team many pellets together in rectangular arrays to create *practical* thermoelectric modules. These devices can pump appreciable amounts of heat, and with their series electrical connection, are suitable for commonly available DC power supplies. The most common TE modules in use connect 254 alternating P and N type pellets and use 12 to 16 VDC supply and draw 4 to 5 amps.

Thermoelectric modules look like small solid-state devices that can function as heat pumps. A 'typical' unit is a few millimetres square to a few centimetres square. It is a sandwich formed by two ceramic plates with an array of small bismuth telluride cubes ('couples') in between.

Heatsinks are used to either collect heat (in heating mode) or dissipate collected heat into another medium (air, water). Heat must be transferred from the object being cooled (or heated) to the Peltier module and heat must be transferred from the Peltier module to a heatsink. Systems are often designed for pumping heat from both liquids and solids. In the case of solids, they are usually mounted right on the TE device; liquids typically circulate through a heat exchanger (usually fabricated from an aluminium or copper block) which is attached to the Peltier unit. Occasionally, circulating liquids are also used on the hot side of TE cooling systems to effectively dissipate all of the heat.

Peltier device cooling and heating speeds can change temperatures extremely quickly, but to avoid damage from thermal expansion the rate of change is controlled to about 1°C per second. It is theoretically possible to get a temperature difference across a Peltier module of 75°C although it has been stated that in practice this is not achieved. In practice the results are about half this figure.

If a TE module is to be used to cool anywhere near freezing then water condensation must be considered. Ever-present water vapour begins to drop out of the air at the 'dew point'. This will result in the TE module, and what it is being used to cool, to get wet. Moisture inside of the TE module will cause corrosion and can result in a short-circuit. A solution to this problem is to operate the TE module in a vacuum (best) or a dry nitrogen atmosphere.

System control
Varying the power supply is often used. Pulse width modulation can be used, but a frequency above 2 kHz is recommended and the voltage applied must not exceed the recommended maximum voltage (T_{max}).

Peltier devices are best suited to smaller cooling applications. They can be stacked to achieve lower temperatures, but are not very 'efficient' as coolers due to the heating effect of the current flowing as well as drawing a great deal of current but act as very good heaters. This disadvantage can be offset by the advantages of no moving parts, no Freon refrigerant, no noise, no vibration, very small size, long life, and the capability of precision temperature control.

Automotive application – Amerigon's proprietary CCS system
Vehicle cabin air is drawn into the cushion and back TE modules and, based on inputs from individual seat controls and from temperature sensors, the unit will either add heat or remove heat to the air flow.

The basis of the system is the Peltier circuit. The Peltier circuit, heatsink (heat exchanger) and fan assembly are mounted as one module. Air is used as a medium to move heat around the seat through the perforated seat layers. Conditioned air is ported to the top surface of the foam through channels which evenly distribute the conditioned air over the surface. Breathable trim covers allow the conditioned air to pass through to the occupant. When the Peltier device is cooling, heat is generated on the opposite side of the device which must be removed to allow the temperature differential to exist.

This heat is pumped into the cabin space and is labelled on Figure 1.66 as 'waste heat'. This will create an extra load on the A/C system if being operated to cool the interior space.

When voltage is applied to the Peltier module in one direction one side of the Peltier device will be hot and the other cool due to the direction of the charge carriers creating a ΔT across the Peltier device. Switching polarity of the circuit creates the opposite effect.

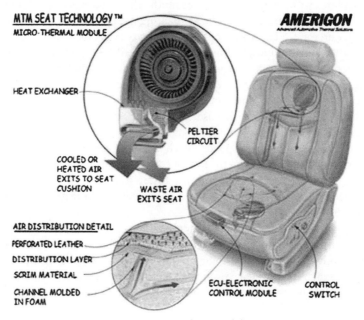

Figure 1.66 Amerigon's Peltier module CCS system
(courtesy of Amerigon Corporation)

Figure 1.67 TE module showing Peltier couples and copper tabs stacked in series
(courtesy of Amerigon Corporation)

Amerigon's Peltier module proprietary CCS system allows occupants to select seat temperatures to promote comfort and reduce driver fatigue through the use of a solid-state heat pump combined with an active, microprocessor controlled temperature management system to vary heating and cooling capacity.

Amerigon states that it is the first to have successfully packaged this technology for use in automotive seating applications.

1.6 The air-conditioning system

System activation

The signal to activate the air-conditioning system comes from the occupant(s). Activation is completed by the onboard Electronic Control Unit (ECU). The ECU has a number of inputs which send electronic signals based on sensed conditions, e.g. temperatures, pressures, speeds, positions. Based on this information the ECU will either activate or deactivate (if already operating) the system. If the system does not activate then a fault in the form of a code will be stored in the computer and on some systems a light will be activated to tell the driver a fault exists. Advanced systems may use telematics to send a signal to a call centre. Operators at the centre will advise the customer of the required action, i.e. urgency on visiting a dealership, often via telephone.

Activation of the air-conditioning system is achieved under some or all of the following conditions:

- The outside air temperature is above 9°C.
- The engine has been running for more than 5 seconds.
- The temperature of the evaporator is above 4°C (no ice forming over the surface).
- The engine coolant temperature is approximately between 40°C and 105°C.
- The vehicle is not rapidly accelerating or the engine is under high load (overtaking etc.).
- The air-conditioning activation button has been selected and the interior fan is on.

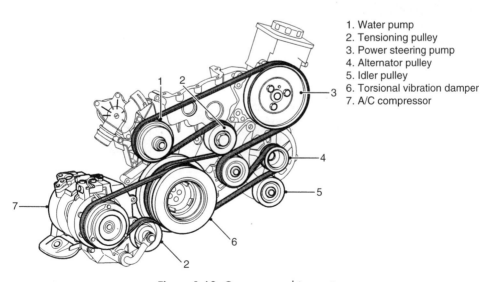

1. Water pump
2. Tensioning pulley
3. Power steering pump
4. Alternator pulley
5. Idler pulley
6. Torsional vibration damper
7. A/C compressor

Figure 1.68 Compressor drive system
(courtesy of Rover Group PLC)

- The sensors in the air-conditioning system have acknowledged that the system is under pressure assuming that a quantity of refrigerant exists inside the system and that it has not leaked out to the atmosphere (sensed by either pressure switches or sensors).
- No fault codes exist in the ECU.

Simple electronic circuit

Upon activation current flows from the vehicle battery through fuses, switches fitted to the A/C system and often an A/C relay to a magnetic clutch. The A/C relay is generally controlled by an onboard computer, which makes the ultimate decision to allow the A/C system to be activated based on system integrity, that is, the system has no faults and conditions are right for system activation. The clutch is positioned behind the compressor pulley and once activated will make a physical connection between the compressor pulley, which is driven by the engine, and the internal pumping elements (Fig. 1.68).

1.7 The expansion valve system

The air-conditioning system works on a continuous cycle (Fig. 1.69). A compressor receives low pressure heat laden refrigerant vapour from the evaporator. The compressor pressurises

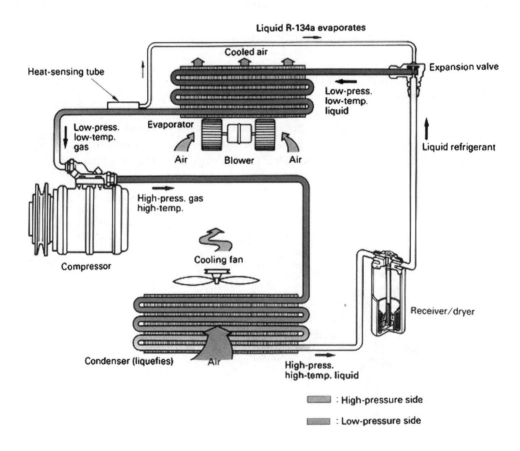

Figure 1.69 Expansion valve system
(with the agreement of Toyota (GB) PLC). This figure is reproduced in the colour plate section.

the refrigerant from 30 psi to approximately 213 psi depending on system demand. This increases the temperature from approximately 0 to 80°C. At this temperature and pressure the refrigerant is above its boiling point of approximately 57°C. The compressor discharges super-heated refrigerant vapour to the condenser.

The refrigerant flows into the condenser. The condenser has numerous cooling fins in which the vapour is pumped. In the condenser the high pressure vapour condenses into a high pressure liquid. This is achieved by reducing the temperature from, for example, 80°C to below 57°C which is the refrigerant's boiling point. This is achieved by forcing air over the surface of the condenser enabling heat to transfer from the refrigerant to the outside air thus reducing its temperature (subcooled). Only refrigerant in the form of a high pressure subcooled liquid leaves the bottom of the condenser outlet.

The subcooled liquid refrigerant flows into the receiver drier which stores, dries and filters the liquid refrigerant.

The subcooled liquid refrigerant then flows from the receiver drier to the expansion valve which then changes the refrigerant into low pressure, low temperature liquid/vapour. This is achieved by lowering the pressure using a variable orifice. The orifice has high pressure one side (from the receiver drier) and low pressure the other (evaporator and compressor) and allows a small quantity of refrigerant to flow through it. The sudden drop in pressure and temperature causes some of the refrigerant to vaporise which is called a flash gas. The low pressure low temperature liquid/vapour then flows to the evaporator where the heat is transferred from its surface to the refrigerant through vaporisation. The heat comes from either inside (recycled air) or outside (fresh intake of air) the vehicle and is blown over the evaporator's surface. Once the refrigerant has completely vaporised and reached its saturation point it should still be able to carry more heat. The refrigerant continues to flow through the remainder of the evaporator coils absorbing more heat and becoming slightly superheated.

The low pressure low temperature slightly superheated vapour refrigerant flows to the compressor and the cycle repeats itself.

> Note – temperatures are approximate and are dependent on refrigerant type, system load, pressure and temperature relationship.

1.8 The fixed orifice valve system (cycling clutch orifice tube)

The air-conditioning system works on a continuous cycle (Fig. 1.70). A compressor receives low pressure heat laden refrigerant vapour from the evaporator. The compressor pressurises the refrigerant from 30 psi to approximately 213 psi depending on system demand. This increases the temperature from approximately 0 to 80°C. At this temperature and pressure the refrigerant is above its boiling point of approximately 57°C. The compressor discharges super-heated refrigerant vapour to the condenser.

The refrigerant flows into the condenser. The condenser has numerous cooling fins in which the vapour is pumped. In the condenser the high pressure vapour condenses into a high pressure liquid. This is achieved by reducing the temperature from, for example, 80°C to below 57°C which is the refrigerant's boiling point. This is achieved by forcing air over the surface of the condenser enabling heat to transfer from the refrigerant to the outside air thus reducing its temperature (subcooled). Only refrigerant in the form of a high pressure subcooled liquid leaves the bottom of the condenser outlet.

The liquid refrigerant then passes through a fixed orifice tube – a tube with a fixed cross-sectional area allowing only a metered quantity of liquid refrigerant to pass through it.

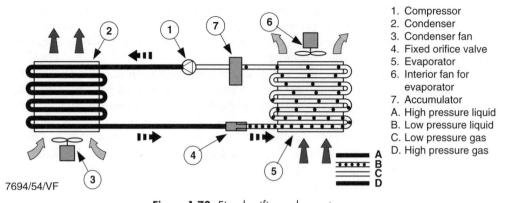

1. Compressor
2. Condenser
3. Condenser fan
4. Fixed orifice valve
5. Evaporator
6. Interior fan for evaporator
7. Accumulator
A. High pressure liquid
B. Low pressure liquid
C. Low pressure gas
D. High pressure gas

7694/54/VF

Figure 1.70 Fixed orifice valve system
(reproduced with the kind permission of Ford Motor Company Limited)

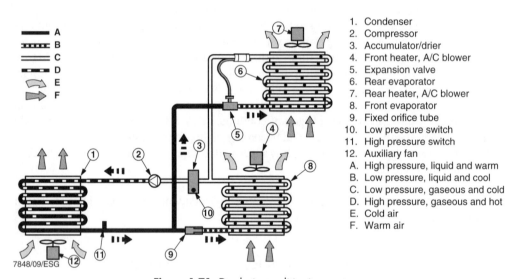

1. Condenser
2. Compressor
3. Accumulator/drier
4. Front heater, A/C blower
5. Expansion valve
6. Rear evaporator
7. Rear heater, A/C blower
8. Front evaporator
9. Fixed orifice tube
10. Low pressure switch
11. High pressure switch
12. Auxiliary fan
A. High pressure, liquid and warm
B. Low pressure, liquid and cool
C. Low pressure, gaseous and cold
D. High pressure, gaseous and hot
E. Cold air
F. Warm air

7848/09/ESG

Figure 1.71 Dual air-conditioning system
(reproduced with the kind permission of Ford Motor Company Limited)

The low pressure, low temperature liquid is then forced to expand rapidly due to the increase in volume of the evaporator. This drop in pressure causes the refrigerant to boil (vaporise) and absorb large amounts of heat energy which is transferred from the air flowing over the evaporator's fins to the evaporator's surface and thus to the refrigerant through vaporisation. After the heat is removed from the air it is directed to the vehicle's interior.

The low pressure, low temperature liquid/vapour refrigerant flows from the evaporator to the top of the accumulator which acts as a drier and storage device and separates any liquid from vapour to protect the compressor (compressors can only pressurise vapour). The large surface area of the accumulator also assists in any final evaporation of liquid refrigerant. The saturated vapour and a tiny percentage of liquid refrigerant to carry oil from the oil bleed leave the accumulator from the top and flow under low pressure to the compressor (suction side) and the cycle repeats itself.

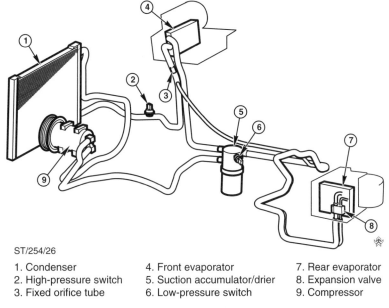

ST/254/26

1. Condenser   4. Front evaporator   7. Rear evaporator
2. High-pressure switch   5. Suction accumulator/drier   8. Expansion valve
3. Fixed orifice tube   6. Low-pressure switch   9. Compressor

Figure 1.72 The components of a dual A/C system

Note – temperatures are approximate and are dependent on refrigerant type, system load, pressure and temperature relationship.

1.9 Dual air-conditioning

This system (Figs 1.71 and 1.72) often combines the use of a fixed orifice valve and an expansion valve. The primary operation of the system shown in Figure 1.71 is a fixed orifice system with additional outlets and inlets which feed an additional expansion valve and evaporator to aid additional cooling. The additional cooling is often fitted in the back of a large multi-passenger vehicle given two temperature control zones – front zone and rear zone. The outlet (8) from the primary system to the additional expansion valve/evaporator is fed from the inlet (7) to the fixed orifice valve. Refrigerant splits and flows from the outlet of the primary system to an expansion valve and evaporator. Their function is identical to the normal expansion valve system. Refrigerant vapour flows from the evaporator in the secondary system to the accumulator (4) then (3) in the primary system. The accumulator ensures that hot liquid refrigerant enters the compressor from both systems.

2 Air-conditioning components

The aim of this chapter is to:

● Enable the reader to understand the operation of the individual components of the A/C system.

2.1 The compressor

The function of the compressor is to compress and circulate superheated refrigerant vapour around a closed loop system (any liquid or dirt will damage the compressor). Compressors vary in design, size, weight, rotational speed and direction and displacement. Also compressors can be mechanically or electrically driven. Some compressors are variable displacement and some are fixed. The compressor uses 80% of the energy required to operate an air-conditioning system. This means that the type of compressor used in the system will determine the overall efficiency of the system. This is particularly important for fuel economy and pollution which is monitored through government regulation.

Operation

The compressor is driven by an engine driven pulley system (Fig. 1.68). At the front of the compressor is a magnetic clutch which when given power engages the compressor. The compressor draws in refrigerant vapour from the suction side which is the outlet of the accumulator (fixed orifice valve system) or the outlet of the evaporator (expansion valve system). Because the refrigerant that left the evaporator/accumulator is a vapour it can no longer absorb heat energy and act as a cooler.

Figure 2.1 Compressor
(courtesy of Sanden UK)

During the compression of the refrigerant inside the compressor the pressure and temperature rapidly increase. An ideal system will increase the pressure from 200 to 2250 kPa (29 to 326 psi). The temperature increase can be as much as 0°C–110°C. When the air-conditioning system is running the suction pressure is between 120 and 300 kPa (17.5 and 43.5 psi), when the system is under high load the pressure and temperature of the refrigerant can reach as high as 2800 kPa (406 psi) and 125°C.

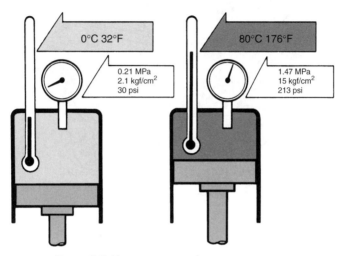

Figure 2.2 Temperature and pressure increase (with the agreement of Toyota (GB) PLC)

The compressor can only compress refrigerant vapour. Any liquid or dirt allowed to enter the compressor will cause damage.

The boiling point of refrigerant at 326 psi is 57°C so the refrigerant will remain in a gaseous state until the refrigerant gives off enough heat to drop below 57°C. To do this the refrigerant flows from the outlet of the compressor to the condenser.

Variable capacity compressors

Variable capacity compressors can vary the volume of refrigerant depending on the demands of the system. The demands of the system are sensed by the pressure of the refrigerant coming out of the evaporator. The demand is the amount of heat transferred to the refrigerant. This increase in temperature will affect the pressure that will eventually enter the inlet of the compressor housing. The minimum displacement of a variable type compressor is about 10 cm^3 and is not zero because the refrigerant carries the lubricant for the compressor which would cause damage if no refrigerant was present inside the compressor while it was operating. Variable capacity compressors greatly reduce the amount of on/off cycles that non-variable capacity compressors are subjected to. This reduces noise from the clicking of the magnetic clutch, increases fuel efficiency through variable loading of the system and reduces the wear of the magnetic clutch plate. Variable types of compressor generally differ from the non-variable type due to the addition of a control valve. The control valve is used to vary the displacement of the compressor to match the demands of the system.

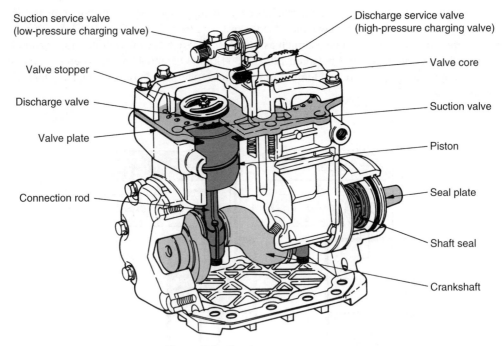

Suction service valve
(low-pressure charging valve)

Discharge service valve
(high-pressure charging valve)

Valve stopper

Valve core

Discharge valve

Suction valve

Valve plate

Piston

Connection rod

Seal plate

Shaft seal

Crankshaft

Figure 2.3 Crank type compressor
(with the agreement of Toyota (GB) PLC)

Types of compressor

There are three main categories of compressor:

1. Reciprocating – crank and axial piston (swash plate).
2. Rotary – vane.
3. Oscillating – scroll type (helix).

Crank type compressor (reciprocating)

Crank type compressors (Fig 2.3) are not generally used in the automotive industry any more. They may have up to two cylinders including 'V' shape configuration. They are driven by the engine pulley system which rotates a crankshaft inside the pump. The crankshaft is connected to a piston via a connecting rod which travels up and down the bore. Above the piston there is a valve assembly to direct the flow of refrigerant.

Pumping operation

The crankshaft and connecting rod convert the pump's motion from rotary (Fig. 2.4) to reciprocating. The piston travels up and down the bore inducing, compressing and discharging the refrigerant. Two valves are fitted per bore: a suction valve and a discharge valve.

On the downward stroke the refrigerant enters the compression chamber through the suction port due to a vacuum being created above the piston and the low pressure of the refrigerant.

On the upward stroke the refrigerant is compressed and an increase in pressure and temperature occurs. When the refrigerant pressure overcomes the force of the discharge valves the high temperature and high pressure (superheated vapour) refrigerant leaves the compression chamber.

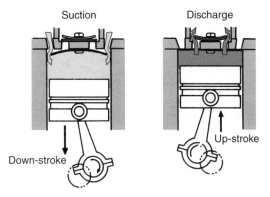

Figure 2.4 Reciprocating piston type compressor
(with the agreement of Toyota (GB) PLC)

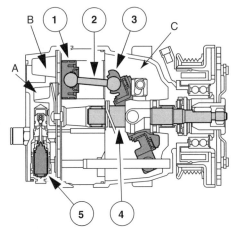

1. Piston (section view showing only one side)
2. Connecting rod
3. Swash or wobble plate
4. Driving shaft
5. Control valve (fitted only to variable
   displacement compressors)
A. High pressure chamber
B. Low pressure chamber
C. Internal pressure chamber

Figure 2.5 Variable displacement axial piston type compressor (swash plate type)
(reproduced with the kind permission of Ford Motor Company Limited)

Axial piston (swash plate) type variable capacity compressor

The axial piston type (Figure 2.5) is one of the most common types of compressor and can be fixed or variable capacity. The pumping cylinders are circumferentially situated around the outside of the drive shaft (4) and parallel to its axis. Each cylinder has a double ended piston with a separate pumping chamber at each end. Each pumping chamber has a set of inlet and discharge reed valves. The inlet reed valves are connected to the inlet port of the compressor via internal drillings and the discharge valves are connected to the discharge port of the compressor via internal drillings.

Pumping operation

A swash plate (3) is attached to the drive shaft. The swash plate is located at an angle. The pistons inside the pumping chamber are connected to the swash plate via swivelling ball joints. The rotation of the swash plate causes the pistons to reciprocate inside their pumping chambers. When the volume above a piston increases the refrigerant flows into the chamber via the inlet reed valves – induction. When the volume above a piston is reducing the pressurised

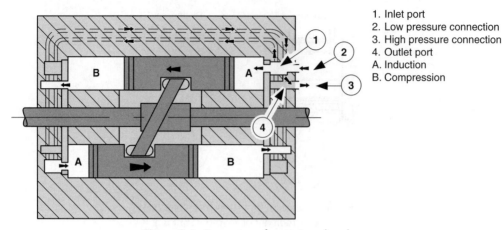

1. Inlet port
2. Low pressure connection
3. High pressure connection
4. Outlet port
A. Induction
B. Compression

Figure 2.6 Operation of pumping chambers
(reproduced with the kind permission of Ford Motor Company Limited)

refrigerant exits the pumping chamber via the pumping chamber discharge reed valves. The position of the swash plate will determine the length of travel by the piston which allows the compressor to vary its output.

Control of the compressor swept volume is regulated by the control valve ((5) in Fig. 2.5), located in the compressor rear end plate. The control valve attempts to keep the compressor low pressure side at a constant pressure, also known as the control point, which is determined by factory settings and cannot be adjusted in service.

The operation of the variable capacity swash plate compressor is based on the creation of three distinctive pressures which are sensed and controlled by a control valve and adjusted to create a mechanical equilibrium within the compressor (this can be seen on Figs 2.7 and 2.8):

A. High pressure (discharge pressure).
B. Low pressure (suction pressure from the evaporator (TX system or accumulator FOV).
C. Compressor internal pressure (generated by the high pressure but only a fraction of it).

The displacement is the difference between the low pressure and the internal pressure, this is called 'delta P (ΔP)'. The change in delta P varies the angle of the swash plate causing the capacity of the compressor to vary.

Control valve operation
The control valve consists of a spring loaded piston and ball valve and a bellows which can change its length according to pressure.

In order to achieve this constant pressure, the control valve changes the pressure in the compressor housing as follows:

Low demand   When the load on the air-conditioning system is minimal the low pressure from the evaporator will be low. This is due to the expansion valve (TX valve) opening only a small amount and allowing only a small amount of refrigerant to flow to the evaporator. This low pressure will flow to the compressor into the suction side and be sensed at point (C), Figure 2.7. The result of this low pressure, which is lower than the control pressure set by the manufacturers, means that the compressor needs to reduce its displacement. The bellows (6) expand and forces the ball valve up which increases the bleed of refrigerant from the high pressure (A) to

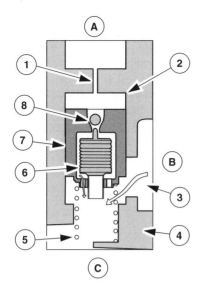

1. Calibrated bore
2. Valve cylinder
3. Bypass
4. Compressor body
5. Spring
6. Bellows
7. Piston
8. Ball valve
A. High pressure
B. Crankcase pressure which is partially compressed refrigerant directed from the high side
C. Low pressure from the suction valve

Figure 2.7 Low cooling demand
(reproduced with the kind permission of Ford Motor Company Limited)

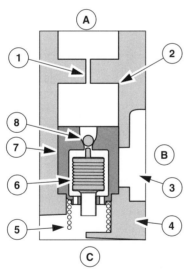

1. Calibrated bore
2. Valve cylinder
3. Bypass
4. Compressor body
5. Spring
6. Bellows
7. Piston
8. Ball valve
A. High pressure
B. Crankcase pressure which is partially compressed refrigerant directed from the high side
C. Low pressure from the suction valve

Figure 2.8 High cooling demand
(reproduced with the kind permission of Ford Motor Company Limited)

the low pressure (C). Because the high pressure (A) comes from the high pressure acting on the pistons inside the compressor the pressure acting on those pistons reduces.

High demand When the load on the air-conditioning system is high, the low pressure (suction side) will be high. This will act at point (C) on Figure 2.8, and will be higher than the control point (set at manufacture). This will cause the bellows to be compressed and the ball valve to be closed. This will stop the bleed from the high side to the low side and will cause the pressure above the valve to increase against spring pressure moving the valve body downwards. The bypass is closed and the entire delivery volume is passed to the compressor high pressure connection. When no refrigerant is added from the high pressure side, the pressure in the crankcase will drop, the displacement increases and the compressor works at a higher capacity. This is achieved through linkages between the swash plate and shaft, which alter its angle and will continually increase as the pistons and the rocker discs move. The increased angle results in an increased displacement. The increased displacement increases the cooling effect due to increased refrigerant volume.

A weak spring, called a start spring, is fitted on the compressor shaft and is used to return the pistons and the swash plate towards the valves as soon as the compressor stops operating and the pressure difference between the high and low pressure sides evens out. In this position, the compressor delivers 5% of its maximum swept volume and the starting torque is low the next time the compressor starts.

Movement of the swash plate

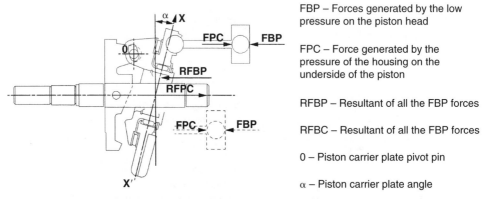

FBP – Forces generated by the low pressure on the piston head

FPC – Force generated by the pressure of the housing on the underside of the piston

RFBP – Resultant of all the FBP forces

RFBC – Resultant of all the FBP forces

0 – Piston carrier plate pivot pin

α – Piston carrier plate angle

Figure 2.9 Force diagram
(reproduced with kind permission of Peugeot)

Low load   If the low pressure decreases the low pressure force FBP decreases and therefore the low pressure refrigerant force RFBP decreases. If the internal pressure of the refrigerant force is constant, this becomes greater than the low pressure refrigerant force RFBP. At the

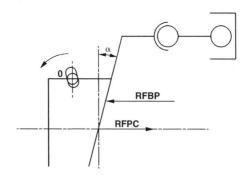

Figure 2.10 Force diagram showing increase in angle
(with kind permission of Peugeot)

point the RFPC is applied it will cause the piston carrier plate to rotate anti-clockwise around pin 0 – the pivot point. This causes the α-piston carrier plate angle to decrease. The compressor positions itself to minimum capacity to reduce the volume of refrigerant output.

High load   When the air passing through the evaporator is hot, humid and in large quantities the refrigerant load is said to be high. This means there will be a high volume of refrigerant at a significant low pressure value. When the low pressure value increases the FBP increases and the RFBP collectively increases. At the point where the RFBP is applied, which is under the piston carrier point, this will rotate clockwise around 0 – the pivot point. This causes the α-piston carrier plate angle to increase.

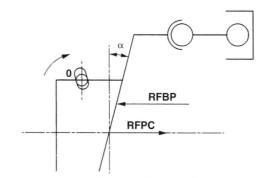

Figure 2.11 Force diagram showing decrease in angle
(with kind permission of Peugeot)

The compressor positions itself to maximum capacity to output as much refrigerant as possible to fulfil the system's requirements.

As previously discussed the displacement of the swash plate compressor is determined by the angle of the swash plate. This is a function of the relationship between the high and low pressures inside the compressor which is controlled by a regulating pattern. The graph will vary from one type of compressor to another.

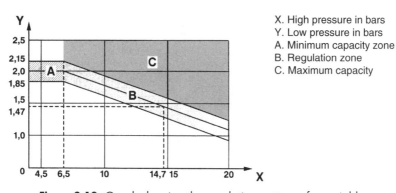

Figure 2.12 Graph showing the regulating pattern of a variable
displacement swash plate compressor
(with kind permission of Peugeot)

Vane type compressor

The vane type compressor is compact and has low frictional losses. It is quiet and has few moving parts. It is a rotary compressor using rotating vanes to increase the flow of refrigerant. There are two types of vane type compressor:

1. Through vane.
2. Eccentric vane.

Through vane type rotor

The through vane type has two vanes mounted at right angles to each other in slots in a rotor housing. As the rotor rotates the vanes slide radially to maintain contact with the surface, this is maintained through centrifugal force. The circulating refrigerant oil with the centrifugal force seals the moving parts relative to one another.

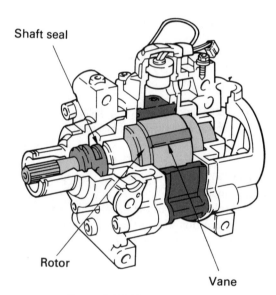

Shaft seal

Rotor

Vane

Figure 2.13 Vane type compressor
(with the agreement of Toyota (GB) PLC)

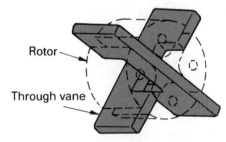

Rotor

Through vane

Figure 2.14 Through vane rotor
(with the agreement of Toyota (GB) PLC)

Pumping operation   It is important to note that the vane type compressor has three compression spaces. When the refrigerant is discharged, a space is available on the suction side of the compressor and is filled with refrigerant. At the same time the compression space of the compressor will have refrigerant being pressurised. This means that suction, compression and discharge are continually occurring during every full rotation.

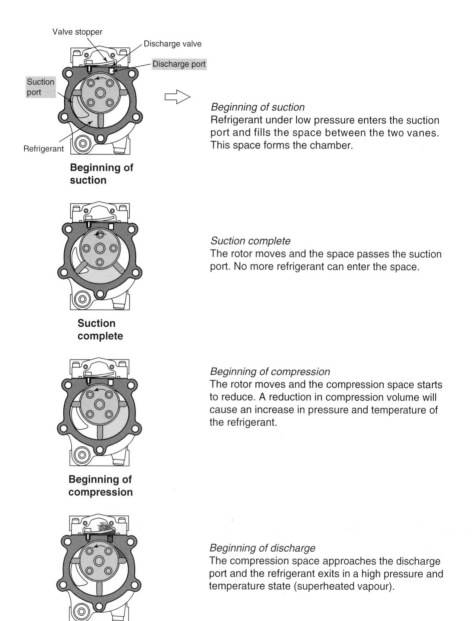

Beginning of suction

Beginning of suction
Refrigerant under low pressure enters the suction port and fills the space between the two vanes. This space forms the chamber.

Suction complete

Suction complete
The rotor moves and the space passes the suction port. No more refrigerant can enter the space.

Beginning of compression

Beginning of compression
The rotor moves and the compression space starts to reduce. A reduction in compression volume will cause an increase in pressure and temperature of the refrigerant.

Beginning of discharge

Beginning of discharge
The compression space approaches the discharge port and the refrigerant exits in a high pressure and temperature state (superheated vapour).

Figure 2.15 Pumping operation of a through vane type rotor
(with the agreement of Toyota (GB) PLC)

Eccentric vane
The eccentric vane (Fig. 2.16) works in a similar manner to the through vane except the vanes are organised separately and are not mounted at right angles to each other. The rotor inside the vane compressor rotates on an eccentric which is used to increase and decrease the volume within the compressor.

Scroll (helix) type compressor
The scroll compressor (Fig. 2.17) consists of two helices with one lying within the other. They are both mounted in a cylindrical housing. One helix is fixed and the other is attached to the drive shaft of the compressor. The driven helix does not rotate itself but does orbit the other helix. The two helices through movement create crescent-shaped compression chambers.

A. The volume within the compressor increases with the movement of the driven helix. This allows refrigerant to enter the compression chamber. The shape of the helix and its movement mean that no physical inlet valve is required. Induction or the suction phase ends when the position of the helix is at the bottom of its eccentric movement.

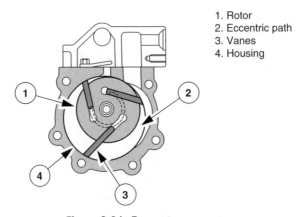

1. Rotor
2. Eccentric path
3. Vanes
4. Housing

Figure 2.16 Eccentric vane rotor
(reproduced with the kind permission of Ford Motor Company Limited)

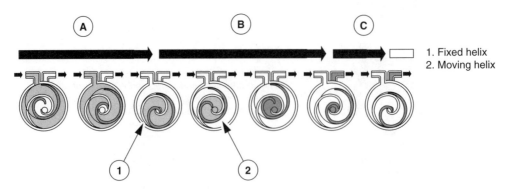

1. Fixed helix
2. Moving helix

Figure 2.17 Scroll compressor operation
(reproduced with the kind permission of Ford Motor Company Limited)

B. Compression occurs by trapping the refrigerant in the centre of the helix and then reducing the volume. This reduction will increase the pressure and temperature of the refrigerant to the required state.

C. Discharge occurs at the centre of the helix where a valve is positioned to ensure that no refrigerant is allowed to flow back into the compressor when not being operated.

During operation all gas spaces are in various states of compression which results in a continuous inlet and outlet flow. The compressor has few moving parts which means less wear.

Electric compressor

The electric compressor has to be used due to the absence of a mechanical drive mechanism for the compressor. Electric compressors are generally used on vehicles with hybrid or electric power units. Hybrid vehicles have both a small engine, generally diesel, and an electric motor powered by a battery unit. The engine intermittently runs in the event that the electric power unit energy level becomes low or high torque output is required. Electric powered vehicles have no engine units and generally take the form of fuel cell vehicles. A fuel cell, which is an inversion of the process of the electrolysis of water, is used to produce electrical energy. Fuel cells do not store energy; they produce it and replace the energy used by the batteries which provide the primary source to drive the electric motor on the vehicle. These technologies will over the next 10 years change the way in which we view, design, manufacture, maintain and repair our vehicles.

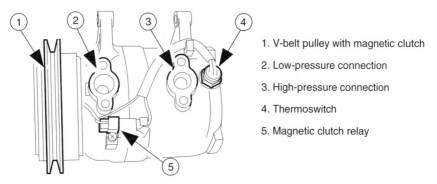

1. V-belt pulley with magnetic clutch

2. Low-pressure connection

3. High-pressure connection

4. Thermoswitch

5. Magnetic clutch relay

Figure 2.18 Vane compressor thermo switch (operates between 140 and 150°C)

Pressure relief valve

If the load within the system is excessive or a blockage occurs and the pressure of the system is excessively high then a pressure relief valve fitted to the compressor and often the accumulator opens and reduces the system pressure. The valve is only used if other safety devices fail. The valve reduces the possibility of a burst pipe or fractured evaporator or condenser due to excessive pressures in the system. The valve operates at approximately 3.5–4.0 MPa.

Thermo cut-out

Some compressors have a thermo cut-out (Fig. 2.19), an electrical switch that works on the principle of a bimetallic strip. If the temperature of the refrigerant in the compressor exceeds a safe limit (approx. 150°C) then the bimetallic strip bends and breaks the electrical connection to the compressor clutch. This shuts the system down.

Table 2.1 Specification of the electric compressor (courtesy of Sanden UK)

Motor	Proto.F653DC Brushless
Displacement (cc)	20.1
Max Speed (rpm)	12 000
Cooling Capacity (kW)	6.5
Rated Voltage (V)	345
Refrigerant/Oil	HFC134a/SE-10
Outer Diameter (mm)	120
Length (mm)	194
Weight (kg)	6.0

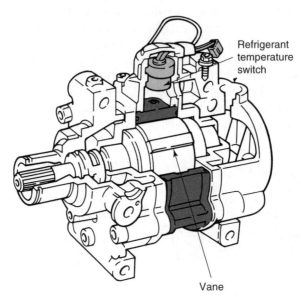

Refrigerant
temperature
switch

Vane

Figure 2.19 Compressor thermo cut-out
(with the agreement of Toyota (GB) PLC)

Compressor magnetic clutch

The compressor is driven by the engine crankshaft via a pulley system. In Figure 2.21 the pulley system provides permanent drive to the multi groove drive belt (4). This means once the engine is started the multi groove drive belt pulley is rotating. There is an air gap between the multi groove drive belt pulley and the drive plate (1). When the engine is running the compressor is stationary until the A/C button is selected and electrical power flows through the clutch field coil generating an electrical magnetic field which attracts the drive plate towards the multi groove drive belt pulley (Fig. 2.20). The drive plate which is attached to the compressor drive shaft is pulled towards and held against the multi groove drive belt pulley system. The clutch is now held together as one unit and the compressor's rotational speed matches the engine speed. When the A/C system is either being cycled or is no longer required the current is switched off and the magnetic force created in the clutch field coil depletes (Fig. 2.21). The drive plate disengages through the help of return springs and the compressor stops.

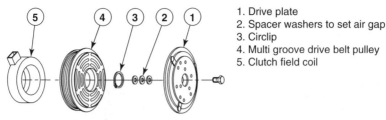

1. Drive plate
2. Spacer washers to set air gap
3. Circlip
4. Multi groove drive belt pulley
5. Clutch field coil

Figure 2.20 Compressor clutch pack
(reproduced with the kind permission of Ford Motor Company Limited)

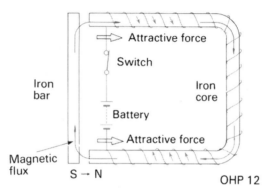

Figure 2.21 Current and magnetic flux flowing through a clutch pack
(with the agreement of Toyota (GB) PLC)

Classification of compressor clutches
Compressor clutches are classified according to their shape (Fig. 2.22).

- F type and G type – crank type compressors.
- R type and P type – swash plate and through vane type compressors.

Clamping diodes
When an electromagnetic actuator is de-energised a voltage spike can be produced within the controlling circuit. This is due to the collapse of the magnetic field. This is often viewed on oscilloscope waveforms and referred to as sawtooth signals. Clamping diodes are used to filter out the voltage spikes and protect driver circuits. If a clamping diode becomes faulty the electrical system may become unstable due to excessive electrical interference. A clamping diode reduces the peak voltage of a signal used to switch any electromagnetic type actuator. For more information (see section 3.3).

Externally regulated compressor (ECC only)
This is a compressor with an externally regulated compressor stroke. The swash plate angle is varied by a Pulse Width Modulated (PWM) controlled valve which is regulated by the Electronic Climate Control (ECC). This enables the compressor to be precisely regulated between minimum and maximum output. For this reason, the magnetic clutch can be dispensed with. Compressor oil supply is guaranteed, even when the compressor is operating at minimum output.

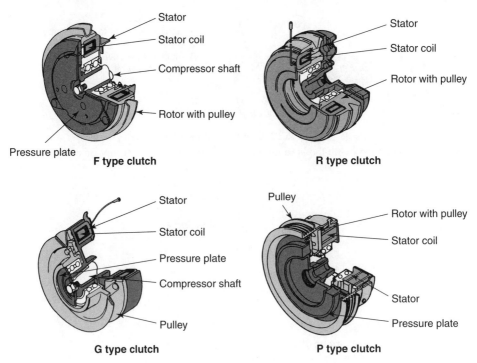

Figure 2.22 Clutch pack configurations
(with the agreement of Toyota (GB) PLC)

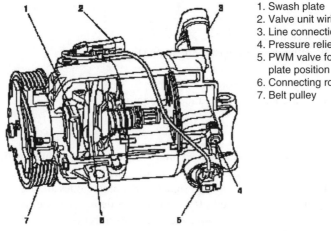

1. Swash plate
2. Valve unit wiring harness multiplug
3. Line connections
4. Pressure relief valve
5. PWM valve for adjusting the swash plate position
6. Connecting rod plate
7. Belt pulley

Figure 2.23 Externally regulated swash plate compressor
(courtesy of Vauxhall Motor Company)

Advantages:

- Reduced overall weight – *weight saving thanks to omission of magnetic clutch.*
- Minimum fuel consumption thanks to precise control of compressor output according to respective requirements.
- No operating jolts when delivery quantity increases.

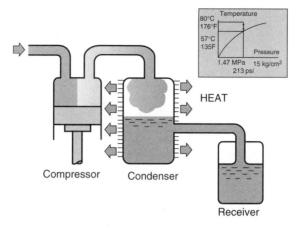

Figure 2.24 Simplified diagram of condenser operation
(with the agreement of Toyota (GB) PLC)

Compressor oil

Compressor oil is necessary for the lubrication of the moving parts of the compressor. The compressor oil lubricates the compressor by dissolving in the refrigerant and circulating throughout the refrigeration circuit. For this reason the recommended oil and quantity must be used. Because the refrigerant oil for R12 is mineral oil and R134a is synthetic PAG oil they must not be mixed. Only the correct oil must be used matching the compressor type and refrigerant.

2.2 The condenser

The function of the condenser is to act as a heat exchanger to dispel the heat energy contained in the refrigerant. Superheated vapour enters the condenser at the top and subcooled liquid leaves the condenser at the bottom. The condenser must be highly efficient but as compact as possible.

The pressure and temperature has been raised by the compressor. There is a need to lower the temperature of the heat laden refrigerant to change it back into a liquid enabling it to act as a cooler again later in the system. To do this the refrigerant flows into the condenser as a vapour and gives off heat to the surrounding area and most of the refrigerant (depending on system load) condenses back into a liquid which then flows into the receiver/drier.

The condenser is located at the front of the vehicle where strong air flow through its core can be achieved when the vehicle is in motion. To aid the removal of heat when the vehicle is stationary or at a low speed the condenser is fitted with a single or double fan system. Shrouds are often used to direct the air flow over the surface of the condenser.

Condenser design

The ideal condenser should have no pressure drop between the inlet and the outlet. Condensers are generally made from aluminium to prevent any chemical reaction between the metal and refrigerant/oil mixture. They are generally constructed with tubes and fins. Tubes to carry the

Figure 2.25 Location of a condenser

refrigerant and fins to increase the surface area in contact with the outside air. Their shapes vary and include:

Serpentine fin type

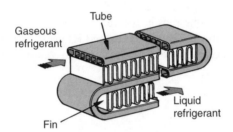

Figure 2.26 Serpentine fin type (a flat tube condenser) (with the agreement of Toyota (GB) PLC)

Tube and plate type

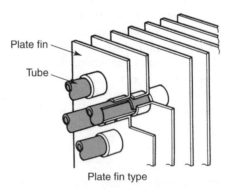

Figure 2.27 Flat extruded aluminium with multiple tubes within its structure (with the agreement of Toyota (GB) PLC)

Tube and plate and fin types have been used for a number of years and can be seen on R12 systems but are not generally used on R134a systems. This type of condenser can also be back-flushed to remove any foreign particles within the system.

Parallel flow type (a flat tube condenser)

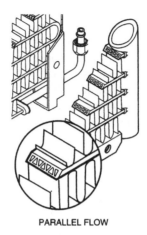

PARALLEL FLOW

Figure 2.28 Parallel flow condenser with flat tube

Parallel flow condensers are very efficient, the condenser breaks up the flow into tiny streams enabling it to transfer heat more rapidly. This type of condenser cannot be flushed and if it becomes blocked can only be replaced.

Condenser refrigerant flow

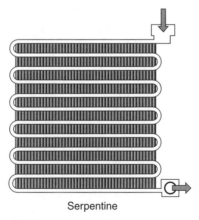

Serpentine

Figure 2.29 Serpentine flow from top to bottom

The flow of refrigerant is either serpentine (Fig. 2.29) or parallel flow (Fig. 2.30). Serpentine flows through the tube(s) evenly eventually condensing while following the same path.

Parallel flow allows the path of the refrigerant to go vertically as well as horizontally across the condenser. Parallel flow is considered to be the more efficient layout. The key to the design is the header tanks/manifolds fitted to the sides of the core allowing the flow to break up into small streams.

Dual condenser

This layout includes a condenser with an integrated multi-flow condenser (Fig. 2.31) and a gas/liquid separator (modulator) – a subcool cycle. In simple terms these are two condensers stacked on top of each other with a receiver drier called a modulator between the two. They are generally used in vehicles with large internal space and cooling requirements. Refrigerant flows through the first parallel flow condenser and then into the modulator as a liquid. Any gaseous refrigerant which the first condenser was unable to condense will travel with the liquid refrigerant to the subcooling portion of the condenser to ensure only liquid refrigerant flows to the FOV (Fixed Orifice Valve) or receiver-drier.

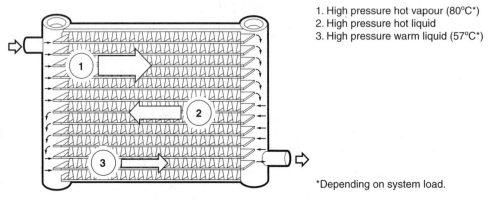

1. High pressure hot vapour (80°C*)
2. High pressure hot liquid
3. High pressure warm liquid (57°C*)

*Depending on system load.

Figure 2.30 A parallel flow condenser
(reproduced with the kind permission of Ford Motor Company Limited)

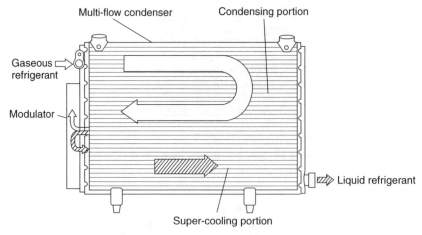

Figure 2.31 Parallel flow with subcondenser (dual condenser)
(with the agreement of Toyota (GB) PLC)

The condenser is the point within the A/C system which is used to remove the unwanted heat. This makes it very important in the overall efficiency of the system. As the latent heat of condensation is transferred to the air stream, the refrigerant vapour makes the necessary change into liquid.

2.3 The receiver-drier/accumulator

The receiver-drier

A receiver-drier (Fig. 2.32) is used when a thermostatic expansion valve metering device is used and is positioned between the condenser and the thermostatic expansion valve.

The function of the receiver drier is as follows:

1. To ensure the system is free from dirt preventing any excessive wear or premature failure of components.
2. To remove moisture from the refrigerant ensuring no ice can form on any components within the system which may cause a blockage and to ensure no internal corrosion can form.
3. To act as a temporary reservoir to supply the system under varying load conditions.
4. To allow only liquid refrigerant to flow to the expansion valve.
5. To act as a point for diagnostics (sight glass sometimes fitted).

Operation

Refrigerant entering the receiver-drier in an ideal system will be in a liquid state. If the system is under heavy load the condenser may have not been efficient enough to completely condense the refrigerant. This means a small amount of vapour may be present. Liquid and vapour can enter the receiver through the inlet where it will separate. Liquid will fall to the bottom of the receiver while vapour will rise to the top. The outlet is connected to a receiver tube internally which has a pickup point at the bottom of the receiver where the filter is positioned.

The refrigerant flows through the desiccant and filter to get to the outlet pickup tube. This ensures that only liquid refrigerant flows to the expansion valve.

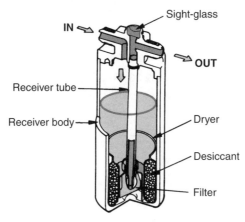

Figure 2.32 Receiver-drier
(with the agreement of Toyota (GB) PLC)

Sight glass

A sight glass is not used on R134a systems because the refrigerant has a cloudy appearance in its normal condition. (For more information see section 4.5.)

Fusible plug (pressure relief valve)

If adequate cooling of the condenser is not provided or the cooling load becomes excessive a fusible plug will release the excessive pressure to reduce the possibility of a burst pipe. The fusible plug contains metal which has a melting point of approximately 100–110°C. Fusible plugs or 'melt bolts' are used generally on older models or as fail safe measures on new ones.

The main safety device for high pressure control is the high pressure switch (for more information, see section 2.7 for control switches).

Pressure relief valve

The problem with a fusible plug is that once the metal melts you lose the full contents of the refrigerant. This is harmful to the environment and can cause damage to vehicle components. To reduce the pressure in the system the pressure relief valve ejects only enough refrigerant to reduce the pressure in the system. Pressure relief valves are fitted to compressors and sometimes to receiver-driers.

Accumulator

An accumulator is used when an FOV metering device is used. The accumulator is fitted between the evaporator and the compressor.

The function of the accumulator is as follows:

1. To ensure that the refrigerant leaves the accumulator as a vapour and not a liquid state for the compressor to induce.

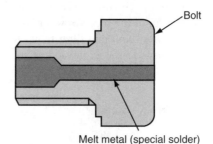

Figure 2.33 Fusible plug
(with the agreement of Toyota (GB) PLC)

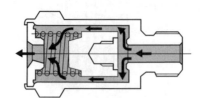

Figure 2.34 Pressure relief valve showing ejection route
(with the agreement of Toyota (GB) PLC)

2. To ensure it is free from dirt to stop any excessive wear or premature failure of components.
3. To remove moisture ensuring no ice can form on any components within the system which may cause a blockage and to ensure no internal corrosion can form.
4. To act as a temporary reservoir to supply the system under varying load conditions.
5. To add lubricating oil for system components like the compressor.
6. Often to house the low pressure switch/sensor.

Operation

The refrigerant enters the accumulator from the evaporator in liquid/vapour form. It enters through the inlet (3) creating a vortex and flowing around the cap (4). The refrigerant passes through the desiccant where it is cleaned and moisture is removed. The vapour collects under the cap (4) where it is extracted through the outlet. During extraction it passes through a U tube in which it is mixed with oil from the small bleed hole (6). This bleed hole allows very small quantities of liquid refrigerant mixed with lubricating oil (3%) to flow with the refrigerant vapour to the compressor. Because the liquid refrigerant vapour is in such small quantities there is no danger of compressor damage.

> Note – a desiccant is situated inside the receiver to absorb moisture and filter particles. Zeolite is used for R134a and silica gel is used for R12.

The receiver-drier or accumulator must be replaced under any of the following circumstances:

- If the A/C system is opened to moisture for more than 3 hours then the receiver/accumulator must be replaced.
- Under service conditions – every two years.
- Compressor seizure or any possibility of foreign matter in the system.
- Excessive moisture in system causing icing.

> Note – often the receiver/accumulator also contains a contrast medium (dye) which dissolves in the compressor lubricant. If there is a leak in the refrigerant circuit, the compressor lubricant, together with the contrast medium, will escape and can be detected with a special UV lamp.

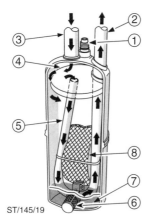

1. Connection for low pressure switch
2. Outlet to the compressor
3. Inlet from the evaporator
4. Cap
5. U-tube with bleed hole
6. Filter screen
7. Refrigerant oil
8. Desiccant

ST/145/19

Figure 2.35 Accumulator
(reproduced with the kind permi.ssion of Ford Motor Company Limited)

2.4 The expansion valve/fixed orifice valve

To control the amount of refrigerant volume flowing through the evaporator a metering device must be used. The function of the metering device is as follows:

- To separate the high pressure and low pressure side of the system.
- To meter the volume of refrigerant and hence the cooling capacity of the evaporator.
- To ensure that there is superheated refrigerant exiting the evaporator.

Currently there are two main categories of metering device used, a Thermostatic Expansion Valve (TXV or TEV) and a Fixed Orifice Valve (FOV). The pressure drop across the evaporator is used to determine which type of valve is the most appropriate. Simple air-conditioning systems will generally use only one of these metering devices. Dual air-conditioning systems may use both a TXV and an FOV within the system.

Superheat

Superheat within an expansion system is very important. It ensures that the entire liquid refrigerant inside the evaporator has been vaporised from a liquid. It is generally measured by the difference in the boiling point of the refrigerant between the inlet and the outlet of the evaporator which can be as high as a 10°C difference. Evaporators, depending on their design characteristics, operate within different amounts of superheat and expansion valves are matched against this by adjusting the superheat valve spring tension.

Thermostatic expansion valve

There are also some variations in the design of a TXV.

- Internally equalised thermostatic expansion valve.
- Externally equalised expansion valve.
- Box or H-valve type.

Operation of externally equalised expansion valve

The externally equalised expansion valve (Fig. 2.36) has the benefit of having refrigerant pressure from the outlet of the evaporator acting directly on the underside of the diaphragm of the TXV. This arrangement overcomes the problem of sensing the pressure drop across the evaporator. The TXV has a pintle valve which is controlled by a diaphragm.

There are three pressures acting on the diaphragm of a TXV valve:

1. Refrigerant within or exiting the evaporator applies pressure under the diaphragm (P_e).
2. Spring pressure (called superheat spring) applying pressure against a pintle valve (ball and seat). This pressure is applied under the diaphragm (P_s).
3. Pressure from the expansion of the liquid within the heat sensing bulb via the capillary tube (P_f) which is above the diaphragm.

The amount of refrigerant that is allowed to flow into the evaporator is determined by the vertical movement of the diaphragm and valve. This is controlled by the difference in pressure above the diaphragm P_f and below the diaphragm which is the sum of P_e and P_s.

When the load on the system is high and additional cooling is required the temperature exiting the evaporator will be high.

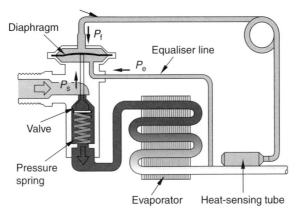

Figure 2.36 Externally equalised expansion valve
(with the agreement of Toyota (GB) PLC)

This high temperature will be transferred to the heat sensing tube (remote bulb) which contains an inert liquid (different to the refrigerant) which expands within the bulb increasing in pressure due to its fixed volume. The increase in pressure is transferred via the capillary tube to the top of the diaphragm enabling the diaphragm to overcome the combined pressure of the superheat spring (P_s) and the refrigerant exiting (P_e). The valve opens further and allows an increase in volume of refrigerant to flow through the evaporator to cope with the increase in load. This will lower the temperature of the refrigerant exiting the evaporator. This reduction in temperature will be transferred to the heat sensing tube and cause the liquid inside to contract. This will reduce the pressure inside the capillary tube and diaphragm. This reduction in force P_f above the diaphragm will allow the superheat sensing spring pressure P_s and refrigerant exiting the evaporator pressure P_e to force the diaphragm upwards reducing the orifice size and thus reducing the volume of refrigerant.

This continual adjustment and balancing of forces controls the volume of refrigerant to ensure that a superheated refrigerant exists within and exiting the evaporator. The manufacturers adjust expansion valves to ensure they operate under superheat conditions. Factory settings must not be tampered with and the correct TXV must always be used as a replacement for a faulty one.

Some expansion valves have a small V cut into the valve seat to ensure that if the valve is closed a small quantity of refrigerant may still flow around the system if a fault occurs or they are seized closed.

Operation of internally equalised expansion valve

The operation of the internally equalised expansion valve is almost the same as the externally equalised expansion valve. The difference is the loss of the benefit of sensing the pressure of the refrigerant as it leaves the outlet of the evaporator. The pressure is sensed on the inlet of the evaporator or just before, using an internal drilling inside the housing of the TXV where refrigerant entering can apply pressure on the underside of the diaphragm. This means that the pressure drop across the evaporator is unknown with this type of valve.

Box or H-valve type

The box type expansion valve (Fig. 2.37) includes the pressure sensing and temperature sensing functions of the externally equalised TXV but has no external tubes (capillary or pressure

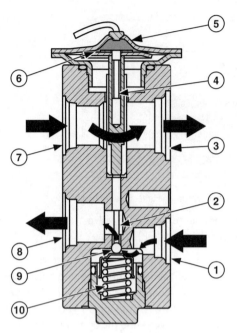

1. Valve inlet from receiver drier
2. Valve slide
3. Valve outlet to compressor
4. Temperature sensitive element
5. Diaphragm head (filled with liquid)
6. Pressure sensitive diaphragm
7. Valve inlet from the evaporator
8. Valve outlet to the evaporator
9. Ball valve
10. Superheat spring

Figure 2.37 Box type expansion valve
(reproduced with the kind permission of Ford Motor Company Limited)

sensing). This is achieved by using two passages – one entering the evaporator and one exiting the evaporator.

Liquid refrigerant enters the block valve housing (1). The orifice is very small and there is a large pressure drop on the other side of the ball valve (8). The liquid and small amount of vapour refrigerant (flash gas) enters the evaporator. The liquid/vapour will boil due to the drop in pressure and the vapour will become saturated. The saturated vapour will continue to flow through the evaporator becoming superheated. The superheated refrigerant will enter the box type expansion valve at point (7). The valve position is controlled by the temperature and pressure of the superheated vapour entering it from the evaporator at point (7). If the temperature of the refrigerant is high due to a high cooling demand then this additional heat will be transferred to the sensing element and diaphragm head. The liquid will expand and apply pressure downwards on the ball valve and superheat spring enlarging the orifice and allowing an increased volume of refrigerant to flow through the evaporator. The increased volume of refrigerant will provide additional cooling capacity and the refrigerant temperature entering the valve at point (7) should reduce. When this occurs the sensing element will transfer the reduced temperature to the diaphragm head which will then contract and reduce the pressure applied to the ball valve and superheat spring. The pressure of the superheated vapour is sensed directly under the diaphragm head via internal drillings. If the pressure is high then the diaphragm head flexes upward reducing the pressure applied to the ball valve and superheat spring. This causes the orifice to reduce in size thus reducing the volume and pressure of the refrigerant flowing into the evaporator. If the pressure applied to the diaphragm is low then the diaphragm will flex downward and apply additional pressure to the ball valve and superheat spring. This will increase the size of the orifice and allow a larger volume of refrigerant at a higher pressure to flow through the evaporator.

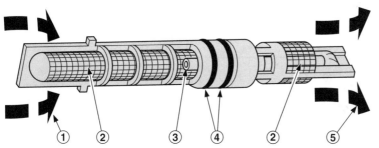

1. Inlet on high
   pressure side
2. Filter screen stop
   contaminants
3. Fixed orifice
   made of brass
4. O-ring to stop
   refrigerant passing
   the valve
5. Outlet to evaporator
6. Outlet on low
   pressure side

Figure 2.38 Fixed orifice valve (FOV)
(reproduced with the kind permission of Ford Motor Company Limited)

Fixed orifice valve (FOV)

The fixed orifice valve (Fig. 2.38) is positioned inside the high pressure line between the condenser and the evaporator. This can often be seen when looking at a system with an FOV fitted due to the increased diameter of the aluminium tube where the valve is situated. This is also where you can sense the change in temperature from hot to cold.

The volume of refrigerant flowing through the orifice is determined by the pressure of the refrigerant and the size of the orifice. The orifice is fixed so the only way to control the volume is to vary the pressure of the refrigerant (1.5–2.9 bar low pressure side). This type of system tends to cycle (switch on/off) the compressor on a regular basis to vary the volume/pressure of refrigerant matched against the load on the system. The compressor is cycled by a cycling switch located in the low side of the system. A way to improve this system is to fit a variable orifice valve or smart valve or a variable displacement compressor.

As previously stated the function of the FOV is to divide the system creating a high pressure and low pressure side and to meter refrigerant into the evaporator.

Liquid refrigerant flows from the condenser at high pressure to the inlet of the FOV. The refrigerant travels through filter screens to remove any foreign particles and then to the calibrated fixed diameter tube inside the plastic valve body. Dependent on the pressure of the refrigerant a small volume flows through the orifice into the evaporator inlet. This changes the refrigerant into a low pressure liquid ready to boil off inside the evaporator. The orifice size of the FOV is matched to deliver the correct volume of refrigerant under maximum cooling loads. This causes problems under light loads when the evaporator is in danger of becoming flooded due to not enough heat available to vaporise the entire refrigerant inside the evaporator. This is the reason why an accumulator must be fitted between the outlet of the evaporator and the compressor ensuring no liquid refrigerant flows to the evaporator, except through the oil bleed inside the accumulator which is used to deliver oil to the compressor (5%).

The FOV comes in a range of orifice sizes (1.19 mm–1.70 mm) which are colour coded to enable them to be matched to the manufactured vehicle (see Table 2.2). Special tools are required to remove the valve (Chapter 5).

Variable orifice valve (Smart VOV)

The fixed orifice valve (Fig. 2.39) is limited due to its design which often leads to poor performance and excessive cycling of the compressor (if fixed displacement). To improve the performance a VOV can be fitted which can respond to the change in pressure of the refrigerant and thus vary its orifice size to compensate. It is important for the correct operation of the

Table 2.2 Colour and orifice size chart

Colour code	Orifice size (diameter)
Black/Blue	1.70 mm
Red	1.57 mm
Orange	1.45 mm
Brown	1.35 mm
Green	1.19 mm

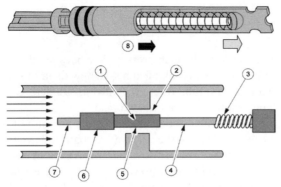

1. Metering pin with different diameters
2. Valve aperture
3. Spring
4. Aperture approx. 1.6 mm, e.g. when driving on highways and motorways
5. Aperture approx. 1.3 mm, e.g. when driving in town
6. Aperture approx. 1.1 mm at idle
7. Aperture approx. 1.4 mm
8. Installation orientation

Figure 2.39 Variable orifice valve
(reproduced with the kind permission of Ford Motor Company Limited)

A/C system to create the correct pressure distribution throughout the system. This must be matched to the load on the system. This will promote the correct condenser and evaporator performance.

The valve assembly contains two ports, a fixed and a variable port. The fixed port is matched against the required refrigerant flow for high vehicle speed. The variable port reacts to the temperature applied to it from the refrigerant exiting the condenser. The temperature is sensed by a bimetallic spring which expands and contracts with changes in refrigerant temperature. At idle and low compressor speeds the orifice size can be reduced. This creates a larger pressure differentiation across the valve and reduces the volume of refrigerant and evaporator flooding. When either the output of the compressor due to vehicle speed or the external load changes the orifice size can respond to this change. The main benefits which are published by aftermarket companies that sell VOV state that the valve is suited to drivers who spend a great deal of time idling in traffic or moving slowly. Performance is improved through reduced compressor loading and greater cooling. There are also published data on improved fuel economy due to reduced compressor loading. Currently VOV are not in widespread use by Original Equipment Manufacturers (OEM). Whenever working on an A/C system the OEM must always be contacted for advice when deviating/changing the original set-up of a system (replacing an FOV with a VOV).

The pressure in the high pressure line pushes the metering pin against a spring with a force which depends on the type of driving (idle, town or motorway/highway). The aperture of the valve is increased or decreased by the different diameters of the metering pin and thus matched to the driving situation.

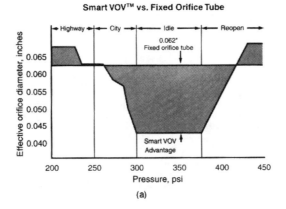

Figure 2.40 Performance chart of a VOV: (a) changes in diameter of VOV compared to FOV under different driving conditions; (b) lower temperature of discharged air from the air vent over a greater variation in driving conditions
(courtesy of Nartron Corporation)

2.5 The evaporator

The evaporator is very similar in construction to a condenser. An evaporator will have a serpentine, tube and fin or parallel type construction. The function of an evaporator is to provide a large surface area to allow the warm often humid air to flow through it releasing its heat energy to the refrigerant inside.

The refrigerant by this time will have just had a large pressure and temperature (Fig. 2.41) drop coming through the expansion/fixed orifice valve causing it to want to boil and just requiring the heat energy to do so. The evaporator absorbs the heat energy from the air flowing over its surface. The energy is transferred and the refrigerant reaches saturation point. At this point the refrigerant can still absorb a small amount of heat energy. The refrigerant will do so and become superheated. The superheated refrigerant will then flow to the compressor (TXV system) or accumulator (FOV system).

The evaporator is extremely cold at this stage and any moisture in the air flowing through the evaporator will adhere to the evaporator's surface. The water droplets on the surface help

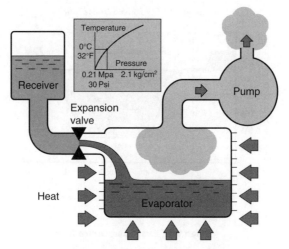

Figure 2.41 Operation of an evaporator
(with the agreement of Toyota (GB) PLC)

Figure 2.42 Location of an evaporator inside the heater box

clean the incoming air by trapping dirt and foreign particles. The humidity content is also reduced so cleaner drier air is delivered to the interior of the vehicle. This improves the comfort level especially in high humid conditions and allows perspiration to evaporate more quickly. The moisture drips off the evaporator's surface into a drain which directs it to the outside of the vehicle via a duct.

Dehumidified air is very effective for window defogging due to a large number of passengers in a vehicle and/or humid conditions.

Design

The design of an evaporator is based on the size, shape, number of tubes and fins and the number of rows. This is to maximise the flow rate and surface area. The evaporator is tested for the maximum amount of heat and moisture which can be removed by the evaporator within a given period.

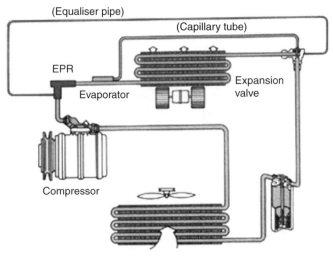

Figure 2.43 EPR (Evaporator Pressure Regulator) valve
(with the agreement of Toyota (GB) PLC)

2.6 Anti-frosting devices

The evaporator pressure regulator (Figs 2.43 and 2.44) is mounted between the outlet of the evaporator and the compressor inlet (suction side). The valve regulates the pressure inside the evaporator to prevent icing. If the pressure drops below a certain threshold (196 kPa (28.4 psi)) then the valve closes to restrict the flow of refrigerant and increase the pressure inside the evaporator. This is to stop the evaporator temperature from reaching 0°C due to the relationship between temperature and pressure.

When the cooling load is high the vapour pressure of the refrigerant in the evaporator is high. The valve fully opens and the refrigerant flows unobstructed to the compressor. The valve operation is based on a spring bellows which expands and contracts with changes in refrigerant pressure. This device virtually eliminates the need for the compressor to cycle on and off to regulate the temperature of the evaporator.

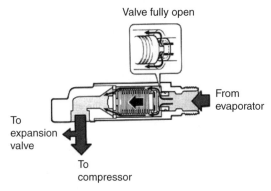

Figure 2.44 EPR valve detail
(with the agreement of Toyota (GB) PLC)

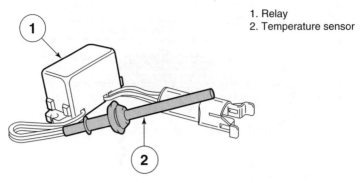

1. Relay
2. Temperature sensor

Figure 2.45 De-ice switch
(reproduced with the kind permission of Ford Motor Company Limited)

The only problem with this type of valve is the possibility of oil starvation due to the reduced flow of refrigerant.

De-ice switch (TXV system)

A de-ice switch (Fig. 2.45) is a temperature sensor and relay built as one unit. The temperature sensor is fitted to the evaporator's fins and measures the temperature of the evaporator surface using an NTC type temperature sensitive resistor. This sensor sends the information to the relay in the form of a voltage drop and when approaching the freezing point of water (0°C) the current to the compressor clutch is interrupted to increase the pressure in the evaporator and avoid the surface water freezing. With a system threshold of 1°C at the surface, the relay will turn the compressor off. Once the surface increases to 2.5°C, the compressor, via the relay, will be switched on again.

Evaporator sensor

An evaporator sensor is used on a system which generally uses electronic control as opposed to electrical control. These systems use the temperature sensor to feed a voltage reading to a control module which is programmed to use the data and compare it to stored data in its memory. The sensor is generally an NTC type sensor which means it has a negative temperature coefficient. This means that with an increase in temperature the resistance of the sensor will reduce. This will affect the current flowing through the sensor and the voltage across the sensor. The module can apply this data and when the corresponding voltage is sensed, associated with 1°C, the module will disengage the compressor clutch via a relay to stop the evaporator from freezing. The compressor clutch will be re-engaged when the temperature of the evaporator reaches 2.5°C. The sensor is generally fitted to the evaporator fins for direct measurement. Advanced vehicle electronic systems may use a multiplex wiring system to transfer the information digitally (see Chapter 3, section 3.2).

Compressor cycling switch (FOV system)

If the temperature in the evaporator approaches 0°C then icing will occur on the surface of the evaporator due to the water droplets which form. This will reduce the volume of air flowing through the evaporator and reduce the efficiency of the system. To prevent this a compressor cycling switch (Fig. 2.46) is fitted to the accumulator to deactivate the compressor clutch when a specific pressure is reached. For example, when 1.5 bar is reached on the low pressure side of

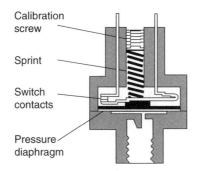

Figure 2.46 Compressor cycling switch

the system the switch contacts will open and directly or indirectly interrupt the current flowing to the compressor clutch. Once the pressure rises to 2.9 bar then the switch contacts will close again and the compressor will re-engage. Often the cycling switch is also used as a low pressure switch in case there is a system leak and the refrigerant escapes. In this case the switch will open at 1.5 bar and disengage the compressor clutch ensuring the compressor is not damaged due to no refrigerant flow (refrigerant carries the lubricant to lubricate the compressor).

2.7 Basic control switches

Pressure switching devices

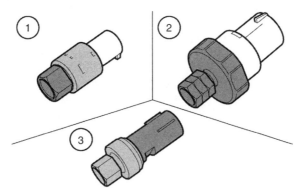

Figure 2.47 Low (1) dual (2) and high (3) pressure switches
(reproduced with the kind permission of Ford Motor Company Limited)

Low pressure switch
Reacts to the pressure in the low side (suction side) of the system, generally the accumulator and disengages the compressor clutch if the pressure drops below approximately 1.5 bar. On an FOV system the low pressure switch and cycling switch are often the same device.

Dual pressure – high pressure switch and condenser fan switch
Two pressure-sensitive switches are contained in the high pressure switch. One of these switches acts as a safety switch to prevent excessive system pressure. The second switch

switches the condenser auxiliary fan on to its second setting at approximately 20.7 bar and off again at 17.2 bar. This switching process improves the performance of the system in cases of excessive heat.

High pressure switch

A single normally closed pressure switch de-energises the compressor if excessive high pressure exists within the A/C system, approximately 30–35 bar. The switch is normally positioned in the high side of the system.

Trinary switch

Three pressure sensitive switches are integrated into a trinary switch. A low pressure switch which creates an open circuit thus removing the current flowing to the compressor if the system pressure drops below approximately 1.4 bar. This could be caused by a refrigerant leak or natural discharge over a number of years. A high pressure switch operates at a pressure of approximately 30 bar which again removes the A/C compressor's current in the event of a system blockage anywhere in the system. The third switch is used for high speed operation of the condenser fan aiding the removal of heat. This operates at approximately 18 bar. The switch is positioned on the high pressure side on TXV controlled systems. If FOV is used then a cycling switch is incorporated into the low side with a dual pressure switch on the high side. Modern A/C systems are replacing switches with a single pressure sensor.

Pressure sensors

The pressure sensor contains two metal plated ceramic discs mounted in close proximity. The disc located closest to the pressure connection is thinner and bends when subjected to pressure. By this means, capacitance between the metal plating of the discs is changed based on the pressure. A circuit integrated in the sensor converts the capacitance to an analogue voltage (see Table 2.3).

For more information, see Section 3.2.

Table 2.3 Results from a pressure sensor

Bar	Voltage
2	0.4
10	1.6
15	2.35
20	3.1
25	3.85

3 Air-conditioning electrical and electronic control

This is the largest chapter of the book and is due to the complexity of understanding which exists to be able to comprehend and predict the operation of an electrical circuit and electronic control. The chapter is broken down into basic electrical theory, circuit operation, sensors and actuators, testing and three case studies. Readers can be selective with the text depending on electrical experience; if unsure start at the beginning.

Fundamental to understanding and predicting circuit operation is applying the simple rules of electrical theory, knowledge of sensors and actuators, ability to use the correct test equipment to gain the right results and understanding how to read wiring schematics. A number of wiring diagrams have been broken down into simple electrical circuits with meters and are presented using Crocodile Clips to aid their explanation. This software can be purchased via the internet (www.crocodileclips.com) and can help the learner experiment with electrical circuits.

The aim of this chapter is to:

- understand basic electrical theory;
- understand the electrical laws behind the operation of parallel, series and compound circuits;
- be able to read a wiring diagram and predict circuit operation;
- understand the underpinning principles behind sensors, actuators and electronic control;
- allow the learner to apply the knowledge gained from the last two chapters in understanding the operation of the systems and appropriate methods of testing.

3.1 Electrical principles

Anything which occupies space and has mass is called *matter*. Matter is composed of small particles called atoms. Atoms combine to form molecules. One molecule of water contains two hydrogen atoms and one oxygen atom; this combination is indicated by the chemical symbol H_2O. In 1808 John Dalton noted that matter consists of individual atoms which are identical and unchangeable. Humphry Davy and Michael Faraday proved that electricity and matter are closely related. Their research into electrolysis helped to establish that electricity is atomic in character and that the atoms of electricity are a part of atoms of matter.

Examination of a hydrogen atom under a powerful microscope shows that it consists of a positively charged *proton* at the nucleus (centre) and one *electron*, which orbits around the proton at high speed. The negatively charged electron is the lightest particle known at this time: its mass is only about 0.0005 of the mass of the hydrogen atom. The complete atom is uncharged because the charge on the nucleus is balanced by the charge on the electron.

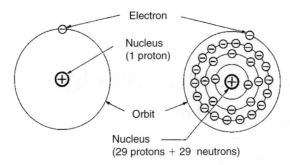

Figure 3.1 Hydrogen and copper atoms (with the agreement of Toyota (GB) PLC)

Other materials have different combinations of electrons and protons. A copper atom has 29 electrons; these move in four different orbits around the atom's nucleus. The central region consists of 29 protons and 29 neutrons. The neutrons have no electrical charge.

If an electron is removed from the atom, then the atom and its material become positively charged. If an electron is added to the atom then it and the material will become negatively charged.

The difference in electric charge creates a potential difference or voltage. The size of the voltage is directly proportional to the charge difference. The charge on a electron is tiny, so to create a useful voltage a very large number of charges have to be separated.

The magnitude of the force produced between two electrically charged bodies was studied by the French scientist Coulomb in 1775. To honour his work, the SI unit of electrical charge is called the *coulomb* (C) and 1 C is the charge equivalent to about 6×10^{-18} electrons.

That is 6 million million million electrons (6 000 000 000 000 000 000). The charge on a single electron is about 1.6×10^{-19} C, or 0.00000000000000000016C.

The actual voltage created by a particular charge separation depends on the area over which the separated charges are spread, the separation distance and the electrical properties of the material between the charges.

Moving charges apart to create a potential difference requires energy because each negative electron is being pulled away from a positive nucleus against the force of attraction. This is similar to the way that energy is needed to lift an object, like a bottle of water, up in the air against the attractive force of gravity (which you can think of as a mass separation). We know that energy can't be destroyed – so what happens to the energy used in creating the charge separation? The answer is that it is transformed into electrical potential energy. Again this is very much like the way that the energy used to lift the water upwards against the pull of gravity is transformed into gravitational potential energy; if the water is poured out the potential energy is transformed into kinetic energy (energy of movement) as the water accelerates downwards. This kinetic energy can, in turn, be harnessed to do useful work, for example in a hydroelectric power station. The amount of energy released depends on the difference in height between the start and end of the drop and the amount of water. In a similar way the potential energy of separated electrons can be released to do useful work. If the electrons are allowed to return to the positively charged atoms from which they were separated, their potential energy will be released and can be harnessed. To do this requires an electric circuit made of a conductor that allows the electrons to return. To summarise, separated charges create an electric potential difference (a voltage). The size of this potential difference is measured in volts. When measuring the size of an electric potential difference you are measuring the difference in electric potential between two points. Separating charges requires energy. When separated charges are reunited, the energy is released and can be harnessed for useful work.

There are a number of ways that charges can be separated:

1. Electrostatics.
2. Electromagnetism.
3. Chemistry.
4. Photoelectric effect.
5. Piezoelectric effect.

> Note – for more information on electromagnetism and the photoelectric and piezoelectric effects see section 3.2.

Electrostatics

When different types of materials rub against each other, electrons can be moved from one material to another. This causes a charge separation, but one that isn't generally useful; the amounts of separated charge are generally too small to do anything practical with. However, engineers have recently been investigating whether the rubbing of your clothes as you walk could be used to generate enough usable electricity to charge a mobile phone.

For example: when a glass rod is rubbed with a silk cloth the surfaces of both the glass rod and the silk cloth become charged with electricity. The charges within the glass rod and silk cloth do not move unless the glass rod and silk cloth are brought closer together or are connected by a conducting substance. If two glass rods are hung on threads and both rubbed with a silk cloth and both rods moved closer to one another they will repel each other. If one rod is brought near to the silk cloth then the rod and cloth will attract each other.

Electromagnetism

When an electron is moved through a magnetic field a force is created that pushes the electron. An electron in a conducting wire is pushed to one end of the wire, so creating a voltage along the wire. If you can arrange for the wire to keep moving in the magnetic field then you will have a continuous force on the electrons; as you release the potential energy by allowing electrons to flow back through a circuit and do useful work, more electrons will take their place and you will have a permanent potential difference. An electrical generator is, put simply, a device that keeps wires moving through a magnetic field to produce a permanent and usable potential difference.

Chemistry

Some chemical reactions can be used to separate charges. A chemical cell (often called a battery – though usually a battery is a collection of cells) contains chemicals that react together; as the potential difference caused by the reaction rises it suppresses the reaction, for example in common batteries the reaction stops occurring at a voltage of 1.5 V. When the battery is used the voltage drops slightly and the reaction starts up to maintain the voltage – until all of the chemicals have reacted and the battery is 'dead'. Fuel cells use a rather different chemical technology to generate charge separation. In a basic fuel cell, hydrogen and oxygen are combined to create water; this is a chemical reaction that releases energy and the cell is designed to use this energy to separate charges creating a voltage of about 0.7 V. The main difference between a fuel cell and a battery is that, when the hydrogen in a fuel cell runs out, the cell can be refuelled by adding more hydrogen (the oxygen used comes from the air).

Photoelectric

Certain materials release electrons when light falls on them. This is called the photoelectric effect and can be used to separate charges. Although first explained by Albert Einstein a hundred years ago, it is only recently that the cost of 'solar cells' based on the closely related photovoltaic effect has become low enough to make them a realistic cost-effective way of generating electricity. See section 3.2 for examples of photoelectric sensors used to measure light intensity (sun load sensor).

Piezoelectric

When you apply pressure to some crystalline materials you get a charge separation within the crystal. The resulting voltage across the crystal can be very high. This effect is commonly used to create a spark to light a flame in hobs and gas fires and also in 'crystal' microphones. In most situations it would be impractical to generate a constant source of separated charges using this effect as this would mean constant rapid pressurising of the crystal. However, as with charge separation through rubbing, engineers have recently been investigating whether the constantly changing pressure in your shoes as you walk could be used to generate enough usable electricity to charge the battery in a mobile phone or MP3 player. See section 3.2 for examples of piezoelectric sensors used for refrigerant pressure sensing.

Conductors and conductivity

Copper is a very good conductor so this metal is used as a material for cabling. The electrons are not tightly bonded to their nucleus and drift at random from one atom to another. Materials that have a number of free electrons make good conductors of electricity because little effort is needed to persuade charge separation.

Conductivity takes a somewhat different form when it comes to semiconductor material. For electronic applications, semiconductor materials are grown into crystalline structures which are given conductive properties by virtue of the impurities (or dopants) which are added. In their purest form (i.e. without dopants), the base semiconductor materials form crystalline lattices which become very stable by sharing electrons among the constituent atoms. In this pure state, the material is not very conductive.

Insulators

Insulation materials have no loosely bound electrons, so movement of electrons from one atom to the next is very difficult.

Semiconductors

A semiconductor is a material having an electrical resistance higher than conductors like copper but lower than that of insulators like glass or rubber. A semiconductor has a range of properties which can be used within the electronics industry. The two most common semiconducting materials are germanium and silicon. In their pure state they are not suitable for practical use as semiconductors. For this reason they must be doped, which is achieved through adding impurities to enhance their effectiveness.

N type
An N type semiconductor consists of a silicon (Si) or germanium (Ge) base or 'substrate' which has been doped with a slight amount of arsenic (As), antimony (Sb) or phosphorus (P)

in order to provide it with many free electrons (i.e. electrons which can easily move through the silicon or germanium to provide electrical current).

P type

A P type semiconductor, on the other hand, consists of a silicon or germanium substrate that has been doped with gallium (Ga), indium (In), or aluminium (Al) to provide 'holes', which can be thought of as 'missing' electrons, and hence as positive charges flowing in a direction opposite that of free electrons.

Movement of electrons

All electrons have a potential energy, given a suitable medium in which to exist they move freely from one energy level to another. This movement from one energy level to another is called *current flow*. Using conventional flow (+ to −), electrical energy is considered to move from a point of high potential to a point of lower potential. Unfortunately it was found that electron flow is the opposite direction since the negatively charged electron is attracted to the positive potential. This text uses the conventional current flow of positive to negative to aid the understanding of electronic principles. A battery and generator are both capable of producing a difference in potential between two points. The electrical force that gives this increase in potential difference at the source is called the *electromotive force*. The terminals of a battery and generator are called positive (+) and negative (−) and these relate to the higher potential and lower potential respectively. Moving charges apart to create a potential difference requires energy because each negative electron is being pulled away from a positive nucleus against the force of attraction. Separating a coulomb of charge to create 1 V requires 1 joule of energy. This is a small amount of energy when compared to kW/h = 3.6 million joules of energy which is the measurement used in domestic homes.

The unit of potential is the *volt*, named after the Italian scientist Volta:

1 volt = 1 joule per coulomb

Summary of potential difference (volt)

Separated charges create an electric potential difference (a voltage). The size of the potential difference is measured in volts. When measuring the size of an electric potential difference you are measuring the difference in electrical potential of two points.

Electromotive force (emf)

An emf is the driving influence which causes the current to flow. The emf is not actually a force but is related to the energy expended during the passing of a unit charge through the source. The emf is related to energy conversion. As a charge passes through a source of electrical energy, work is done on them. The emf of the source is the work done per coulomb on the charges. A typical car battery has an emf of 12 volts which means that 12 joules of work is done on each coulomb which passes through the battery.

Ampere (A)

The ampere is the unit of electric flow and is the rate of electron movement along a conductor. A coulomb is the quantity of electrons so when one coulomb passes a given point in one second the current is one ampere:

1 ampere = 1 coulomb per second

Various standards have been used in the past to define the ampere; nowadays it is defined in terms of the force between conductors. If two parallel conductors are placed a given distance apart, when current is passed through the conductors, a force is set up which is proportional to the current.

Watt (W)

The watt is a unit of power and applies to all branches of science. It is equivalent to work done at the rate of one joule per second. (One joule is the product of the force, in newtons, and the distance, in metres, $1\,J = 1\,Nm$.)

A power of $1\,W$ is developed when a current of $1\,A$ flows under the 'pressure' or potential difference (pd) of $1\,V$:

$$\text{Power} = \frac{\text{energy supplied (joules)}}{\text{time (seconds)}}$$

$$= \frac{\text{voltage} \times \text{current} \times \text{time}}{\text{time}}$$

$$\text{Watts} = \text{volts} \times \text{amperes}$$

Voltage drop

This represents the energy used by free electrons while engaged in current flow. Passive devices such as resistors create voltage drops in a circuit because they absorb some of the energy and dissipate it in the form of heat. The volt drop across a series circuit is equal to the emf of the circuit.

Ohm's law

In 1826 Ohm discovered that the length of wire in a circuit affected the flow of current. He found that as the length was increased, the current flow decreased and from these findings he concluded that:

Under constant temperature conditions, the current in a conductor is directly proportional to the pd between its ends.

This statement is known as *Ohm's law*.

Resistance

From Ohm's law the relationship between potential difference (V) and current (I) is expressed:

$$R = \frac{V}{I}$$

In this case, R is a constant which changes only when the length, cross-sectional area or temperature of the conductor is altered. Evidence shows that the value of the constant is related to the conductor's opposition to current flow, so R is called the resistance and given the unit name of ohm:

The ohm is the resistance of a conductor through which a current of one ampere flows when a potential difference of one volt is across it.

Rearranging the expression $V/I = R$ is often called Ohm's law. Rearranged it gives:

$V = IR$ or volts = amperes $\times$ ohms

If two of these values are known, the third can be calculated, so this expression has a number of practical uses. The resistance of a conductor is affected by its material, temperature and dimensions. The SI unit of resistivity of a material is the ohm-metre.

The rule can be expressed using the following equation:

$$R = \rho \frac{I}{A}$$

R = resistivity in ohms of material
L = length of the current path through the conductor in metres
A = cross-sectional area of the conductor in metres squared
ρ = constant of proportionality between R and I/A

Most metals increase their resistivity when the temperature is raised, so these metals are said to have a *positive temperature coefficient*. Conversely if the resistivity decreases with an increase in temperature then the material has a *negative temperature coefficient*.

Circuit resistors

Motor vehicle circuits normally consist of a number of resistors/electrical consumers controlled by switches and connected to an electrical supply. Consumers can take many forms; they can be a lamp, motor, solenoid or be a part of some other energy-consuming device.

A basic understanding of circuit behaviour may be helped if the effect of resistors on voltage and current flow is considered. A temperature sensing circuit will be discussed to aid the explanation of series resistors/consumers. The temperature sensing circuit is used to sense cabin temperature; the temperature of the air inside the vehicle. This is measured by an air temperature sensor and monitored by an onboard computer (climate electronic control unit). The information is used to control cooling rate by adjusting air distribution and blower speed.

Electrical consumers may be connected in series, in parallel, or a combination of both. The diagram in Figure 3.2 shows the temperature sensing circuit in series. The circuit is a typical temperature sensing circuit used by most manufacturers. Figure 3.3 shows the circuit modelled using Crocodile Clips software.

Circuit operation (Fig. 3.3)

The Semi Automatic Climate Control (SATC) module will have a power supply from the battery and the ignition switch. The SATC A58 has internal circuitry that can reduce the power supply from 12 volts (14 V when engine is running) to 5 volts. Using DIN standards this changes the wire code from 30 (battery feed) to 8 (sensor signal). This allows differentiation between a battery supply of 12/14 V and a fixed 5 V supply from a computer. The current then flows to the fixed resistor situated inside the SATC. The resistor causes a potential difference to occur across itself. This potential difference depends on the current flowing through the resistor and the number of consumers it is in series with. Because electrical circuits are dynamic the other resistor in the circuit will affect the current flow. The current will reduce through the circuit and two potential differences will be created across both consumers – fixed resistor inside the SATC module and the temperature sensitive resistor called a thermistor. Although the thermistor changes with temperature, when calculating the current and pd across each resistor the value is fixed as illustrated $+25°C$ is $500\,\Omega$. For a full explanation on the operation of a thermistor see section 3.2.

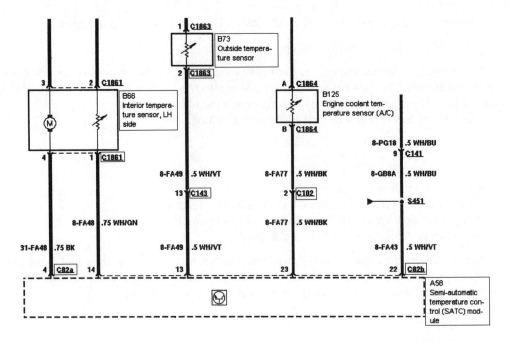

Figure 3.2 A wiring diagram of a temperature sensing circuit on an SATC system Ford Cougar (reproduced with the kind permission of Ford Motor Company Limited)

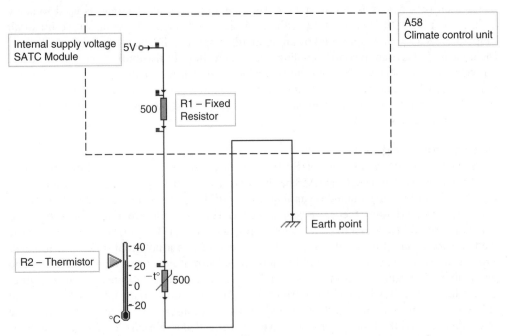

Figure 3.3 Simplified diagram of a temperature sensing circuit to aid the explanation of consumers in series (courtesy of Crocodile Clips)

Resistors in series

Placing two resistors in series (Figure 3.3) means that the *full current* must pass through each resistor in turn. When they are connected in this end-to-end manner, the resistance of the two resistors is the sum of their values, so:

$$R = R1 + R2$$

$$R = 500 + 500 = 1000\,\Omega \text{ or } 1\,k\Omega$$

Assuming the resistance of the cables is negligible, by applying Ohm's law the current flow can be calculated:

$$V = IR$$

$$I = \frac{V}{R} = \frac{5\,V}{1000\,\Omega} = 0.005 \text{ ampere or } 5\,mA$$

In this case, a current of 0.005 amps will pass through each resistor and around the whole circuit.
Inserting an additional resistor of 500 Ω in series with the other two would give a total resistance of:

$$R = R1 + R2 + R3$$

$$R = 500 + 500 + 500 = 1500\,\Omega \text{ or } 1.5\,k\Omega$$

And a current flow of:

$$V = IR$$

$$I = \frac{V}{R} = \frac{5\,V}{1500\,\Omega} = 0.00333 \text{ ampere or } 3.3\,mA$$

Voltage distribution (resistors in series)

Figure 3.4 shows two resistors in series with two ammeters and two voltmeters, positioned to measure the current and pd.

It will be seen that the ammeter is fitted in series with the resistors. This means that all current flowing in the circuit must pass through the ammeter, no matter where the meter is inserted in the circuit.

Energy is expended driving the *current* through a *resistor* so this causes the *potential to drop*. The voltage drop V1 and V2 (decrease in pd) when the current passes through R1 and R2 can be found by applying Ohm's law:

$$V = I \times R1 = 0.005 \text{ amps} \times 500 \text{ ohms} = 2.5 \text{ volts}$$

A voltmeter connected to R1 will measure a volt drop. In this case it will register 2.5 volts so the pd across R2 will also be 2.5 volts as well because the total voltage is 5 V.

$$5\,V \text{ (Internal supply)} - 2.5\,V\ (R1) = 2.5\,V\ (R2)$$

This is because V1 and V2 will always add up to the voltage supply in a series circuit, 5 volts in this case.

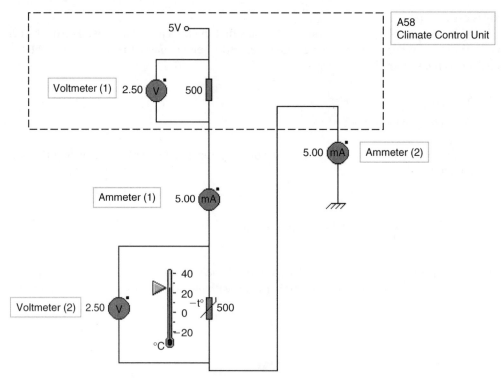

Figure 3.4 Voltage distribution
(courtesy of Crocodile Clips)

By moving the voltmeter around the circuit, the voltage distribution can be determined:

Voltmeter position	Supply	Potential difference (V)
		5
V1	0.005 A × 500 Ω	2.5
V2	0.005 A × 500 Ω	2.5

Rules of a series circuit

In a series circuit the resistance is added together to calculate the total resistance; this is due to the constant current flow through the resistors, unlike parallel circuits where the current flow splits due to branches. The volt drop across resistors in a series circuit will always add up to the supply voltage. The current flow in a series circuit is constant throughout the whole circuit.

What happens if the temperature of the thermistor changes? Figure 3.5 shows the thermistor has now changed its resistance due to a change in its temperature from 25°C to 35°C, this could be due to the vehicle being stationary (parked) for a period and has picked up heat due to solar radiation (UV through the glass area). This increase in temperature has caused the resistance to reduce from 500 Ω to 333 Ω. This means that the material of the sensor is NTC – a negative temperature coefficient, as the temperature increases the resistance reduces.

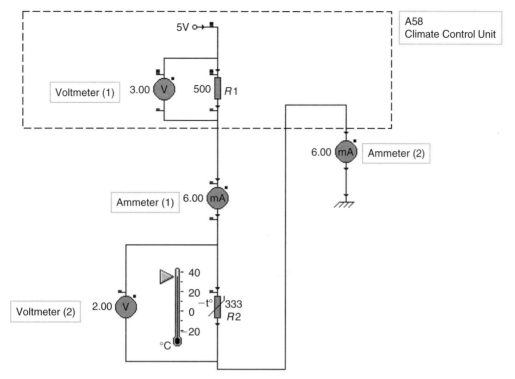

Figure 3.5 Change in circuit values
(courtesy of Crocodile Clips)

It will be seen that the ammeter fitted shows an increase in current flow due to the overall reduction in the total resistance of the circuit.

Ohm's law:

$R = R1 + R2$

$R = 500\,\Omega + 333\,\Omega = 833\,\Omega$

And a total current flow of:

$V = IR$

$I = \dfrac{V}{R} = \dfrac{5\,\text{V}}{833\,\Omega} = 0.006$ ampere or 6 mA

Voltage drop across $R1$:

$V = I \times R1 = 0.006$ amps $\times$ 500 ohms $= 3.0$ volts

A voltmeter (2) connected as shown will measure a volt drop. In this case it will register 3.0 volts. The pd across $R2$ will be 2.0 volts because the total voltage is 5 V:

$5\,\text{V} - 3\,\text{V} = 2\,\text{V}$

This is because $V1$ and $V2$ will always add up to the voltage supply in a series circuit, 5 volts in this case.

By moving the voltmeter around the circuit, the voltage distribution can be determined:

Voltmeter position	Supply	Potential difference (V)
		5
$V1$	$0.006\,\text{A} \times 500\,\Omega$	3.0
$V2$	$0.006\,\text{A} \times 333\,\Omega$	2.0

As the temperature acting on the thermistor changes the component's internal resistance changes and so does the potential difference across it. This causes the pd across the fixed resistor to change also. The SATC senses this change and can alter the fan speed and flap position to maintain a certain cabin temperature.

Resistors in parallel

Connecting resistors in parallel ensures that the pd applied to each resistor is the same. Current flowing through the ammeter is shared between the two resistors and the amount of current flowing through each resistor will depend on the resistance rating of each resistor. To aid the explanation of parallel resistance an A/C heater control module circuit will be used.

Lighting circuit for the A/C display

Figure 3.6 shows the heater control module A128. C41 is the connector code which is referenced so you can obtain a pictorial representation of the connector and its wiring connections. As with all DIN standard diagrams the power enters the top of the diagram and the grounds (earth) are based at the bottom. Standard symbols are used for bulbs, LEDs (Light Emitting Diodes) etc. The heater control module is a part of the facia panel and includes switches for the A/C system and heater control. These switches must be illuminated so they can be seen in the event of poor light conditions, at night time and to tell the operator the A/C has been switched on.

C41 pin 4 has the code *29S-LA28 .35 OG/GN*. Using DIN standards, *29* indicates voltage supplied at all times overload protected. *S* indicates 'switched', this generally means that voltage will be supplied all the time once it has been switched on and it will have a fuse in the circuit. *LA* is the code for the 'Switches and Instruments'. *.35* is the cross-sectional area, which is comparatively low, so it is a low current carrying wire. *OG* is orange, which is the functional base colour for 29 voltage at all times and *GN* is green, which is the tracer colour for identification purposes only. This means that the wire will be live when the headlight/sidelights of the vehicle are switched on (Switches and Instruments). It is overload protected (fused) and supplies the heater control module with power for illumination via bulbs and LEDs. For more information on wiring identification, see section 3.7.

Only the lamps will be included in this explanation due to their being positioned in parallel with each other. Lamps generally have a hot and cold resistance value. The lamps when cold will have a low resistance value, for example $3\,\Omega$. When the filament gets hot inside the bulb its resistance value increases to approximately $100\,\Omega$. This is known as a PTC (Positive Temperature Coefficient), this is because the resistance increases with an increase in temperature. To calculate the resistance of the circuit we will use the hot resistance value of the lamp and not the cold. If the hot resistance of the lamp is unknown then it can be calculated using Ohm's law by knowing the voltage applied to the lamp and the current flowing through it.

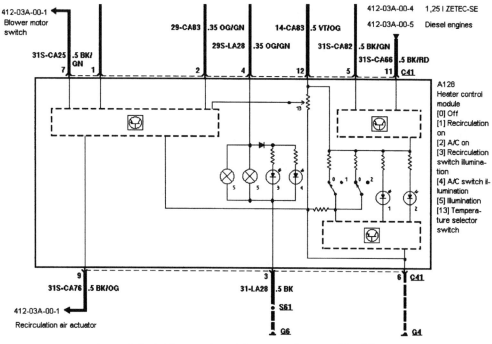

Figure 3.6 A/C control module wiring diagram (Ford Fiesta)
(reproduced with the kind permission of Ford Motor Company Limited)

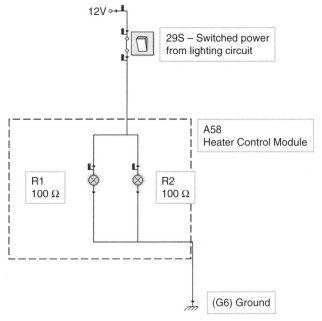

Figure 3.7 Simplified diagram of heater control module illumination
to aid explanation of conductors in parallel
(courtesy of Crocodile Clips)

Resistors/consumers in parallel (Figs 3.7 and 3.8)

Applying Ohm's law to find the current flow in each lamp:

$$I = \frac{V}{R}$$

Current flow through $R1$ = $\dfrac{12\,\text{V}}{100\,\Omega}$ = 0.12 amperes

Current flow through $R2$ = $\dfrac{12\,\text{V}}{100\,\Omega}$ = 0.12 amperes

Total current flow through the whole circuit (both branches) = 0.12 A + 0.12 A = 0.24 A

When calculated in this way, the current through each branch circuit can be found easily. Also it is possible to find the total resistance of the circuit. Applying Ohm's law:

$$R_T = \frac{V}{I}$$

$$R_T = \frac{12\,\text{V}}{0.24\,\text{A}} = 50\,\Omega$$

Note – this method does not work with compound circuits.

The *equivalent resistance* of a number of resistors $R1$, $R2$, $R3$ can also be found by applying either formula 3.1 or 3.2:

$$R_T = \frac{1}{R1} + \frac{1}{R2} = \frac{1}{100} + \frac{1}{100} = \frac{100}{2} = 50\,\Omega \qquad (3.1)$$

or

$$R_T = \frac{R1 \times R2}{R1 + R2} = \frac{100 \times 100}{100 + 100} = \frac{10\,000}{200} = 50\,\Omega \qquad (3.2)$$

Formula 3.1 works with any amount of resistors in parallel including resistors in compound circuits.

Formula 3.2 works with two resistors in parallel including resistors in a compound circuit.

Rules of a parallel circuit

The potential difference (voltage drop) across each consumer is the same value as the supply voltage, for example if the supply is 12 V then every consumer in the circuit will have a potential difference of 12 V (theoretically). The total current is equal to the sum of all the branch currents, i.e. if the current splits into two branches then the total current of the circuit is the sum of the two branch currents.

Compound circuit (series/parallel circuit)

This circuit uses resistors, or consumer devices, connected so that some parts are in series and other parts are in parallel (Figure 3.9).

When calculating the current flow in these circuits, it is imagined that the parallel resistors are replaced by a single resistor of equivalent value so as to produce a series circuit.

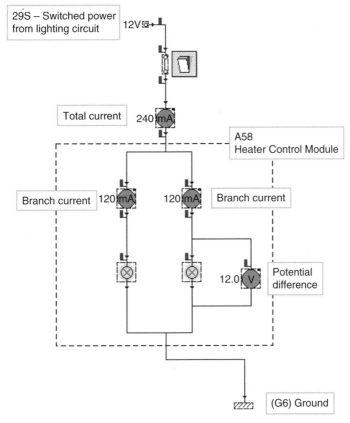

Figure 3.8 Simplified diagram of heater control module – circuit 1
(courtesy of Crocodile Clips)

Not all resistance values are known so they must be calculated. The resistance of the diode and LED is calculated by knowing the voltage drop across it and the current flowing through it:

$$R5 = \frac{0.727\,\text{V}}{0.056\,\text{A}} = 13\,\Omega$$

$R2$ and $R4 = 330\,\Omega$ (the resistors in series with the LEDs)

$$R1 \text{ and } R3 = \frac{2.03\,\text{V}}{0.028\,\text{A}} = 72.5\,\Omega \text{ each}$$

So Figure 3.9 has changed to Figure 3.10.

Important note – because $R2$ and $R3$, $R4$ and $R5$ are in series they must be added together before they can be considered to be placed in parallel together.

The total resistance of the circuit is found by:

$$\begin{aligned}
R_\text{T} &= \frac{1}{R1} + \frac{1}{R2} \\
R_\text{T} &= \frac{1}{402\,\Omega} + \frac{1}{402\,\Omega} = 201\,\Omega
\end{aligned} \tag{3.3}$$

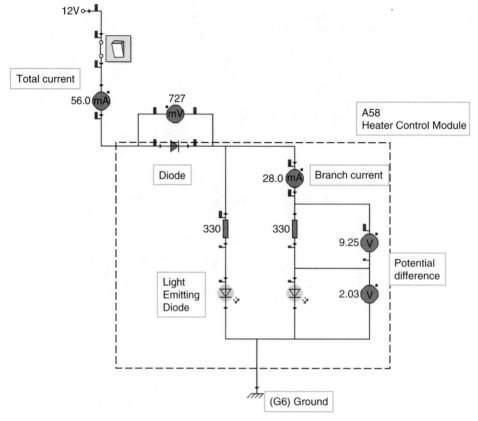

Figure 3.9 Simplified diagram of heater control module – circuit 2
(courtesy of Crocodile Clips)

201 Ω is the resistance of the parallel circuit so you must now add the series part of the circuit for the total resistance R_T:

$$R_T = 201\,\Omega + 13\,\Omega = 214\,\Omega$$

Total current flow in the circuit is calculated by $I = \dfrac{V}{R}$

$$I = \frac{12\,\text{V}}{214\,\Omega} = 0.056\,\text{amps}$$

We must remember that amps split when going through a parallel circuit, 0.056 amps is the total amp flow $I = I1 + I2$.

Consideration of the current flow through $R1$ and $R2$ shows that 0.056 A is the total current flowing through both resistors. This current divides according to the resistor values – the higher the value, the smaller the current:

$$\text{Current flow through } R1 = \frac{R2}{R1 \times R2} \times I$$

This result may be verified by adding the two branches together $I1 + I2 = I$.

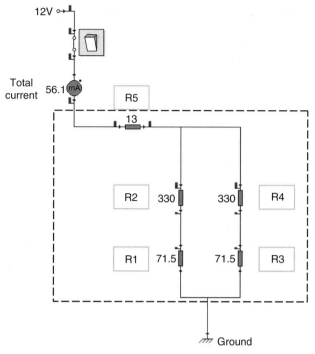

Figure 3.10 Simplified diagram of heater control module – circuit 3
(courtesy of Crocodile Clips)

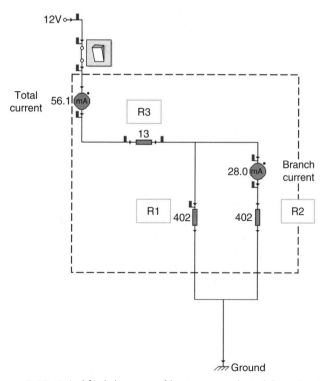

Figure 3.11 Simplified diagram of heater control module – circuit 4
(courtesy of Crocodile Clips)

Another method to calculate the current flow through the branches $I1 + I2$ is to calculate the volt drop across each resistor starting with the resistors in parallel. Once the volt drop and the resistance value are known the current can be calculated.

Resistance calculation using circuit equivalents

If a network of resistors ends with just two terminals as shown in Figure 3.12 then an equivalence resistor can be placed in the circuit to represent a combination of two previous resistance values.

Figure 3.12: Because the two 400 Ω resistors are in parallel the current passing through the resistors will divide and because the values are equal the current will divide equally. The equivalence resistance therefore is half the single value due to half the current flow, i.e. 200 Ω.

Figure 3.13: You can see that a 200 Ω resistor has replaced the two 400 Ω resistors. The 300 Ω and 200 Ω resistors are in series so the resistance can be added.

Figure 3.14: The two resistors have now been added and form two 500 Ω resistors in parallel. This means that the current will split and form two paths. The equivalent resistance is half due to the equal split.

Figure 3.15: Finally, the 250 Ω resistor represents the equivalent resistance of the whole circuit replacing the 500 Ω, 300 Ω, 400 Ω and 400 Ω resistors.

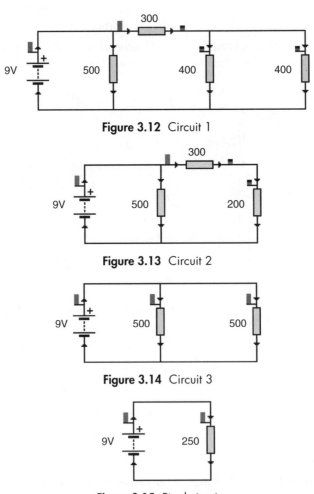

Figure 3.12 Circuit 1

Figure 3.13 Circuit 2

Figure 3.14 Circuit 3

Figure 3.15 Final circuit

3.2 Sensors and actuators

This section describes a range of sensors and actuators used in common A/C systems. It describes the underpinning theory associated with the sensors and actuators and provides practical examples of their application, and also includes wiring diagrams and measured data in various forms representing the output of a sensor or control signal used for an actuator. The sensors and actuators included are as follows:

Sensors:

- Temperature – NTC and PTC.
- Sun load – photovoltaic and solar cells.
- Pressure – capacitive, strain gauge, piezoelectric.
- Position – linear and rotary potentiometer.
- Speed – Hall effect and inductive.
- Humidity – capacitive.
- Pollution – metal oxide semiconductor (MOS).

Actuators:

- Solenoids – relays, coolant valves.
- Motors – permanent magnet.
- Stepper motors – DC permanent magnet, variable reluctance, hybrid.

Temperature sensor

Theory of operation
A temperature sensor uses direct measurement. It is a temperature dependent, non-linear semi-conductor which is termed 'thermistor'. It comes in various packages and is fitted in a range of environments. Temperature sensors are generally NTC or PTC type semiconductor material.

NTC temperature sensor

These are generally made from different oxides of metals such as iron, cobalt, nickel copper and zinc. The sensor is extensively used in temperature measurement. The NTC (Negative

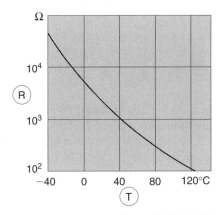

R – Resistance in ohms

T – Temperature in Celsius

Figure 3.16 NTC graph
(reproduced with the kind permission of Ford Motor Company Limited).

Temperature Coefficient) thermistor, decreases in resistance as it increases in temperature. The sensors are generally arranged in a potential divider circuit with a fixed resistor. The fixed resistor will be inside the control module to allow the module to measure a variation in voltage or current, see Figure 3.2.

Sensor monitoring locations:

1. Engine coolant temperature sensor – direct measurement of coolant temperature.
2. Ambient temperature sensor – fitted near the condenser/front bumper.
3. Evaporator temperature sensor – fitted to the evaporator to measure its surface temperature informing a module to control the compressor in case icing occurs.
4. Cabin temperature sensor – integral to the A/C controls panel and/or fitted to the air ducting.

°C (F°)	kohm	Volt
0 (32)	28	3.7
10 (50)	18	3.2
20 (68)	12	2.7
30 (86)	8.3	2.3
40 (104)	5.8	1.8
50 (122)	4.1	1.5
60 (140)	3.0	1.2
70 (158)	2.2	0.9
80 (176)	1.6	0.7
90 (194)	1.2	0.5

Figure 3.17 NTC sensor voltage and resistance values at varying temperatures, used for comparison

Sensor failure

If a temperature sensor like the exterior temperature sensor fails, then often the A/C module will apply a fixed temperature value, e.g. 10°C. If the interior temperature sensor fails then a temperature of approximately 24°C is fixed. This allows the system to operate with a fault.

An example application is an air vent outlet and interior temperature sensor with integrated fan (Figure 3.18)

Sensor monitoring location:

1. Built into the heater controls or
2. Fitted to the air ducts.

The sensor unit has a small electric fan (Fig. 3.18) to prevent distortion of the temperature due to heat build-up. The outlet and cabin temperature sensors are the most important values for the A/C control unit. A comparison is made between the cabin temperature and the desired temperature in order to decide if the mixed-air temperature should be increased or decreased. The cabin temperature is adjusted with respect to the outside temperature so that the temperature experienced matches the set temperature. As the difference between selected temperature and adjusted temperature increases, the interior fan speed will increase. When the ignition is turned off, the suction fan on some systems may continue to operate for up to 10 minutes. This decreases the risk of incorrect temperature value when restarting after a short stop.

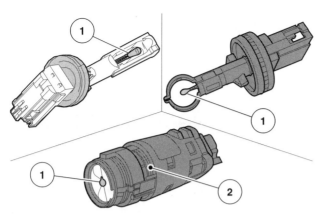

1. NTC temperature sensor
2. Fan body

Figure 3.18 NTC sensor integrated fan assembly
(reproduced with the kind permission of Ford Motor Company Limited)

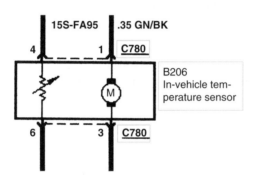

Figure 3.19 Sensor wiring diagram
(reproduced with the kind permission of Ford Motor Company Limited)

Sensor wiring diagram

The motor is powered by the A/C control module (Fig. 3.19). Some systems will have the unit built inside the module while other systems have a separate unit near the centre vent area to accurately measure interior temperature. The motor will be powered by a 12 V feed and the sensor by a 5 V feed. The sensor will be a part of a potential divider circuit enabling the measurement of a volt drop across a fixed resistor situated inside the module.

The technician is able to measure the volt drop across the sensor by placing a voltmeter or oscilloscope (trend measurement) across pins 4 and 6.

Waveform

The plot in Fig. 3.20 shows the volt drop across an NTC sensor. As the temperature applied to the sensor by the A/C system increases the resistance reduces with a corresponding reduction in volt drop. A DC analogue signal can also be measured using the Min and Max selection of an oscilloscope. This enables the measurement of the total variation in voltage to be known which can be compared to a known value for analysis.

Signal checks:

1. Measuring the temperature of the air flowing past the sensor to carry out a comparison should be carried out.

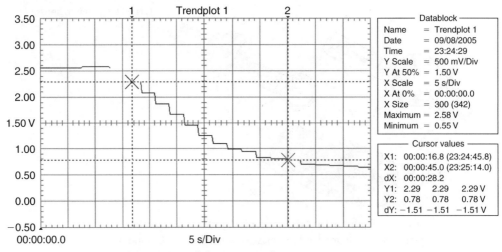

Figure 3.20 Left face vent outlet temperature sensor
(courtesy of Fluke)

2. The waveform should correspond to an NTC or PTC graph (Figs 3.16 and 3.22).
3. The peak voltage should be referenced to the specification of the sensor.
4. Voltage transitions should be steady and reflect a change in temperature.

The measured temperature of the face vent outlet temperature at the start of the waveform is 2.58 V at 4°C and 0.55 V at 60°C.

Data logger

TIS 2000 - Snapshot Upload/Display [(5) 2005, Vectra-C / Signum, ECC (Electronic Climate Control)]

File Applications Sessions Configuration View Snapshot Options Playback Help

Left Floor Outlet Temperature	21.2 °C	Left Interior Outlet Temperature	2.71 V
Left Floor Outlet Temperature	1.90 V	Right Interior Outlet Temperature	6.4 °C
Right Floor Outlet Temperature	20.4 °C	Right Interior Outlet Temperature	2.74 V
Right Floor Outlet Temperature	1.94 V	Engine Coolant Temperature	87.0 °C
Left Interior Outlet Temperature	7.0 °C		

Figure 3.21 NTC ambient temperature reading using serial tester Tech 2
(courtesy of Vauxhall Motor Company)

PTC temperature sensor/element

These are generally made from barium titanate. A characteristic of a positive temperature coefficient sensor/element is that there is an increase in resistance due to an increase in temperature. The increase in temperature can also be caused by the current flowing through the sensor. A PTC element can be used as a protection device, overload protection due to a large heating effect from the current flow. When the current flows and the element increases in temperature its resistance also increases, this has a counter effect and reduces the current flow. This characteristic prevents the PTC element from overheating.

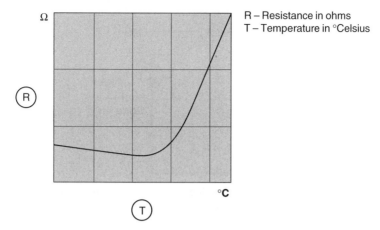

R – Resistance in ohms
T – Temperature in °Celsius

Figure 3.22 PTC graph
(reproduced with the kind permission of Ford Motor Company Limited)

Specification	Unit	Output
Power supply +	V	12
Max power consumption, jammed motor	mA	700–800
Normal power consumption at rated voltage	mA	200

Figure 3.23 Specification of a recirculation motor with PTC sensors

PTC material can also be used in electric heating systems, see Chapter 1, section 1.2.

Example recirculation flap controlled by PTC sensors
The air recirculation flap on some A/C systems is operated by a DC motor. The flap has only two positions (100% open or 0% closed). When the motor has turned the flap to one of the end positions, the current passing through the motor winding is limited by two PTC resistors built into the motor. Once the motor reaches its maximum position the current will increase and heat the PTC sensors. The sensors increase in temperature and reduce the current flow due to a corresponding increase in resistance.

Sensor monitoring location:

1. Fitted inside the recirculation motor which is attached to the recirculation flap.

Sensor failure
If a temperature sensor like the exterior temperature sensor fails, then often the A/C module works with a fixed temperature value, e.g. 10°C. If the interior temperature sensor fails then a temperature of approximately 24°C is fixed. This allows the system to operate.

Sun load (photovoltaic diode and solar cells)

Theory of operation
The sun load sensor (Fig. 3.26) can be a photoelectric diode which is designed to exploit the photovoltaic effect and measures sun intensity. If a voltage in reverse bias is placed across the diode and light is directed on it, a reverse current will flow (photovoltaic current). Exposing the diode to

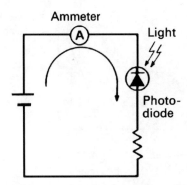

Figure 3.24 Electrical circuit showing diode and load resistor conducting in reverse bias (with the agreement of Toyota (GB) PLC)

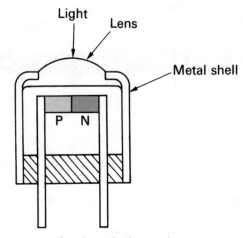

Figure 3.25 Construction of a photo diode *P* and *N* type semiconductor material (with the agreement of Toyota (GB) PLC)

light energy produces more electron hole pairs (free charge carriers) which pass the junction of the diode and increase current flow. The amperage of this current is proportional to light intensity. Sensor monitoring location:

1. The sensor is located above the instrument panel near the windshield.

Explanation of P and N type material (Fig. 3.25)
N type
Once impurities are added to a base material their conductive properties are radically affected. For example, if we have a crystal formed primarily of silicon (which has four valence electrons), but with arsenic impurities (having five valence electrons) added, we end up with 'free electrons' which do not fit into the crystalline structure. These electrons are loosely bound. When a voltage is applied, they can be easily set in motion to allow electrical current to pass. The loosely bound electrons are considered the charge carriers in this 'negatively doped' material, which is referred to as N type material. The electron flow in an N type material is from *negative to positive*. This is due to the electrons being repelled by the negative pole and attracted by the positive pole of the power supply.

P type
It is also possible to form a more conductive crystal by adding impurities which have one less valence electron. For example, if indium impurities (which have three valence electrons) are used in combination with silicon, this creates a crystalline structure which has 'holes' in it, that

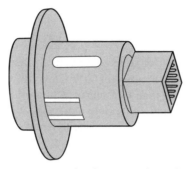

Figure 3.26 Sun load sensor (photo diode)
(reproduced with the kind permission of Ford Motor Company Limited)

State	Unit	Output
Light	kohm	0–1
Dark	kohm	>4.5

Figure 3.27 Example resistance value

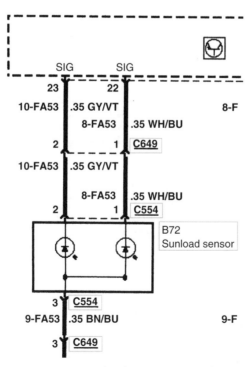

Figure 3.28 Sun load sensor wiring schematic
(reproduced with the kind permission of Ford Motor Company Limited)

is, places within the crystal where an electron would normally be found if the material was pure. These so-called 'holes' make it easier to allow electrons to flow through the material with the application of a voltage. In this case, 'holes' are considered to be the charge carriers in this 'positively doped' conductor, which is referred to as P material. Positive charge carriers are repelled by the positive pole of the DC supply and attracted to the negative pole; thus 'hole' current flows in a direction *opposite to that of electron flow*.

Sensor failure

If the sun load sensor fails, the A/C module works with a value corresponding to darkness.

Photo diode can be tested using a powerful 60 W light. This will alter the light intensity and vary the sensor's output. If more than one sensor is housed within the sun load sensor, for example dual zone A/C with left and right photo diodes, the light is placed towards one side to test the sensor and more towards the opposite side to test the other sensor. This can be monitored via a scope, diagnostic tester or volt/current meter. An amp clamp (low current sensing clamp) could be fitted to measure the current flow, access permitting.

Wiring diagram

The sensor (Fig. 3.28) receives a 5 V signal from the A/C module pins 1 and 2. The A/C module will also have resistors in series with both the left and right sensor. This will provide a potential divider circuit. The division will be based on the amount of light falling on the diode causing it to conduct in a reverse bias direction. Pin 3 is the sensor ground (earth).

A typical sensor that works on the photo diode principle has the following output:

    Dark signal voltage: 4.6 volts measured at pin 22/23
    Light signal voltage: 0.4 volts measured at pin 22/23

Data logger

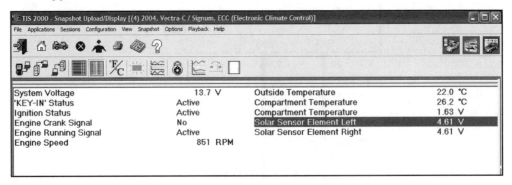

Figure 3.29 Double photo diode uploaded from a Tech 2 diagnostic tester
(dual zone climate system)
(courtesy of Vauxhall Motor Company)

Sun load sensor using solar cells

A solar cells converts light energy into electrical energy. The sensing element contains a PN junction, so the charge is carried separately in its electric field before proceeding to the metal contacts on the semiconductor's surface (see page 118 for an explanation of P and N type material. A DC electric (photoelectric voltage) is produced across the terminals; the electrical potential is between 0.5 and 1.2 V depending on the semiconductor material being used. Silicon is the most common material for solar cells.

State	Unit	Output
Dark	V	0
Illuminated with approx. 60 W bulb	V	approx. 0.5

Figure 3.30 Technical data from a solar cell sensor

Sun load sensor (infrared)

Theory of operation

Infrared radiation lies in the optical waveband and forms part of the electromagnetic spectrum. The infrared range is adjoined by visible light of long wavelengths. Every warm body releases infrared radiation. The higher the body's temperature the more energy is released in the form of infrared radiation. This makes the measurement of heat source and intensity possible with infrared sensors. The infrared sun load sensor is situated on the top of the dashboard and contains five infrared-sensitive elements: left, right, front, rear and top. It is supplied with power in the form of battery pulses from the climate control module. Voltage from the five sensor elements is sent consecutively to the climate control module. For the control module to be able to detect which sensor element is reporting, the transmission is preceded by the pulse pattern 5 V–0 V–5 V. The control module synchronises the solar sensor pulse transmission by sending short ground pulses at 25 Hz from pin 16. With the voltage from the five sensor elements, the control module calculates the solar intensity, azimuth and height. The values are used to calculate the current temperature at head height for the front seat passenger and driver.

Pressure sensor – capacitive type

Theory of operation

The pressure sensor contains two metal plated ceramic discs mounted in close proximity. The disc located closest to the pressure connection is thinner and bends when subjected to pressure. By this means, capacitance between the metal plating of the discs is changed based on the pressure. A circuit integrated in the sensor converts the capacitance to an analogue voltage. Capacitive measurements are based on the principle of a capacitor with the physical property of storing electrical charge. Its ability to store a charge is based on the distance between the two plates, size of the plates and the material it is made from. The distance is the main variable which determines the plates' charge difference. If the plates are far apart then this creates a low charge difference (Figure 3.31). If the plates are close together then the charge difference will increase proportionally (Figure 3.32). Linear pressure sensors allow for greater fan control.

Sensor monitoring locations:

1. A/C high pressure sensor is fitted to the A/C pipe work or receiver-drier.

Waveform

The trend plot (Fig. 3.33) shows the A/C system off, producing a voltage reading of 1.3 volts and a pressure of approximately 5 bar. The A/C was then switched on under light load; the voltage increased proportionally with an increase in pressure. The A/C stabilised at a voltage of approximately 2 volts which is approximately 12 bar. The load on the system was increased by setting the interior temperature to the lowest value on the climate control system (low <16°C) and increasing the blower speed to maximum. This caused the sensor voltage to peak at 2.2 volts, approximately 13.5 bar.

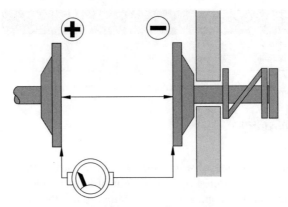

Figure 3.31 Low charge rate
(reproduced with the kind permission of Ford Motor Company Limited)

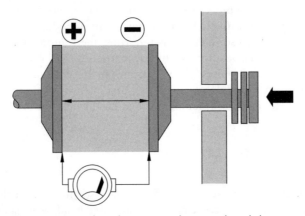

Figure 3.32 Higher charge rate due to reduced distance
(reproduced with the kind permission of Ford Motor Company Limited)

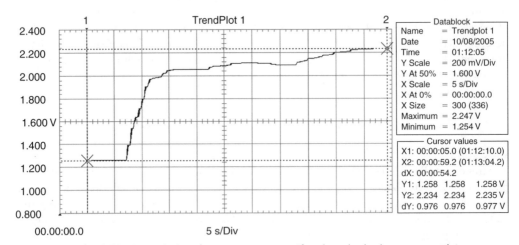

Figure 3.33 A trend plot of a pressure sensor (fitted on the high pressure side)
(courtesy of Fluke)

Pressure (bar)	Pd (volts)
2	0.4
10	1.6
15	2.35
20	3.1
25	3.85

Figure 3.34 Results from a pressure sensor

Signal checks:

1. Monitor a steady change in voltage level directly proportional to a change in pressure.
2. Check for glitches in the signal (drops to zero or rise to reference voltage).
3. Volt drop on the reference voltage and ground signal should not be greater than 400 mV (see Power-to-power test and Earth-to-earth test under section 3.3).

Wiring diagram

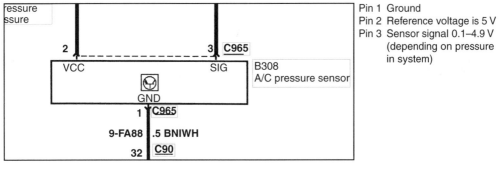

Figure 3.35 A/C pressure sensor

Data logger

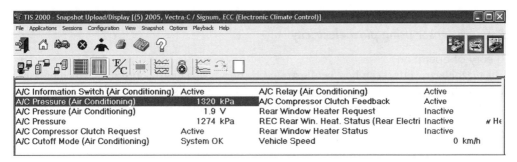

Figure 3.36 A/C pressure sensor voltage and interpreted pressure
uploaded from a Tech 2 diagnostic tester
(courtesy of Vauxhall Motor Company)

Pressure sensor using strain gauge

Theory of operation

A micro-machined membrane sensor with a strain gauge is used to measure pressure. A strain gauge is a group of resistors printed onto a membrane in the form of a bridge circuit (Wheatstone bridge). The bridge circuit Figure 3.37 shows four resistors (two series resistors in parallel with each other). The potentials at points U_M are equal when the bridge is balanced (both sets of parallel resistors have the same potential difference across them). When pressure is applied to the sensor's membrane the resulting mechanical force changes the electrical resistance of some of the resistors. The resistors are arranged so that two resistors increase in length and two resistors decrease in length due to deformation. This causes the bridge to become unbalanced giving a potential difference output across point U_M. The output of the potential difference is proportional to the deformation of the membrane. It is important to remember that resistance is proportional to length and inversely proportional to cross-sectional area. By changing the shape of the resistors the bridge becomes unbalanced. The sensor and the hybrid circuitry for signal processing are located together in a single housing.

Sensor monitoring locations:

1. A/C high pressure sensor is fitted to the A/C pipe work or receiver-drier.

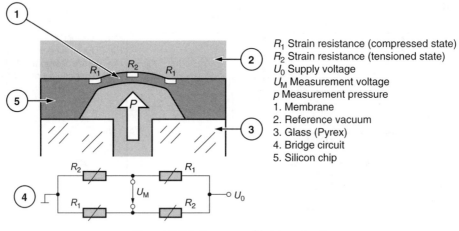

R_1 Strain resistance (compressed state)
R_2 Strain resistance (tensioned state)
U_0 Supply voltage
U_M Measurement voltage
p Measurement pressure
1. Membrane
2. Reference vacuum
3. Glass (Pyrex)
4. Bridge circuit
5. Silicon chip

Figure 3.37 Sensor and bridge circuit
(reproduced with the kind permission of Ford Motor Company Limited)

Pressure sensor using piezoelectricity

Theory of operation

The piezoelectric effect can best be illustrated by means of a quartz crystal on which pressure is exerted. The quartz crystal is electrically neutral in its rest state, that is, the positively and negatively charged atoms (ions) are in balance. External pressure exerted on a quartz crystal causes the crystal's lattice to deform. This results in ion displacement. An electric voltage is generated as a consequence. The direct piezo effect is primarily utilised in sensors. Today's technologies use high performance piezoceramic materials instead of quartz crystals. As sensors, piezoceramics convert a force acting upon them into an electrical signal when the ceramic material is compressed against its high rigidity. Owing to dielectric displacement (dielectric = electrically non-conductive), surface charges are generated and an electric field builds up. This field can be registered as a (measurable) electrical voltage via electrodes. In the case of sensors, mechanical energy is converted into electrical energy by means of a force acting on a piezoelectric body.

Sensor monitoring locations:

1. A/C high pressure sensor is fitted to the A/C pipe work or receiver-drier.

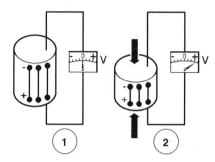

Figure 3.38 Piezoceramic material under pressure creating a voltage (reproduced with the kind permission of Ford Motor Company Limited)

Angle sensors (potentiometer)

Theory of operation

An angle sensor simply measures the angular rotation of a component (e.g. air distributor door or throttle valve). Figure 3.39 shows a sliding contact and a control track. A reference voltage for the sliding contact is supplied via a contact track. This contact track has a constant, low ohmic resistance from start to end. The sliding-contact position sensor is actuated by the movement of the distribution door (which is actuated by a motor), the resistance of the variable resistance track changes along its length. The sensor track operates on the principle that resistance is proportional to length and inversely proportional to cross-sectional area. This means if you double the length of a conductor the resistance will double and if you halve the cross-sectional area of a conductor the resistance will double. The slider moves across the track measuring the voltage at different points.

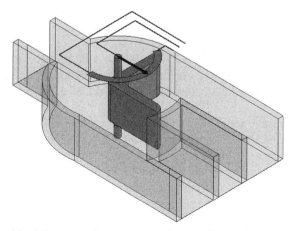

Figure 3.39 Simplified diagram of a rotary distribution door with potentiometer feedback

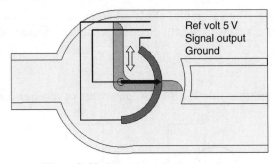

Figure 3.40 Potentiometer connections

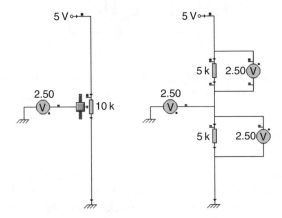

Figure 3.41 Shows a linear potentiometer on the left and the
equivalent potential divider circuit on the right
(courtesy of Crocodile Clips)

The rotary distribution door has a DC permanent magnet motor (not shown) with a potentiometer which provides closed loop feedback to verify its position.

A reference voltage and ground is applied to the resistive track. This creates a 5 V volt drop across the resistor. The signal output sliding contact makes contact with the surface of the resistive track and divides (potential divider circuit) the circuit creating two volt drops across its length (Fig. 3.41).

If the sliding contact position sensor is in the middle of the resistive track, 50% of travel has occurred on the rotary door. Then the volt drop theoretically should be equal at 2.5 volts. This is due to the resistor track being divided into two halves of equal proportion so equal volt drops exist.

If the rotary distribution door is closed, which represents a 0% movement ratio, then the output of the sensor would be very low at around 0.1 V due to a low resistance between the reference voltage and the output. If the rotary door was in the fully open position, 100% movement ratio, then the volt drop would be high due to the high resistance.

A formula to calculate voltage output (V_{out}) of a rotary sensor:

$$V_{out} = \frac{\varnothing}{300} \times V_S$$

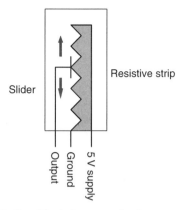

Figure 3.42 Simplified diagram of a linear potentiometer

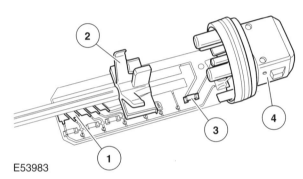

1. Slide path (resistance track)
2. Slider
3. Resistor (fixed)
4. Electrical connector

E53983

Figure 3.43 Linear potentiometer
(reproduced with the kind permission of Ford Motor Company Limited)

V_s – voltage supply
Ø – angle of rotation
300 – total available angle of rotation

$$\text{e.g. } V_{\text{out}} = \frac{150°}{300°} \times 5\,\text{V} = 2.5\,\text{V}$$

Pressure sensor sliding contact potentiometer (linear)
A linear potentiometer (Fig. 3.43) operates on the same principle as a rotary potentiometer except the movement is linear and not rotary, hence its name. Pressure from the refrigerant is applied to a membrane that deforms and transfers this movement to the potentiometer slider (1). This varies the linear movement and the sensor's output.

Sensor monitoring locations:

1. A/C pressure sensor (linear measurement) is fitted to the A/C pipe work Schrader valve or receiver drier housing (high pressure side of A/C system).
2. Throttle position (angle measurement) is found in the throttle housing, accelerator pedal position.
3. Heater door position feedback is attached to heater door motor or mechanism.

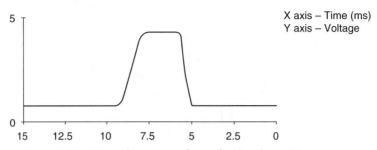

Figure 3.44 DC analogue waveform of a throttle position sensor (reproduced with the kind permission of Ford Motor Company Limited)

Waveform

The waveform represents an increase in voltage output with a corresponding increase in angular movement.

Example specification:
Voltage output approximately 0.8 V at rest position, throttle valve closed
Voltage output approximately 4.8 V at Wide Open Throttle (WOT) position

Resistance of the sensor in the rest position – 1.1 kΩ
Resistance of the sensor in fully open position – 4.4 kΩ

Signal checks:

1. Monitor a steady change in voltage level directly proportional to a change in rotary or linear motion.
2. Check for glitches in the signal (drops to zero or rise to reference voltage).
3. Volt drop on the reference voltage and ground signal should not be greater than 400 mV (see Power-to-power test and Earth-to-earth test under section 3.3).
4. Min and max values indicate maximum angular or linear motion.
5. Signal noise may indicate worn or faulty wiper contact.

Speed sensor (Hall effect)

Theory of operation

If a current flows through an electrical conductor positioned at right angles (90°) to a magnetic field, the charge carriers (electrons) are deflected (Lorentz force). The Hall effect (Figs 3.45 and 3.46) is generated by means of a semiconductor plate (Hall plate) which receives a defined voltage (U). Application of the supply voltage U results in an evenly distributed electron flow over the entire surface of the Hall plate. As a result, a magnetic field builds up around the Hall plate. The evenly distributed electron flow leads to charge equalisation ($U_H = 0\,\text{V}$) on both sides of the Hall plate.

> Note – changes in the magnetic field lead to corresponding changes in electron flow. If the north pole of a permanent magnet meets the north pole of a Hall plate magnetic field, the field moves away from the permanent magnet.

As a result, the electrons (negatively charged particles) driven by the longitudinal potential are suddenly deflected vertically with respect to the current's direction of flow, away from the permanent magnet (repulsion of electron flow). The resulting charge difference between the two sides of the Hall plate gives rise to a Hall voltage(U_H). If the south pole of a permanent magnet

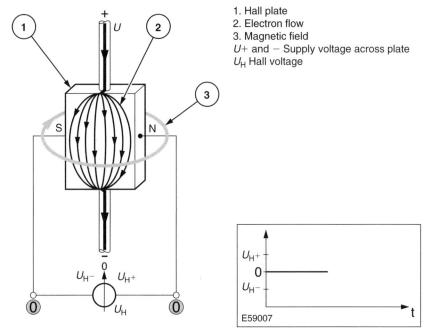

1. Hall plate
2. Electron flow
3. Magnetic field
$U+$ and $-$ Supply voltage across plate
U_H Hall voltage

Figure 3.45 Hall effect with charge equalisation,
waveform shows zero output ($U_H = 0$)
(reproduced with the kind permission of Ford Motor Company Limited)

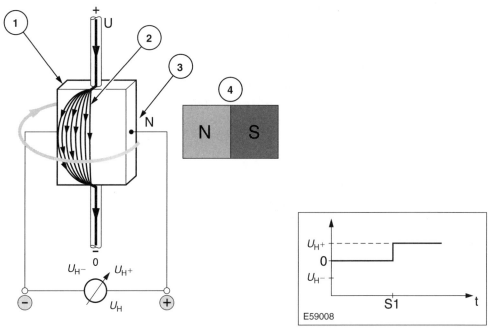

Figure 3.46 Hall effect unequalised, waveform shows
voltage output (U_H+)
(reproduced with the kind permission of Ford Motor Company Limited)

meets the north pole of a Hall plate magnetic field, the field moves toward the permanent magnet. As a result, the electrons (negatively charged particles) driven by the longitudinal potential are suddenly deflected vertically with respect to the current's direction of flow, toward the permanent magnet (attraction of electron flow). The sudden changes in electron flow correspondingly change the polarity of the Hall voltage (from positive to negative or vice versa). Hall voltages are generally very low. Lying in the millivolt range, these voltages must be processed appropriately. Sensor technology usually makes use of Integrated Circuits (ICs) to process Hall voltages and output them as square-wave signals to the terminal device (e.g. PCM). The square-wave signals can be made visible with the aid of an oscilloscope or tested with an LED.

> Note – The Hall plate magnetic field can also be deflected by moving an iron element (e.g. a ferrous pulse wheel) toward it. In this case, there is no alternation of electron flow between the sides of the Hall plate. The magnetic field and electron flow are always displaced in just one direction: from charge equalisation to charge difference (0 signal edge/ high signal edge).

Sensor Monitoring locations:

1. Interior blower speed feedback is fitted on blower inside heater assembly.
2. Vehicle speed sensor is fitted on the transmission housing output shaft.

Example vehicle speed sensor

Vehicle speed is an important factor used to assist in calculating the cooling rate of the interior due to increased natural flow rates (ram air). Often an increase in vehicle speed will cause the control module in automatic mode to reduce the blower speed (automatic climate control system). The accuracy of the exterior temperature sensor is also adversely affected by the ram air. For this reason, the A/C module calculates the exterior temperature from the measurement of the exterior temperature sensor and vehicle speed. The vehicle speed is detected by the Vehicle Speed Sensor (VSS), which also supports other systems (speedometer, PCM (Powertrain Control Module), suspension, braking). It is designed as a Hall sensor and sends a digital signal to the A/C module in the form of a square-wave signal.

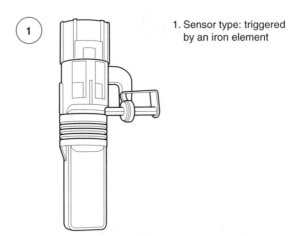

1. Sensor type: triggered
   by an iron element

Figure 3.47 Hall effect vehicle speed sensor
(reproduced with the kind permission of Ford Motor Company Limited)

Waveform

The signal trace depends on the installed sensor and
the transmission

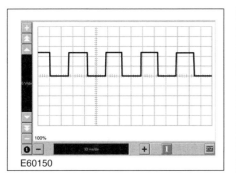

E60150

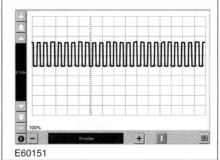

E60151

Signal trace of a VSS signal (Hall) at constant speed   Signal trace of a VSS signal (inductive) at constant speed
(approx. 30 km/h)   (approx. 130 km/h)

Figure 3.48 A frequency modulated signal from a Hall effect
vehicle speed sensor

Signal checks:

1. Monitor a steady change in frequency which is proportional to a change in vehicle speed.
2. Check for glitches in the signal (drops to zero or rises to reference voltage).
3. Volt drop on the reference voltage and ground signal should not be greater than 400 mV.
4. Peak–peak voltages should be the same and equal reference voltage, allow 400 mV difference (see Power-to-power test and Earth-to-earth test under section 3.3).
5. The lower horizontal lines should almost reach zero, allow for 400 mV difference.
6. Signal transitions should be straight and vertical.
7. Check for any background interference.

Inductive type speed sensor

Theory of operation
Figure 3.49 shows a permanent magnet with north and south poles. An electrical conductor is positioned between the north and south poles. If the conductor is moved in the direction of the arrow, it intersects with the permanent magnet's field lines. Charges inside the conductor are displaced in this process. Free electrons move to one end of the conductor. Correspondingly, a shortage of electrons occurs at the other end. If you can arrange for the conductor to keep moving in the magnetic field then you will have a continuous force of electrons available to do work. The resulting potential between the conductor's ends is termed induction voltage.

   Sensor monitoring locations:

1. Compressor speed sensor – measures the rotational speed of the pulley or main compressor shaft.
2. Blower motor speed sensor – measures the rotational speed of the blower motor for feedback control.

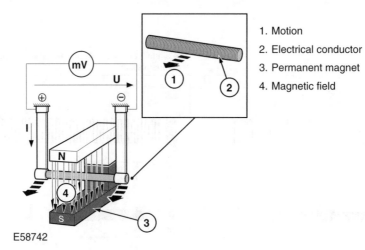

1. Motion
2. Electrical conductor
3. Permanent magnet
4. Magnetic field

E58742

Figure 3.49 Creating an induced voltage

Example speed sensor

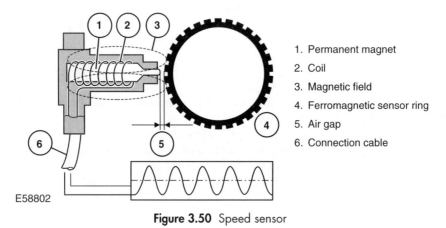

1. Permanent magnet
2. Coil
3. Magnetic field
4. Ferromagnetic sensor ring
5. Air gap
6. Connection cable

E58802

Figure 3.50 Speed sensor

The inductive speed sensor contains a permanent magnet and soft permeable pole pin surrounded by a coil. The speed sensor is mounted so that its front face is a defined distance from the sensor ring. The rotation of the sensor ring induces a voltage proportional to the periodic variation in magnetic flux. The variation in magnetic flux is caused by the movement of the ferromagnetic sensor ring. When the magnetic flux increases or decreases an emf is induced into the coil windings. As a tooth of the ring approaches the sensor a positive emf is generated in the coil due to the lines of flux being cut in the magnetic field. When the tooth is aligned with the sensor there is no change in magnetic flux so the emf is zero. As the tooth rotates away from the sensor it again breaks the lines of magnetic flux which generates a negative emf. The changes in the magnetic flux *induce* an *alternating voltage* in the inductive sensor coil. A uniform tooth pattern will create a near sinusoidal voltage curve.

Waveform

Voltages generated by induction constantly alternate in amplitude and polarity (Fig. 3.51). Accordingly, they are also termed alternating voltage. Alternating voltage rises from 0 V to its positive peak value (amplitude), then drops back via the 0 V level to its negative peak value,

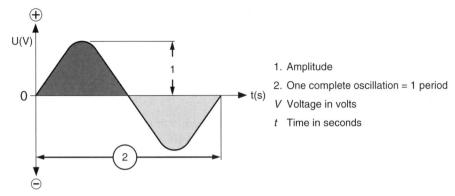

Figure 3.51 Voltage curve

rises again to its positive peak value etc. The number of complete alternations (periods) per second is termed the voltage frequency.

The air gap of this particular type of sensor is crucial to its operation. Because air is not very permeable if the gap is too large the amplitude of the output is very low or zero.

Humidity sensor

Theory of operation

Humidity sensors determine relative air humidity using capacitive measurement technology. For this principle, the sensor element is built out of a film capacitor on different substrates (glass, ceramic etc.). The dielectric is a polymer which absorbs or releases water proportional to the relative environmental humidity, and thus changes the capacitance of the capacitor, which is measured by an onboard electronic circuit (see Pressure sensor – capacitive type (above) for additional information).

A temperature sensor is often used with a humidity sensor. The temperature and humidity sensors together form a single unit, which enables a precise determination of the dewpoint without incurring errors due to temperature gradients between the two sensor elements. The sensors are coupled to an amplification, analogue-to-digital (ND) conversion and serial interface circuit all on the same chip. The integration provides improved signal quality and insensitivity to external disturbances (EMC).

Air quality sensor

Theory of operation

If the vehicle is in an exhaust gas cloud, the air intake process will always be stopped and the system will switch to recirculation mode. This is to prevent the air quality in the passenger compartment from becoming contaminated by the air outside. Recirculation will also be disengaged if the mode was permanently on and the air in the passenger compartment was not being exchanged at all; in this case the system works dynamically. In exceptional cases such as these, the mode of operation ensures that an adequate supply of fresh air is fed into the system.

An Air Quality Sensor (AQS) (Fig. 3.52) is located in the main air inlet duct of the HVAC system. When the threshold for carbon monoxide or nitrogen dioxide is reached, the AQS rapidly communicates to the HVAC system to initiate the air recirculation mode. The Metal Oxide Semiconductor (MOS) sensor consists of a sensing material and a transducer (substrate). Surface reactions at the sensitive layer change its resistivity. The transducer keeps the sensing material at an elevated temperature, and its resistance is measured. Changes in the composition of the ambient atmosphere create a corresponding change in the resistance of the sensing layer, allowing the sensor to detect a wide range of toxic gases even at very low concentrations. The sensing layer is a porous thick film of polycrystalline tin oxide (SnO_2). In normal ambient air, oxygen and water vapour-related gases are absorbed at the surface of the SnO_2 grains. For reducing gases such as CO, a reaction takes place with the preabsorbed oxygen and water vapour-related gases which decreases sensor resistance.

1. Sensor monitoring location: Main air inlet duct of the HVAC system.

The sensor uses integrated signal conditioning electronics mounted with the sensing element on a circuit board. An integral microcontroller monitors the pollution level and creates a Pulse Width Modulated (PWM) or serial output signal in relation to pollution levels.
   Benefits:

* Improved comfort.
* Fast, automatic protection against potentially harmful external pollutants.
* Cabin pollutant concentrations reduced by 20%.
* Occupant discomfort due to odours reduced by 40%.

Figure 3.52 Air quality sensor
(reproduced with the kind permission of Valeo)

Wiring diagram

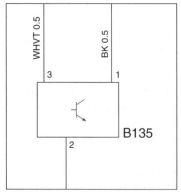

1. Reference voltage of battery supply (12–14V)
2. Ground
3. Signal output depends on level of pollution sensed

Figure 3.53 Air quality sensor wiring schematic
(courtesy of Vauxhall Motor Company)

Data logger

An air quality sensor output of 100% advises the technician that air quality is good. Any value below this figure means that some pollution is present.

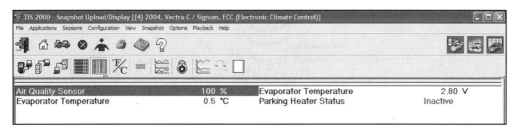

Figure 3.54 Data logger air quality sensor reading
(courtesy of Vauxhall Motor Company)

Actuators

1. Relay.
2. Solenoids.
3. DC permanent magnet motor.
4. Stepper motors – variable reluctance, permanent magnet and hybrid.

Relay

Theory of operation

Current flowing through a conductor like a straight copper wire creates a magnetic field around itself. If the copper wire is twisted (Fig. 3.55) and shaped like a coil and current passed through it then a magnetic flux is created. The relationship between the direction of current flowing through a conductor and the direction of magnetic flux is expressed by 'Ampere's rule of the right-hand screw effect'. If the current is reversed then the magnetic poles will reverse.

If a permeable material (material easily magnetised) is placed under the coil it will become magnetised and attracted by the magnet field generated in the coil. The action of the coil and current is the principle of electromagnetic force. The force created is not strong enough in Fig. 3.55

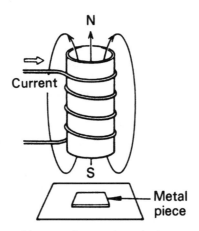

Figure 3.55 Copper cable wound around a cylinder creating a magnetic flux
(with the agreement of Toyota (GB) PLC)

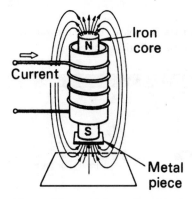

Figure 3.56 Combined lines of force of the coil and bar attract the metal piece
(with the agreement of Toyota (GB) PLC)

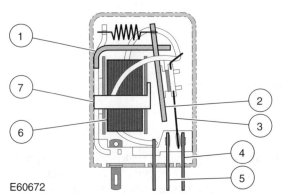

1. Yoke
2. Armature
3. Two-way contact
4. Normally closed contact
5. Normally open contact
6. Relay coil
7. Coil core (permeable material)

E60672

Figure 3.57 Operational parts of a relay
(reproduced with the kind permission of Ford Motor Company Limited)

to move the metal piece. If an iron core is inserted (Fig. 3.56), the number of magnetic fluxes is intensified and the metal is under a greater force. This is because the coil and the bar create lines of magnetic force. This principle explains the operational characteristics of a relay.

Actuator example relay

In Figure 3.57 a current flows through the relay coil (6) and creates a magnetic field, which is magnified by the coil's core (7). This Magnetic Motive Force (MMF) is applied to the armature (2) and it is attracted to the centre of the coil. The force is great enough for the armature to overcome spring force and the armature makes contact with the coil. Contact points are attached to the armature. Once the armature has made contact with the coil the relay contacts are closed from (4) to (5). This creates a closed circuit on the high current side of the relay.

Relays are electrically controlled switches. The function of a relay is to use a low current to operate a high current. The switch inside the relay will be in one of two positions, depending on whether the electromagnetic relay coil is energised or de-energised. In basic relays, there is one input and either one or two outputs. Relays are either Normally Open (NO) or Normally Closed (NC). In either case, the relay switch input (Fig. 3.59a) is always connected to pin 30. Pin 30 not only designates the input to the relay switch, but in accordance with DIN standards, we also know that it's connected to battery positive. The relay outputs on the other side of the relay switch are designated either 87, 87a or 87b. The two remaining relay terminals are connected to the relay coil. Applying current to the coil is what makes the relay close or open. According to DIN standards, pin 85 should be connected to ground (usually controlled by another switch) and pin 86 should be connected to battery positive (usually protected by a fuse). This uses a low current (switched by the A/C module) to operate the relay thus switching on a consumer which uses a large current (condenser fan). This means that the relay must have at least two circuits, a low current and a high current circuit.

The relay in Fig. 3.58 shows two circuits running vertically. The left which obtains its power from fuse F30 (15A) is the low current side (switching side). The right circuit is the high current side (switched side). Generally the switching side is operated by a switch or a module and the switched side operates a high current consumer like the compressor or condenser fan. The diode is used to protect the switching device (generally a control module) from back-emf.

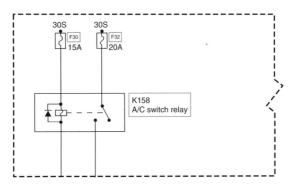

Figure 3.58 A/C relay with internal protection diode
(reproduced with the kind permission of Ford Motor Company Limited)

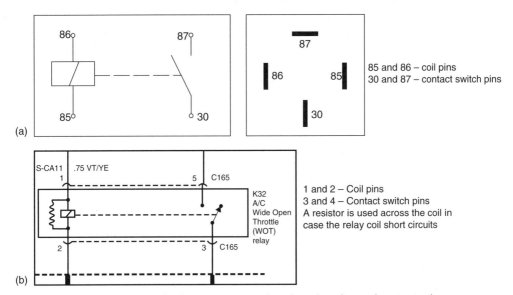

Figure 3.59 (a) DIN standard 72 522 pin codes; (b) A/C relay with ISO pin designation
(reproduced with the kind permission of Ford Motor Company Limited)

Relay pin coding

DIN standard 72 552 pin codes and ISO pin designation

Testing relays

Table 3.1 offers some appropriate tests for a relay.
   Some common relay applications:

- A/C relay.
- WOT (Wide Open Throttle) relay for the compressor clutch.
- Blower motor relay.
- Condensor fan relay.

Actuator locations:

1. Battery junction box, central junction box.

Solenoid

Theory of operation

A solenoid (Fig. 3.60) operates in a similar manner to a relay except when a permeable material like an iron bar is inserted into a coil shaped like a cylinder and current flows through the coil, the bar is pulled to the centre of the coil. The applied current can be DC, AC, or a pulse width modulated control signal.

   A water control solenoid can be used as an example (Figure 3.61). The key to lifting the plunger inside the valve body is the generation of magnetism 'lines of force'. These lines of force need to be strong enough to move the mass of the plunger (the so-called iron bar). They quite often need to be strong enough to overcome a spring force (used to hold the plunger closed) and the force of gravity (plunger is in the vertical position). The more lines of force the stronger the magnetic field or 'flux (Φ)'. The unit for flux is the weber (Wb). The flux density (B) of a magnetic field is the amount of flux (Wb) per unit area perpendicular to the magnetic field. The unit for magnetic flux density is either the weber-per-metre-squared or tesla which has the designated letter (B). Another factor, which determines the strength of a magnetic field, is reluctance. Reluctance is the ratio of magnetic motive force (mmf) to flux in a magnetic conductor. It is the equivalent to electrical resistance and so is proportional to length and inversely proportional to cross-sectional area. The unit of reluctance is the ampere-per-weber and is designated the letter (R).

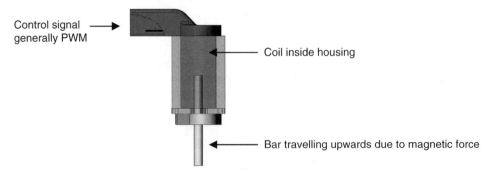

Control signal generally PWM →

— Coil inside housing

← Bar travelling upwards due to magnetic force

Figure 3.60 Control solenoid

Table 3.1 Tests for fault diagnosis of relays

Tester	Notes
Multimeter	*Volt drop tests* This test involves placing the voltmeter leads across the input and output to the relay contacts (pins 87 and 30) (pins 5 and 3). The maximum volt drop should not exceed 0.5 V. *A current test* is the preferred test. A fuse which is in series with the circuit can be removed and the meter placed in series by placing the meter across the fuse terminals. Current measurement is taken to test the relay under load. Relays are current rated so information is available. *A resistance* measurement can be made of the coil and the relay contacts. The relay coil will have a low resistance of approximately 2–3 ohms. *Continuity* To test the sensor contacts (right circuit) the relay would have to be removed from its fuse box and current passed through the switching side. This means that current would pass through the left side circuit to be able to measure using a multimeter the resistance of the right side. The resistance value may not be known so all you would be testing is continuity. *A diode test* can be used to ensure the diode operates correctly inside the relay. The diode should have continuity in one direction and have an infinite resistance in the opposite direction.
Power probe (with built-in LED display)	The use of a *power probe* is ideal to power a relay. Technicians often remove the relay and apply power to the relay female connectors inside the fuse box housing. This is done when they are confident that the relay is faulty and a test is required to check the response of the system. The power probe can also be used to power the relay itself to check audibly or via a meter that the relay contacts are closing. *The LED display* is ideal when checking earth switched pins of the relay. This is done by removing the relay and placing the power probe on the earth switch terminal (controlled by the A/C module). When the A/C module tries to switch the earth path of the relay on your power probe green (green for ground) the LED will illuminate.
Serial test	Relays that send power to a module and are sometimes represented on *data lists*. *Fault codes* can be accessed for power supply faults. *Actuator command* can be used to operate some relays, depending on the ability of the serial tester.
Break-out box testing	Very useful for pinpointing power supply problems. Checked by using either an oscilloscope or multimeter connected to pins on the break-out box which are powered by relays.
Oscilloscope	Very useful for measuring any background interference or electrical noise within the circuit and checking any volt drop going to and across the relay (under load). *Power to power test* This is when the leads of the meter are placed one on the battery terminal and one to the supply of the relay. This is to check that the relay supply voltage is good. The maximum volt drop should not exceed 0.5 V. *Earth to earth test* This is when the leads are placed on the earth of the relay and the earth of the battery. This checks that the earth path is good. The maximum volt drop should not exceed 0.5 V.

The coil creates an mmf, which drives flux left through the plunger, then around the frame of the solenoid then through the air gap and back into the plunger. The reluctance of this path is mostly made up by the air gap. When the plunger is out (valve is closed), the reluctance is quite high. When current is applied to the coil, the plunger moves to the top and the reluctance decreases (due to a smaller air gap). This is an example of forces in magnetic systems; they act to reduce the reluctance, or increase the inductance. Eventually, the plunger will collide with the frame at the top, the air gap will be zero, and the reluctance will be at a minimum.

The main variables for the performance of the solenoid are the following:

1. Plunger shape.
2. Plunger material.
3. Number of windings around the plunger.
4. Supply voltage/current flow through windings.
5. Wiring configuration of the coil.
6. Air gap.

Example water control valve

In water-based heating systems, the heater control valve interrupts the coolant circuit between the engine and the heater core. The valve contains a plunger which closes off the openings between the feed from the engine and the return to the heater core. The solenoid valve is fully open when de-energised.

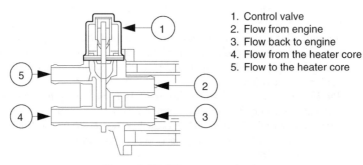

1. Control valve
2. Flow from engine
3. Flow back to engine
4. Flow from the heater core
5. Flow to the heater core

Figure 3.61 Water control valve
(reproduced with the kind permission of Ford Motor Company Limited)

Characteristics	Unit	Value
Resistance	Ω	14–16
Power supply	V	12
Signal type	V	Clock pulse
Frequency		18/min

Figure 3.62 Solenoid specification

During testing, the temperature control should be brought to the centre setting. The valve must open and close regularly, approximately 18 times per minute when the engine is running.

Electric motor permanent magnet

Theory of operation

All conventional electric motors depend for their operation on a conductor such as a wire or a coil creating or operating within a magnetic field. Current flowing through a conductor like a straight copper wire creates a magnetic field around itself. The relationship between the direction of current flowing through a conductor and the direction of magnetic flux is expressed by 'Ampere's rule of the right hand screw effect'. If this conductor is placed between the poles of a magnet and current is passed through it. There will be a reaction between the two magnetic fluxes produced.

The strength of the electromotive force varies in proportion to the density of the magnetic flux, the current flowing through the conductor and the length of the conductor within the magnetic field.

The magnitude of the force varies directly with the strength of the magnetic field and the amount of current flowing in the conductor:

$$F = I \times L \times B$$

F – force (newtons)
I – current (amperes)
L – length (metres)
B – magnetic flux (webers/m^2)

If a conductor formed in the shape of a square coil (Figs 3.63 and 3.64) is placed between the north and south poles of a magnet and a commutator segment is fitted to the end of the coil with brushes to enable an electrical contact, an interesting relationship occurs.

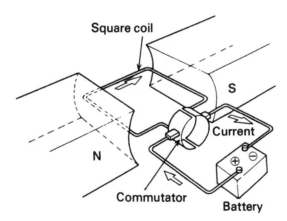

Figure 3.63 Conductor formed in a square placed within a permanent magnet (with the agreement of Toyota (GB) PLC)

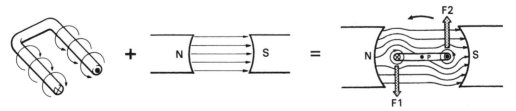

Figure 3.64 Composite magnetic fields and reactive forces $F1$ and $F2$ (with the agreement of Toyota (GB) PLC)

When no current flows through the conductor only one magnetic flux is present, which is created by the permanent magnet (Fig. 3.64). When current flows from the battery to the conductor via the brushes and commutator a magnetic flux is produced. The composition of the magnetic fluxes from the magnets and the conductor creates reactive forces ($F1$ and $F2$). The reactive forces are in different directions due to current flowing forwards on the left side of the square coil and in the opposite direction on the right side of the square coil; this is due to the layout of the conductor within the magnetic field. The direction of the two forces $F1$ and $F2$ adheres to 'Fleming's left hand rule' and causes the conductor to rotate around its axis P. Problems occur when the conductor rotates vertically and counter forces are created that try to reverse the direction of the conductor. As a solution (Fig. 3.65) to this problem individual commutators are fitted to each end of the coil to periodically reverse the current flow whenever rotated 180°; this allows the magnetic forces to be applied in a fixed direction and allows the conductor to fully rotate.

Construction of a permanent magnet motor

Electric motors basically consist of a rotor (moving part) and a stator (stationary part). Generally, the stator comprises a housing with magnets. The brushes and electrical connections are located in the housing cap.

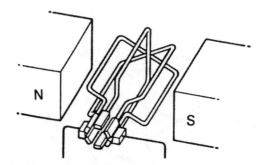

Figure 3.65 Conductors with commutators (armature)
(with the agreement of Toyota (GB) PLC)

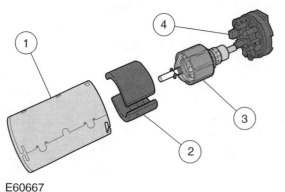

1. Housing (stator)
2. Permanent magnets
3. Rotor (armature)
4. Housing cap with bearing and connections

E60667

Figure 3.66 Permanent magnet motor
(reproduced with the kind permission of Ford Motor Company Limited)

The rotor (Fig. 3.67) consists of the armature and an axle, which are bearing mounted in the housing cap. In electrical engineering, the term 'armature' refers to a moving component; it can rotate or it can move back and forth like the armature in a solenoid.

The standard brush type DC motor found in a car has a wound rotor and a permanent magnet stator (starter motors are an exception as their stators are wound as well). Commutation, switching from one phase to another, is accomplished by incorporating commutator bars on the rotating rotor and stationary brushes in the housing. As the rotor turns, the brushes contact the next phase, allowing the motor to continue to rotate. Commutation is simple and done automatically. Regardless of motor speed, the commutation happens at the right time and no electronic control is required.

So-called brushes (Fig. 3.68) (usually made from graphite) are used to transfer the power via the connections (commutator) of the moving armature.

The brushes are pressed against the commutator (Fig. 3.69) by means of a spring. In the event of excessive power consumption, e.g. due to blocking, bimetal switches (thermoswitches) are used for overload protection. These interrupt the circuit to the electric motor and the contact is only closed again once the motor has cooled down.

Because a permanent magnet motor uses the magnet as a stator the direction of rotation of the motor is determined by the electrical polarity of the supply voltage to the armature. If the

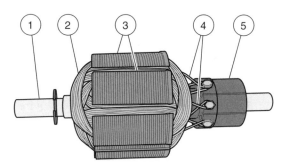

1. Axle
2. Copper coil
3. Iron armature
4. Connection between copper coil and commutator
5. Commutator bars

Figure 3.67 Rotor with armature coil
(reproduced with the kind permission of Ford Motor Company Limited)

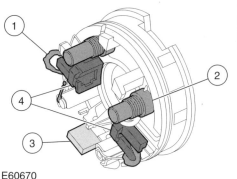

1. Connection between brushes and power supply
2. Spring
3. Thermoswitch (overload protection)
4. Brushes

E60670

Figure 3.68 Housing cap with brushes and connections
(reproduced with the kind permission of Ford Motor Company Limited)

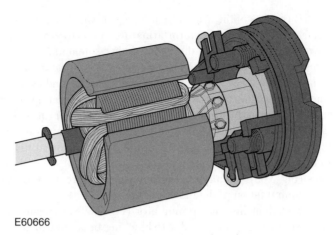

E60666

Figure 3.69 Electric motor without housing
(reproduced with the kind permission of Ford Motor Company Limited)

battery supply is reversed then the motor will run backwards. Permanent magnets also provide constant field strength.

> Note – not all motors use permanent magnets as a stator. Some motors operate in a similar manner to previously described but use a wire wound (called electromagnetic) stator. The stator can be series wound which means it uses the same current as supplied to the armature winding. The back-emf produced by a series wound motor when started is negligible which allows a large current to flow producing high torque This is typical of a starter motor. Most motors used on the A/C system are low torque and require fine control. This results in most motors being either DC permanent magnet or stepper motor type.

Example:

1. DC permanent magnet motor used for heater control doors.

Example DC motor recirculation door
The air recirculation flap is operated by a permanent magnet DC motor. The flap has only two positions. When the motor has turned the flap to one of the end positions, the current passing through the motor winding on some systems can be limited by two PTC resistors built into the motor. The A/C control module will close the output to the motor after a fixed period, e.g. 10 seconds. PTC means Positive Thermal Coefficient and means that the resistance increases with increased heat (the opposite of NTC). This method does not require the motor to have additional sensors for closed loop feedback.
   Location:

1. Recirculation door attached to heater box (see Figure 1.29).

Wiring diagram DC motor
In Figure 3.70 M6 is a recirculation motor controlled by the A/C module. The door position will be either open or closed. Some systems use current sensing for motor positioning. For

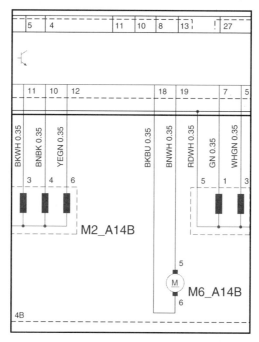

Figure 3.70 DC motor (M6A14B) used for recirculation door control

example, under normal load the motor may draw 200 mA. When the recirculation door reaches the end of its travel the current will increase to 700 or 800 mA. The A/C module can sense this and cut the driving current. The A/C unit reverses output polarity in order to change the direction of the motor's rotation.

Stepper motors

Theory of operation

In essence, step motors are electrical motors that are driven by digital pulses rather than a continuously applied voltage. They are also referred to as Electronic Commutation (EC) motors due to the absence of a commutator and brushes which are generally present in conventional motors. Inherent in this concept is open loop control, wherein a train of pulses translates into a number of shaft revolutions, with each revolution requiring a given number of pulses. Each pulse equals one rotary increment, or step (hence, step motors), which is only a portion of one complete rotation. Therefore, counting pulses can be applied to achieve a desired amount of shaft rotation. The count automatically represents how much movement has been achieved, without the need for feedback information. The precision of step motor controlled motion is determined primarily by the number of steps per revolution; the more steps, the greater the precision. For even higher precision, some step motor drivers divide normal steps into half-steps or micro-steps. With the appropriate logic, step motors can be bi-directional, synchronous, provide rapid acceleration, stopping, and reversal, and will interface easily with other digital mechanisms.

The main advantages of a stepper motor are as follows:

1. No brushes means a maintenance reduction and no brush residue contamination to bearings or the environment.

2. Because there is no brush arcing or brush commutation, brushless motors are much quieter both electrically and audibly.
3. Feedback position is not required because the control system can count steps from a known starting point and calculate position.

Disadvantage:

1. Complex control electronics for commutation.

Range of stepper motors:

1. Variable reluctance.
2. Permanent magnet.
3. Hybrid.

Full-step

In full-step operation, the motor steps through the normal step angle, e.g. 200 steps/revolution motors take 1.8 steps while in half-step operation, 0.9 steps are taken. There are two kinds of full-step modes. Single phase full-step excitation is where the motor is operated with only one phase energised at a time. This mode should only be used where torque and speed performance are not important, e.g. where the motor is operated at a fixed speed and load conditions are well defined. This mode requires the least amount of power from the drive power supply of any of the excitation modes. Dual phase full-step excitation is where the motor is operated with two phases energised at a time. This mode provides good torque and speed performance with a minimum of resonance problems. Dual excitation, provides about 30 to 40% more torque than single excitation, but does require twice the power from the drive power supply.

Half-Step

Half-step excitation is an alternate single and dual phase operation resulting in steps one half the normal step size.

Micro-step

In the micro-step mode, a motor's natural step angle can be divided into much smaller angles. For example, a standard 1.8 degree motor has 200 steps/revolution. If the motor is micro-stepped with a 'divide-by-10', then each micro-step would move the motor 0.18 degrees and there would be 2000 steps/revolution.

Permanent magnet stepper motor

This motor is constructed in almost the opposite manner to a DC permanent magnet motor. The armature becomes a two pole permanent magnet and the stator is wound. Commutation is achieved by electronically controlling the current flowing through the stator. Permanent magnet stepper motors have a permanent magnet rotor with no teeth, and are magnetised perpendicular to the axis. In energising a number of phases in sequence, the rotor rotates as it is attracted to the magnetic poles. Direction of rotation depends on the polarity of the stator when current is applied. Altering the frequency of the pulses to the stator varies the motor's speed of rotation. Increasing the number of stator and rotor poles reduces the step angle. Step angles are generally 90°, 45°, 18°, 15°, 7.5°. Permanent magnet stepper motors have a holding torque (detent torque) when not energised due to using a permanent magnet as a rotor. Rotor direction can be in the opposite direction by changing the sequence of pulses.

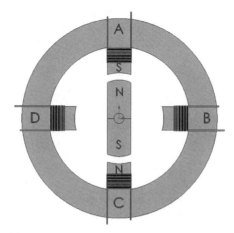

Figure 3.71 Permanent magnet stepper motor

The motor shown in Figure 3.71 will take 90 degree steps as the windings are energised in sequence ABCD. Half-steps can be achieved by these motors. Generally they have step angles of 45 or 90 degrees and step at relatively low rates, but they exhibit high torque and good damping characteristics.

Advantages:

1. High torque compared to variable reluctance.
2. Permanent magnet stepper motors have a holding torque (detent torque) when not energised due to using a permanent magnet as a rotor.

Disadvantages:

1. Complex control electronics for commutation.
2. Fall of performance due to changes in magnetic strength.

Example brushless blower motor

In a permanent magnet BLDC (brushless DC) motor design (Fig. 3.72) the rotor has permanent magnets and the stator is wound. The stator is then commutated electronically using control circuitry which is built into the unit. The circuitry is called an inverter. An inverter is a series of half bridges driving each of the phases of the motor. Typical BLDC designs use three-phase wound stators, so the inverter consists of three half bridges.

Since the electronics are providing commutation for the motor, knowing the rotor speed becomes an issue. This is achieved by a controller that can determine rotor position, either by absolute position sensors or by back emf sensing methods. Most DC motors use a Hall sensor to determine rotor speed and position. This information is often supplied to the bus network or direct to the A/C module as feedback.

Brushless blower motors are equipped with electronics which control the rotational speed. In this case, no separate testing of the electronics is possible, the complete blower must be replaced in the event of a fault. In the case of a blocked pollen filter (air supply), the serial resistor built into the motor may burn out owing to the long-term lack of cooling. The blower motor may also be damaged in the long term owing to insufficient air flow.

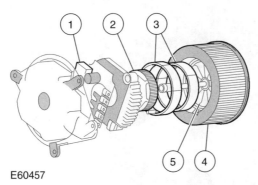

1. Control electronics
2. Stator with windings
3. Rotor with fixed magnets
4. Centrifugal fan
5. Cooling blower motor

E60457

Figure 3.72 Brushless blower motor
(reproduced with the kind permission of Ford Motor Company Limited)

Figure 3.73 Voltage applied to stator coil by control circuitry

Blower setting	Voltage (V)
1	3.5
2	4.5
3	5.2
4	6.5
5	8.3
6	10.7
7 (max)	> 12

Figure 3.74 Shows the average voltage produced by the PWM signal to control the blower motor speed

The electronic control circuitry inside the motor housing applies a voltage using a frequency modulated signal with a variable duty cycle ratio to regulate the motor speed and torque.

Installation position
In the evaporator housing or blower housing.

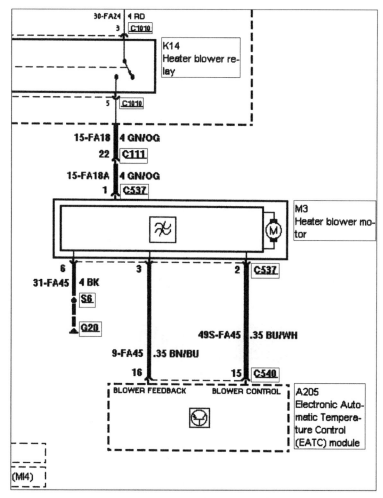

Figure 3.75 Brushless blower motor
(reproduced with the kind permission of Ford Motor Company Limited)

Wiring diagram (Fig. 3.75)

The brushless blower motor (M3) receives its main current via a relay (K14). The motor has its own control electronics built inside the motor housing on a large heatsink (Fig. 3.73). The electronic control unit (A205) sends a speed demand signal to the blower motor. The example provided of a blower motor circuit in Figure 3.75 shows a wiring diagram with a signal sent from pin 15 of the EATC module (A/C module) to pin 2 of the blower motor. The signal is a PWM signal (Fig. 3.76) with a fixed frequency but variable duty cycle ratio to tell the motor circuit what speed is required. The blower motor has a Hall effect speed sensor fitted inside the motor housing which acts as a closed loop feedback signal (Fig. 3.8) which is sent back to the EATC module from pin 3 to pin 16. The signal from a Hall sensor has a fixed duty cycle ratio of 50% and a variable frequency based on speed.

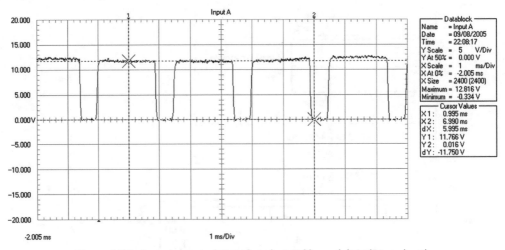

Figure 3.76 Speed demand signal, pulse width modulated signal with
a fixed frequency 400 Hz, duty cycle ratio 76%
(courtesy of Fluke)

Blower speed	Duty cycle ratio (%)	Frequency (Hz)
1	76.5	400
2	70	400
3	63.7	400
4	54.7	400
5	43.7	400
6	30	400
7	26	400

Figure 3.77 Blower control signal (demand signal)

Waveforms fixed frequency variable duty cycle ratio
Signal check:

1. Monitor a steady change in duty cycle ratio which is proportional to a change in blower speed selection (selected by the occupants and communicated to blower control circuitry).
2. Check for glitches in the signal (drops to zero or rises to reference voltage).
3. Volt drop on the reference voltage and ground signal should not be greater than 400 mV.
4. Peak–peak voltages should be the same and equal reference voltage, allow 400 mv difference.
5. The lower horizontal lines should almost reach zero, allow for 400 mv difference.
6. Signal transitions should be straight and vertical.
7. Check for any background interference.

Note – the blower motor waveform is a signal to communicate with the blower electronic control circuitry. If the signal was used to control the blower motor itself then the signal would resemble a sawtooth pattern due to the inductance of the stator coils (see Hybrid stepper motor – Waveform on page 155).

Figure 3.76 shows the duty cycle ratio being manipulated as a signal to the blower motor of the required blower speed set by the occupant (via the control panel or auto A/C mode).

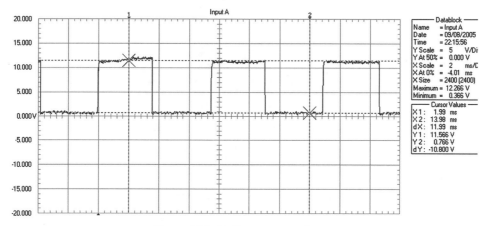

Figure 3.78 Hall sensor signal
(courtesy of Fluke)

Blower speed	Duty cycle ratio (%)	Frequency (Hz)
1	50	76
2	50	133
3	50	160
4	50	190
5	50	230
6	50	274
7	50	317

Figure 3.79 Blower feedback signal (closed loop signal)

Blower feedback (modulated frequency signal)

Figure 3.78 shows a waveform produced from a Hall sensor. Due to the construction of a Hall sensor the duty cycle ratio is fixed and only the frequency varies with a change in blower speed. This signal is used as a feedback signal to confirm the rotational speed of the blower motor to the module it communicates with.

Signal check:

1. Monitor a steady change in frequency which is proportional to a change in blower motor armature speed.
2. Check for glitches in the signal (drops to zero or rises to reference voltage).
3. Volt drop on the reference voltage and ground signal should not be greater than 400 mV.
4. Peak–peak voltages should be the same and equal reference voltage, allow 400 mv difference (see Power-to-power test and Earth-to-earth test under section 3.3).
5. The lower horizontal lines should almost reach zero, allow for 400 mv difference.
6. Signal transitions should be straight and vertical.
7. Check for any background interference.

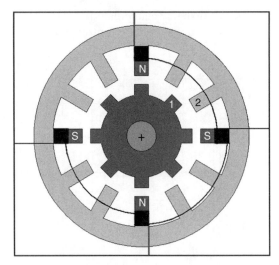

Figure 3.80 Variable reluctance stepper motor

Eventually brushless blower motors will become smart actuators and plug directly into the multiplex wiring system. This means the data stream signals will be sent and not PWM signals.

Variable reluctance

Permeable material rotor – low torque

A variable reluctance stepper motor has a soft iron rotor with radial poles and a stator which is wound. The stator has more poles than the rotor.

The rotor in the centre is made of a permeable material and has eight poles. The stator has 12 poles. The stator is wound with copper cable. Only one phase of the motor is wound on Figure 3.80, normally there would be three-phase windings. To operate the stepper motor the windings would be pulsed in a specific sequence. This example shows four poles of the stator wound in series. When current flows through the four poles of the stator they create a magnetic field. The rotor aligns due to mutual induction to give the shortest magnetic path.

This is the path of minimum reluctance. From this point all that is required is the correct sequence of pole sets to be energised to give the motor its clockwise or anticlockwise motion. To move clockwise, poles 1 and 2 are the closest so these set of four poles should be energised.

These stepper motors operate at high frequency and small step angles. Current in windings does not change direction. To change direction the order of the sequence of steps is changed. Step angles of 15, 7.5, 1.8, 0.45 can be achieved.

Disadvantages of reluctance stepper motors:

1. Complex control electronics for commutation.
2. No holding torque (when not energised).

Example variable stepper motor (Fig. 3.81)

The unipolar stepping motor consists of a rotor (10 poles) and four stator windings. A gearbox transfers the power from the rotor to the lever that controls the damper. The four stepping motor windings receive a common B+ feed from ACC unit pin 20 (K42) and grounds each corresponding winding (pins 11, 31, 10 and 30 in K42) in a specific order thus operating the motor in short pulses. The rotational direction can be changed. When the stepping motor is stationary

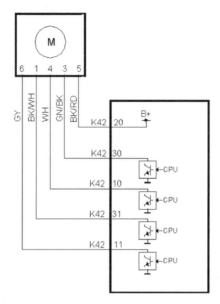

Figure 3.81 Variable reluctance stepper motor wiring diagram
(courtesy of SAAB)

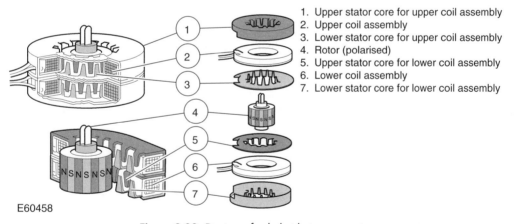

1. Upper stator core for upper coil assembly
2. Upper coil assembly
3. Lower stator core for upper coil assembly
4. Rotor (polarised)
5. Upper stator core for lower coil assembly
6. Lower coil assembly
7. Lower stator core for lower coil assembly

E60458

Figure 3.82 Design of a hybrid stepper motor
(reproduced with the kind permission of Ford Motor Company Limited)

the two windings are continuously ground to lock the motor. The stepping motor does not need to be reconnected to the control module (position sensor), but by sending out a specific number of ground pulses the ACC unit always knows how far the air distribution flap has moved.

Hybrid stepper motor

Stepper motors are used for precise mechanical angular positioning. These motors (Fig. 3.82) feature a rotor made from a magnetic material (e.g. steel) with non-magnetised poles.

The stator consists of a large number of pole pairs and energised windings. The stator is designed in a claw pole configuration with two or four ring coils. Each coil assembly is surrounded by a stator core, which is divided into two parts – the lower and upper stator core. Each

Characteristics	Units	Value
Resistance of windings	Ohm	100
Stepping frequency	Hz	200
Power supply	V	12

Figure 3.83 Example specification of a stepper motor

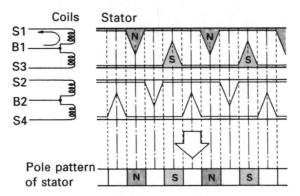

Figure 3.84 Pole pattern of the stator
(with the agreement of Toyota (GB) PLC)

stator core features numerous teeth. These teeth are all offset to one another and are arranged so that they project in the direction of the rotor. The controller cycles the current from one stator pole to the other, deflecting the rotor poles. A torque is generated. If, for instance, four stator cores are installed each with 12 teeth, this means that a total of 48 teeth are available as opposite magnetic poles. As a result, 48 steps per revolution (step angle of 7.5°) are achieved.

Example stepper motor blend door – operation
Current flows through one of the four windings of the cores which is governed by the controlling module (earth switched). Each core is staggered. Each core set has two coil windings wound in opposite directions.

Current is flowing through coil S1 as shown in Figure 3.84. The direction of rotation of the motor is reversed by changing the order in which current is allowed to pass through the four coils.

When the stator coils are energised and the rotor rotates one step the positional relationship shown in Figure 3.85 develops. Since the north pole is attracted to the south pole (opposite poles attract and similar repel) the rotor moves one step.

The N stator claw repels the north pole of the rotor and the south pole of the stator claw attracts it. This allows the rotor to move one step, 11° (1/32 of a revolution).

Wiring diagram
The same wiring schematic as the variable reluctance type (Figure 3.81).

Data logger
Note that the blend and distribution positions are presented as '%'. 0% is closed and 100% is fully open. This is due to the programming of the module and diagnostic software. The live data from the module to control the motor is operated via a sequence of pulses with a fixed

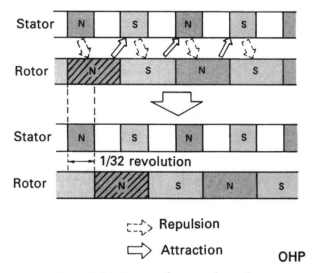

Figure 3.85 Rotor and stator relationship
(with the agreement of Toyota (GB) PLC)

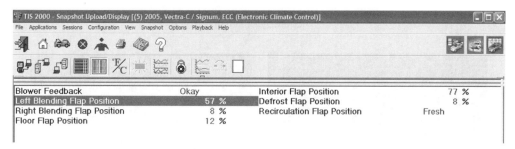

Figure 3.86 Data from the A/C module via data link connector

frequency and duty cycle ratio. A variable ratio is used when speed is important. Blend motors are designed to have accurate positioning and a small angle of movement so the frequency and duty ratio tends to be fixed. Only the sequence varies.

Waveform

Figure 3.87 shows two coils being pulsed in a sequence one after the other to move the motor in one direction. A signal was sent from the A/C control panel (in this example via touch screen display) to change the temperature from 27°C to low (below 16°C) so a sequence of pulses is sent to the coils to move the blend door ensuring less air from the evaporator travels through the heat exchanger and flows directly to the air distribution doors.

Signal check:

1. Check consistent sawtooth pattern (due to inductance reactance), all the signals should be in line.
2. Check for glitches in the signal (drops to zero or rises to reference voltage).
3. Volt drop on the reference voltage and ground signal should not be greater than 400 mV.
4. Maximum peak voltages should be equal to each other.
5. The lower horizontal lines should almost reach zero, allow for 400 mv difference.
6. Check for any background interference.

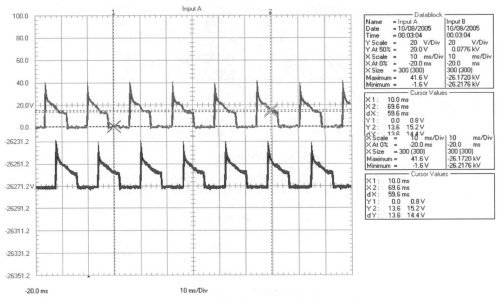

Figure 3.87 Two hybrid motor coils being pulsed
(courtesy of Fluke)

Figure 3.88 Manual control system

Note – the blower motor waveform is a signal to communicate with the blower electronic control circuitry. If the signal was used to control the blower motor itself then the signal would resemble a sawtooth pattern due to the inductance of the stator coils.

A/C modules and displays

Air-conditioning control modules vary depending on the system they are controlling. The module generally incorporates the A/C controls unless the vehicle is of a high specification, which includes satellite navigation, DVD player, telephone system, which requires a graphical interface. High specification vehicles will often have a multi-zone A/C system which requires a great deal of control compared to a manual system. The following examples are provided to allow the reader an appreciation of the differences between a simple and more complex A/C module.

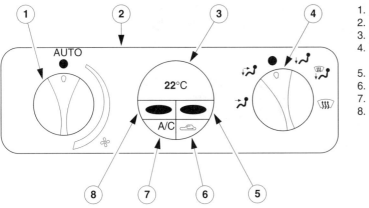

1. Blower control
2. A/C module
3. LCD display
4. Manual selection for air distribution
5. Temperature switch (cold)
6. Recirculation switch
7. A/C switch
8. Temperature switch (hot)

Figure 3.89 Semi-automatic control module
(reproduced with the kind permission of Ford Motor Company Limited)

Module used for manual control (Fig. 3.88)

The manual control system illustrated has a manual blower control selection, manual air distribution, manual temperature control, and switch operated A/C and recirculation. The only information the module must process is based on A/C switch input, recirculation input, and temperature variation for the solenoid operated water control valve. The blend door and air distribution are all Bowden cable operated.

The electronic circuitry required for the operation of such a system is very simple. The module has no memory functions (EPROM), cannot be programmed and is not a part of a multiplex network. This means that the circuitry is designed as an ASIC (Application Specific Integrated Circuit). The output from the unit can be monitored by more electronically advanced modules like the Powertrain Control Module (PCM). These modules (PCM) often make the final decision on whether the A/C compressor clutch will be energised or not.

Semi-automatic systems (Fig. 3.89)

The semi-automatic control system as illustrated in Figure 3.89 has a manual and automatic blower control selection, manual air distribution, manual and automatic temperature control, often a display (LCD), and switch operated A/C and recirculation.

With semi-automatic temperature control, the air distribution (except in 'DEFROST' mode) must be set by the user. In addition, the user can switch on/off automatic operation and set the temperature for a single zone (whole interior space).

The system is often more advanced and can be programmed electronically on and off the vehicle (has EPROM). Semi-automatic modules often incorporate self-test diagnostics (displaying fault codes via the LCD) as well as being able to communicate via multiplex communication networks.

Fully automatic control system (Fig. 3.90)

The fully automatic control system has a manual and automatic blower control selection, air distribution and temperature control. Switch operated A/C and recirculation is built into the module to activate the system although recirculation will be automatically controlled in some A/C modes. The blend door and air distribution and recirculation are motorised.

The module incorporates an interior temperature sensor and fan which can be seen in Figure 3.90 at the bottom of the module on the left hand side behind the grille.

Figure 3.90 A/C control module with LCD
(low specification vehicle)

TFT (Thin Film Transistor)
liquid crystal display

Figure 3.91 Touch screen display (TFT LCD) which communicates with A/C module
(module not built in)

Using advanced graphical displays often requires the A/C module to be a separate unit. Multiplex communication is used to communicate user selection information between the two modules (Fig. 3.91).

Wiring diagram

Figure 3.92 shows a mid-speed CAN network. On the MS CAN bus, the transmission rate is 125 kbit/s. Cabling between two nodes, the touch screen A363 module pins 15 and 5 and the A/C module A205 (EATC) pins 18 and 19. CAN high signal is between pins 5 and 18 and CAN low signal between pins 15 and 19.

This bus is used to transfer all the A/C information the occupants select from the touch screen menu. For example, if auto A/C control is selected by the occupants with a control temperature of 16°C then a signal via the bus is sent to the A/C module allowing it to process the command and compare data against programmed values within its memory. The A/C module will take readings on internal and external air temperature, vehicle speed for natural air flow calculations, blower speed and compare these to the desired temperature. If required the A/C system will then control blower speed, blend door position and air distribution to meet the desired temperature as quickly as possible. All this information will be in the form of a data stream.

Figure 3.93 shows CAN high and low signal (anti-phase).

See section 3.6 for additional information.

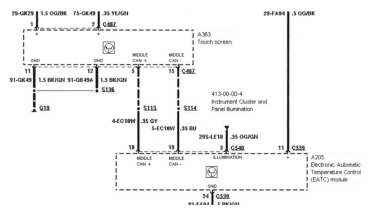

Figure 3.92 Connection between two nodes for A/C control
(reproduced with the kind permission of Ford Motor Company Limited)

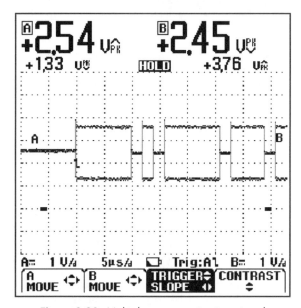

Figure 3.93 Multiplex communication signal
(courtesy of Fluke)

Testing A/C modules

Table 3.2 provides the reader with a number of possible tests which can be used to diagnose system faults on A/C modules.

3.3 Testing sensors and actuators

This section discusses a wide range of test equipment and approaches available when diagnosing faults on A/C systems. Live data from meters, scopes and serial data obtained from scanners are important elements of this section. The differences between when and how to obtain such data is crucial to obtaining information reliably and quickly.

Table 3.2 Tests for fault diagnosis on A/C modules

Tester	Notes
Multimeter	*Volt drop tests* This test involves placing the voltmeter leads across the input from the power supply to the module and the battery positive terminal and checking for acceptable volt drops.
	Ground- to- ground test involves placing the meter leads from the ground terminal of the module and battery ground.
	The maximum volt drop should not exceed 0.5 V.
	A current test can be carried out to check for excessive current due to an internal or external short circuit which is diverting all the power (see Fig. 3.98).
	A resistance test is not applicable
	Continuity test can be carried out with the power isolated and the A/C module disconnected from its harness connector. This test can be used to check for any open circuits in wiring and connectors.
	A diode test is not applicable
Power probe (with built-in LED display)	The use of a power probe is ideal to power an A/C module if used cautiously.
	The power probe can be used to power the A/C module directly. This allows the bypassing of any potentially faulty power supplies. It is also possible to supply a ground signal.
	The LED display is ideal when checking power and earth switched pins on the A/C module. The earth path of the module will power the green LED and the power to the module will illuminate the red LED.
Serial test	A very important test to read any fault codes, check data lists and any possible actuator functions. If the A/C module is faulty then this may prevent you from communicating with this particular system via a serial tester. You should still be able to communicate with other modules in this instance. If you cannot then a problem with power distribution or diagnostic communication is evident.
Break-out box testing	Very useful for pinpointing power supply problems. Checked by using either an oscilloscope or multimeter connected to pins on the break-out box which are powered by relays to supply current to modules.
Oscilloscope	Very useful for measuring any background interference or electrical noise within the circuit and checking any volt drop going to and across the relay (under load).
	Power- to- power test – this is when the leads of the meter are placed one on the battery terminal and one to the supply of the relay. This is to check that the relay supply voltage is acceptable. The maximum volt drop should not exceed 0.5 V. Earth-to-earth test – this is when the leads are placed on the earth of the relay and the earth of the battery. This checks that the earth path is good. The maximum volt drop should not exceed 0.5 V.
Other tests	Use of any self-tests which may be available within the A/C module.

Using a Multimeter

A high impedance meter with a minimum internal resistance of 50 megaohms/volt is required for measuring modern sensors and actuators on automotive systems. A meter with a range of up to at least 50 VDC and as low as 200 mV. VAC should be at least 200 VAC and have resistance ranges between 0–2 megaohms. If the multimeter has a frequency, tachometer, PWM, Duty cycle and a connector for a temperature probe then other data can be obtained. Current measurement is extremely important. Cheap meters only go up to 10 or 20 amps. If high current measurement is required then purchase a quality meter which can measure high currents with a shunt or a compatible amp clamp. The ideal approach is to purchase a digital oscilloscope (Fluke is highly recommended) and amp clamp.

Example specification:

VDC 0–50
VAC 0–200
Resistance 0–2 MΩ
Current 0–500 amps (shunt or amp clamp)
Tachometer 0–10 000 rpm (not essential if you have frequency measurement)
Frequency 0–500 Hz
Dwell 0–90°
Duty Cycle 0–100%
PWM 0–20 ms

Note – when using a multimeter you can only assess the readings obtained by the meter and not the control signal itself. This means you cannot measure any background interference or electrical noise. AC readings will be based on the rms (root mean square) voltage (Urms) which is approximately 70% of the peak value. This means that these meters are not really adequate for measuring AC type vehicle sensors. Oscilloscopes should be used in these circumstances. An oscilloscope will measure the peak (amplitude) and the peak-to-peak value of the waveform including the signal shape, polarity, frequency and any interference.

Wiggle test

A wiggle test is carried out by attaching a meter to a circuit and obtaining a reading. The technician wiggles the cables and connectors to observe variations in the meter reading in the pursuit of verifying a fault. This can also be done using a serial tester and oscilloscope (trend plot). A serial tester/code reader will produce a fault code if a fault is sensed in a circuit during a wiggle test. This allows the technician to use the control module as the diagnostic tester. This test should be a part of a sequence and not relied upon as a singular result.

Load variations

When testing sensors, actuators, modules and cabling a variation in loading on the component will be required. Similar to a wiggle test, where a load is applied by using a force on a component, its cabling etc. Temperature sensors need to be subjected to a real temperature variation. Speed, load, temperature, position, vibration can be used to try to simulate the real environment which the component is *normally* subjected to. While the sensor or actuator is being subjected to these variations a quality tester can be used to view or preferably record the data for analysis.

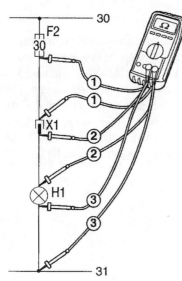

1st measurement:
Ohmmeter indicates less than 2 Ω: circuit OK up to this measurement point.
Indication too large: lead interruption or contact problems between fuse F2 and wiring harness plug X1.
2nd measurement:
Ohmmeter indicates less than 2 Ω: circuit OK up to this measurement point.
Indication too large: lead interruption or contact problems between wiring harness plug X1 and lamp H1.
3rd measurement:
Ohmmeter indicates more than 2 Ω: lead interruption between lamp H1 and ground.

Figure 3.94 Resistance checks
(courtesy of Vauxhall Motor Company)

Resistance (Ω)

Resistance measurement is carried out in parallel, across the component. Ohmmeters use their own power source which means that all power in the circuit being tested must be isolated. If this is not done and the meter is placed in the circuit then the meter can become damaged. A resistance measurement is not a dynamic test. This means that the sensor will not be under load.

Resistance measurement with digital multimeter:

Fault symptom:
Lamp H1 does not illuminate.
Pre-condition:
Battery voltage OK, fuse OK, bulb OK. Set digital multimeter to Ω (resistance), fuse F2 pulled (during the resistance measurement, current must not flow through the lead being measured).

Continuity

A continuity check is based on checking for a complete unbroken electrical path in a circuit, part of a circuit or through a component. A continuity test checks if a circuit is 'continuous' (continuity) and not interrupted (open circuit). A value is not required from the circuit just an indication that the circuit is open or closed.

Ohmmeter

When testing continuity with an ohmmeter the circuit or component being tested must not be a part of a live circuit. All power to the circuit or component must be isolated. The meter is positioned in parallel, across the component. If no continuity exists in the circuit then an infinite sign will appear on the meter.

Voltmeter and ammeter

A voltmeter and ammeter will test the continuity in a *live* circuit. The voltmeter and ammeter are positioned in series and generally any reading above '000' indicates continuity.

Voltage – potential difference (V)

Voltmeters should have high internal resistances and only draw a small amount of current. A volt drop is measured by placing the meter in parallel (across the component) at two points, while the circuit is under load. The voltage at a point in a system can be measured by ensuring that one of the points of the meter has a zero reference (fitted to the battery earth). Meters which have an automatic range will provide the reading as long as it is within the capabilities of the meter. Analogue meters should not really be used on a vehicle system. If they are used to measure voltage then the meter should have an internal resistance of no less than 50 kohms/volt.

Permissible volt drops in a circuit

The volt drop in a circuit depends on the cross-sectional area of a cable, its length and the current flowing through it. The current flowing through the circuit has the greatest effect. For example, when the engine is cranked the battery voltage can reduce to as low as 10.5 V. This is a volt drop of 2 volts due to the high current flow. In some circuits this may be acceptable. In low current circuits up to approximately 2–3 amps a volt drop within the circuit of approximately 0.5 V should not be exceeded.

A power-to-power test is when the leads of the meter are placed one on the vehicle battery positive terminal and one to the supply of a control module or other consumer powered by battery voltage. This is to check that the supply voltage is acceptable. The maximum volt drop between the two points should not exceed 0.5 V. Low current systems should be as low as 0.25 V (sensor circuit, i.e. temperature sensor). If the voltage is exceeded then a resistance between the two points exists, for example faulty relay contacts, poor battery terminal connection, faulty driver circuit inside the module.

The power to a sensor supplied with 5 V from a control unit can also be tested. Place the ground lead of the voltmeter on the module 5 V output pin on the module connector and the positive lead on the power at the sensor. A volt drop of no more than 0.25 V should be expected.

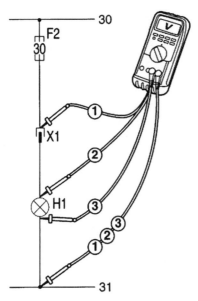

1st measurement:
Voltmeter indicates more than 11.5 V: circuit OK up to this measurement point.
No indication or indication too small: lead interruption or contact problems between fuse F2 and wiring harness plug X1.

2nd measurement:
Voltmeter indicates more than 11.5 V: circuit OK up to this measurement point.
No indication or indication too small: lead interruption or contact problems between wiring harness plug X1 and lamp H1.

3rd measurement:
Voltmeter indicates more than 11.5 V: lead interruption between lamp H1 and ground.

Figure 3.95 Voltage checks
(courtesy of Vauxhall Motor Company)

An earth-to-earth test is when the meter leads are placed on the earth of a component like a sensor and the earth of the battery. This checks that the earth path is good. The maximum volt drop should not exceed 0.5 V. Low current circuits should be as low as 0.25 V (sensor circuit, i.e. temperature sensor). If the voltage is exceeded then a resistance between the two points exists, for example a bad earth or trapped wire.

Voltage measurement with digital multimeter:

Example fault symptom:
Lamp H1 does not illuminate.
Pre-condition:
Battery voltage OK, fuse OK, bulb OK. Set digital multimeter to VDC (direct voltage).

Different meter configuration giving different readings

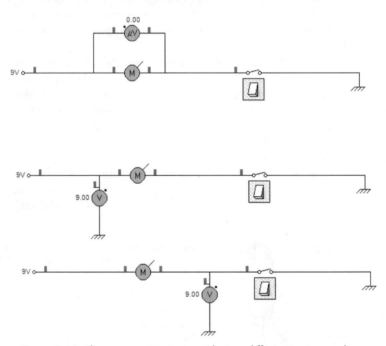

Figure 3.96 Three open circuits provide two different meter readings
(courtesy of Crocodile Clips)

The simulation in Figure 3.96 shows a simple motor circuit in three different configurations. The configurations only change with the position of the voltmeter positions. The simulation is designed to show the different readings obtained by placing the meter in different configurations. In all the configurations the meter is reading the potential difference between the positive and the negative terminal. The difference in the meter readings are due to whether the negative connector is attached to a zero reference or across the component. The circuits in Figure 3.96 are not operational due to the switch being open. This means that the motor is ground switched. By placing the meter across the motor you get a potential difference (pd) of zero due to 9 V being measured both sides of the meter. The two other meters show voltage due to the ground connector being ground referenced (attached to battery earth). If the switch is integral to a control module then voltage will be measured all the way to the pin on the module

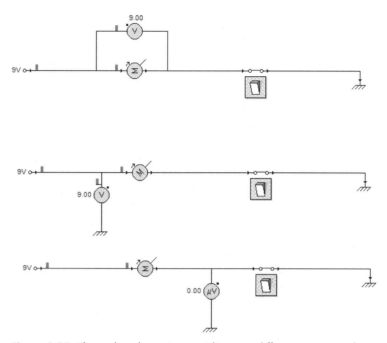

Figure 3.97 Three closed circuits providing two different meter readings
(courtesy of Crocodile Clips)

but no current will flow until the switch inside the module is closed. This can often lead to confusion when trying to understand when current is flowing and when it is not.

The closed circuits have produced two different meter readings and one meter which has not changed. If a component is ground switched like the example in Figure 3.97 then placing the meter across the component will provide a zero voltage when the component is off and supply voltage (minus volt drop caused by current flow) when on. If the meter is placed on the supply side of the component which is ground switched then no change in voltage will be measured during its operation. If the meter is placed on the ground side of the component then during its operation it will change from supply voltage to between zero and 500 mV (volt drop depends on amount of current flowing) also called a ground-to-ground test.

Measuring the pd across the component is always recommended. If just the supply to the component is required then place the multimeter cables on the supply of the component and a zero ground reference (battery ground). A power-to-power check should also be made by placing the cable of the meter onto the supply of the component and the other meter cable on the battery supply (or other voltage supply, i.e. 5 V from ECU). This will provide a reading on the difference of the two voltages. This difference should not be more than about 500 mV.

Measuring AC voltage

Alternating voltage is only produced on a motor vehicle by the alternator which is then rectified into DC before charging the battery. The only other means of measuring AC on a motor vehicle is measuring the output from the sensor. An inductive sensor produces an AC signal which can be used to measure the speed, position and rate of change (velocity) of a variable (e.g. the vehicle speed). The signal will be transmitted to a control module. The best way to measure an AC signal is using an oscilloscope. Inductive signals change in amplitude, frequency, and voltage as well as shape. Obtaining peak-to-peak voltages, frequency, shape and amplitude from a multimeter is beyond its ability.

Measuring current

Measuring current through a circuit is the most important test available to the technician. For example, if the current was measured flowing through the power supply of an A/C module the test would give the technician a great deal of information. To understand how much information which could be obtained with such a test an appreciation of the current path is required. The current flows from the vehicle battery and terminals through cabling, fuses, relays, more cabling and eventually the module connector. Once inside the module the current will supply internal driver and memory circuits, other sensors which the module powers, and eventually go to earth. The earth path includes earth connection to the chassis, then through the chassis and back to the battery. During the test the circuit is under load. Compare this to measuring the potential difference of two points. Current is the preferred measurement but the problem of finding data on current values is not always easy. It is always good practice to take measurements from vehicles that are operating correctly and record data for comparison purposes. The value of potential difference measurement is important but current should always be applied whenever possible.

An ammeter is a low resistance instrument and is very easy to destroy. It is also connected in series with a circuit and is polarity conscious. This can cause problems due to the inability to break the circuit to place the meter in series. Fuseholders are often a very good location for fitting an ammeter. Upon removal of the fuse the ammeter set on the correct meter range can be placed across the pins of the fuse holder to allow all the current of the circuit to flow through the meter. Short circuits, open circuits and low current flow due to a high resistance in a circuit can be measured. Always invest in a good ammeter. Vehicle batteries can supply up to and above 500 amps so if a short circuit to earth exists within a circuit, this will flow through your meter. Shunts are often used to protect ammeters and are worth investing in. Current measurement of a compressor coil can benefit the diagnosis of a faulty coil or a large air gap.

In the case of a lightbulb fitted to a fuseholder of a circuit which is not operating due to being open circuit, if the circuit becomes intermittently closed then the bulb will light up. This is often done while carrying out wiggle tests (wiggling cables and connectors) on electrical cables during faultfinding.

AMP clamp

A current clamp is a passive tester which has sensing circuitry that calculates accurately how much current is flowing through a circuit. It is passive because the circuit being measured can remain complete and not be disconnected like conventional methods of testing currents in series. The clamp connects via a BNC connector to the multimeter leads of an oscilloscope or

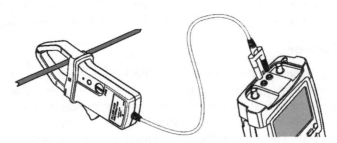

Figure 3.98 Measuring current using an amp clamp
(courtesy of Fluke)

conventional multimeter. The meter is set to voltage and the clamp senses the effects of current flowing through the cable and produces a voltage reading that represents a current flow through conversion, e.g. 100 mV/A so 500 mV means that 5 amps is flowing in the circuit. Amp clamps are a must for modern diagnostic testing.

Measurement of PWM using a voltmeter

If we measure PWM by means of a voltmeter (AC), the instrument will show the average voltage carried by the cable so that a higher pulse ratio will show a higher average voltage reading. Using the voltmeter, we can obtain a rough estimate of the pulse ratio. A pulse ratio of 9% (switched by 13 V) will often give a reading of about 1.2 V (0.09×13 V = 1.17 V). In the case of positive-triggered PWM, connect the red test lead to the cable and the black test lead to a good ground. In the case of negative-triggered PWM, connect the black test lead to the cable and the red test lead to battery positive. Select 'Smooth' on the voltmeter if it has this function.

Power probe

Power probes are useful when a current or a ground signal needs to be applied to a vehicle system. The probe is connected to the vehicle battery and can supply voltage at 12–14 V (if the engine is running) and current up to approximately 20 amps. The probe has Light Emitting Diodes (LEDs) built into the casing which indicate if a wire is live (red LED) or ground (green LED). Power probes also have trip switches in case of an accidental short circuit. If a voltage is absent or weak, additional voltage can be applied to the circuit by pressing a switch on the probe. It must be noted that components can be destroyed through the misuse of a power probe. Power probes should not be used to apply voltage to a control module. Power probes are very useful for testing relay circuits (power and earth points) and upon the removal of a relay the probe can be used to power an A/C compressor clutch providing the fault is electrical and not with the A/C system, i.e. lack of refrigerant (care must be taken or damage to the A/C system can occur due to lack of oil or refrigerant in the system). Power probes can also be used to provide voltage to low current DC motors like recirculation motors.

LED logic probe tester

LED probe testers have two LEDs which are used for a range of tests. One LED is green and the other is red. The red will illuminate whenever it touches a voltage above 0.7 V; the green LED will illuminate whenever it senses a ground signal. The most useful test is a continuity test. The sensor will light when a voltage above 0.7 V is sensed. This tester is very useful for testing pulsed circuits. These circuits include stepper motors, Hall sensors or variable reluctance sensors where the voltage drops to zero and rises above 0.7 V. This causes the LED to switch on and off rapidly (flash) indicating a signal is present. They are also very useful for placing across an A/C cycling switch to provide an indication of when the switch is cycling.

Serial testing (OBD 16- pin connector)

Serial testers are diagnostic testers used to communicate with vehicle systems via the multiplex communication network. This means that the serial tests must be able to 'communicate' via the protocols used throughout all the different manufactured vehicles. It is important to note that the information a serial tester provides is simulated to aid diagnostics and inform the technician on how the system is functioning. The data is not live and is often not 'seen' in the same format as it is presented by the diagnostic tester. An example to aid the explanation is an air distribution door stepper motor. The serial data may be presented as a percentage of opening: 0% closed, 50% half open and 100% fully open. The signal the A/C module sends to the motor is a pulse train used to rotate the door. This means that the serial information will be in a different format than the actual signal which is live and would be sampled using a different process.

Serial testers play a very important role in modern diagnostics. The most common use of a serial tester is for fault code analysis. This allows the technician to see if any faults have been recorded by the modules within the system and logged using a unique fault code which is assigned to every sensor and actuator on the vehicle. Serial testers often provide the technician with the ability to see more simulated information at a glance than say using a dual oscilloscope when only two signals can be monitored at once. The opportunity exists to carry out actuator tests where actuators are energised to witness operation. Serial testers also have important programming functions so that adjustments to a module's memory (data stored in EPROM) can be made allowing the vehicle system to be reprogrammed 'in service'. Reprogramming changes the operational parameters the system puts in place. Advanced serial testers which are dedicated to manufacturers' use, often have comprehensive guided diagnostic capabilities. As vehicles advance the units can be updated by being programmed via intranet or CD ROM and interface with the other knowledge-based systems which the manufacturer uses (technical data systems and data collection).

Serial test – fault code check

While the parameters, or readings, required by OBD II regulations are uniform, auto manufacturers had some latitude in the communications protocol they used to transmit those readings to scanners. Expensive scanner consoles costing thousands of pounds often include decoding software and firmware for all protocols in their units, making them universal. Less expensive units are usually customised for a specific communication protocol. Be sure the scanner you are using suits the protocol of your customer base.

For the retrieval of such codes the diagnostic tester is plugged into the Diagnostic Link Connector (DLC). The diagnostic tester communicates with the system (if networked) or individual module to match the protocols the system is using. If a match cannot be found then a communication error will be displayed. If the communication is successful then information about the vehicle or system will be displayed. A simple menu structure is used on the tester providing a range of facilities for diagnostics. DTC will allow the user access to any stored trouble codes (referred to as historical codes) which should be recorded by the technician and then removed. DLC monitoring is continuous so the vehicle can be taken for a test drive and monitored for any codes that may present themselves. A DLC often only provides an indication that a fault exists within a particular system or associated with a component. It will not tell you exactly what is wrong. Additional testing is often required.

> Note – always check fault code history. If in doubt of system performance then record and clear the fault code history and take the vehicle for a dealer drive cycle test.

Data collection

On some serial testers data collected during a test session may be saved in the instrument for further analysis and also transferred to a PC for future reference. The serial tester is able to perform efficient fault diagnostics on different types of control system, including engine management, ABS, airbag and of course air-conditioning. The serial tester is portable, has a built-in battery supply and may be used for on-road testing.

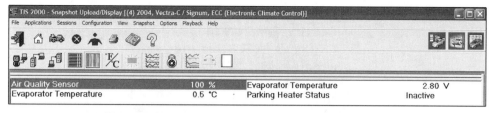

Figure 3.99 Data list uploaded from a serial tester Tech 2
(courtesy of Vauxhall Motor Company)

Sensor	Output	Range
Interior temperature sensor	110 Hz	
A/C panel position status (face vent*)	35%	0–100%
A/C blend door position	25%	0–100%
Defrost door position (windscreen vent)	Closed	Open/closed
Recirculation door position	Open	Open/closed
Fan speed increase	Off	Off/on
Fan speed decrease	Off	Off/on
Temperature increase	Off	Off/on
Temperature decrease	Off	Off/on
Ignition status	On	On/off
Dimmer switch input status	1	1–20 (1 dim 20 light)

* Face vent selected when testing.

Figure 3.100 WDS serial information Mondeo ECAT

Data list

A serial tester reads current data from the vehicle's control unit. The values are updated continuously (small delay).

Data list information can be used to test sensors' wiring. A typical example is using a jumper wire across terminals to check extreme sensor variations. A jumper wire may be placed across the input and output of a sensor to test the wiring of the sensor.

Connect fused jumper wire to: B135 sensor/air quality wiring harness connector (wiring harness side) terminal 3 and terminal 2.

Diagnostic tester data list parameter air quality sensor will read 0%.

Actuators

In most cases, actuators can simply be switched on and off.

Snapshots

All screens that display data received from the vehicle can be saved as 'snapshots'. The snapshots can be downloaded to a PC to create test reports. The saved information is displayed in the same way as when the information was saved.

Programming

Advanced serial testers can be used to program control modules or other components like transponder keys.

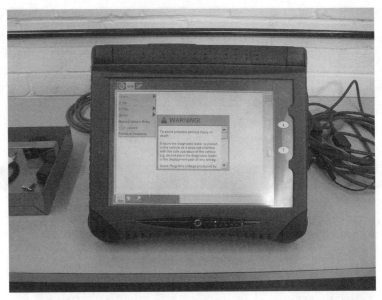

Figure 3.101 An advanced manufacturer specific serial tester (WDS)

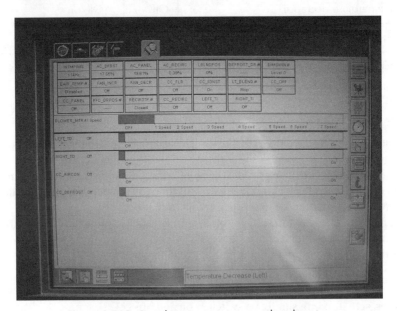

Figure 3.102 Serial tester carrying out data logging

Wiggle test

A wiggle test detects intermittent faults in harnesses and/or connectors. When the test is performed the vehicle module is used to detect and record any faults that present themselves during the test. Hence, it is the intelligence of the module that determines the accuracy of the test. During the test, wiggle multi-plugs and wiring without using excessive force or detaching connectors. Some testers provide an audible beep when a fault has been found in the affected plug or associated wiring.

A/C systems with built-in self-test diagnostics

ATC and SATC systems have the facility to monitor their own inputs and outputs by means of a self-test. The system recognises continuous or sporadic faults and stores them under an appropriate fault code. These fault codes can be read out by activating a diagnostic mode. Diagnostic mode is accessed by sequential operation of a certain key combination on the climate control selection panel. To start self-diagnosis and read out the fault memory, the ignition key must be turned to the 'ON' position and the battery voltage must be between 9 and 16 V. Generally codes are presented in a two-digit format or a flashing LED (blink code). They are presented on the climate control graphics interface.

This is in addition to fault codes accessed via a diagnostic tester by connecting the 16-pin Data Link Connector (DLC) to the vehicle.

Activation of self-diagnosis

Activation can be automatic which means that as soon as a fault is present an LED will flash or message will be displayed on the graphics interface. More common is activation through pressing a sequence of buttons.

For example, briefly press the 'OFF' and 'FOOTWELL' buttons on the A/C controls simultaneously for at least 2 seconds, then press 'AUTO' within 1.5 seconds. The self-diagnosis which then starts lasts a few seconds. An animated display appears during this time. Any faults found are displayed in the form of trouble codes. Example: first of all, '93' flashes for 2 seconds, then '42' flashes for 2 seconds – DTC 9342. If no faults are stored, then all of the segments in the display are actuated. Diagnosis mode can be stopped at any time by pressing any button on the A/C controls. If the 'DEFROST' button is pressed to end the diagnosis mode, all DTCs in the fault memory will be deleted!

Read out stored faults

On the A/C controls briefly press the 'OFF' and 'FOOTWELL' buttons simultaneously for at least 2 seconds, then press 'HEADROOM' within 1.5 seconds. Any stored intermittent faults are output on the graphics display and should be noted for safety reasons. By pressing the 'DEFROST' button, the fault memory is cleared and diagnosis mode is ended. To end diagnosis mode without clearing the DTCs, press any other button on the A/C control.

The above is just one example of a self-test activation procedure. Always refer to manufacturers' specifications.

Break-out box testing (parallel communication)

Whenever possible, break-out boxes should be used for testing electrical signals. A break-out box is a parallel interface which attaches to the existing control module connectors allowing easy connection to wiring without reducing its integrity. In simple terms, remove the connector (female) of a module which may contain hundreds of pins and plug it into the connector of a break-out box (male) and then plug the connector (male) of the break-out into the module. All the electrical signals travelling into and exiting the control module will travel via the break-out box.

As an example, if the module has 104 pins on its connector then a 104 pin break-out box can be attached allowing the technician quick and easy access to testing all of the electrical signals. The pin numbers on the module connector usually match the pin numbers on the break-out box. For example, pin 1 on the module connector should be pin1 on the break-out box.

Some aftermarket suppliers produce generic break-out boxes that fit to a range of manufacturers. They often use overlays on the break-out box pins to match the pin numbers of the module with the numbers on the box.

Figure 3.103 104 pin break-out box
(without serial interface)

Some manufacturers use break-out boxes that can interface with their serial testers which allow the monitoring of live data instead of simulated data via the 16-pin serial connector. Other manufacturers attach oscilloscopes to break-out boxes to monitor, record, and plot live data. Break-out boxes are extremely useful, even small boxes with 2, 3, 4, 5 pins fitted to allow the connection to individual sensors and actuators.

If a break-out box is not available then suitable adaptor leads must be used to take readings from the module connector. Wiring information on connector views and colour codes are required for such a task.

3.4 Oscilloscope waveform sampling

Most oscilloscopes used within the motor vehicle field are handheld allowing portable use.

A modern digital oscilloscope will have a number of functions such as:

- data capture – snapshot, trend, glitch, recording waveforms and set-up functions;
- freeze frame and zoom functions;
- data storage and retrieval, automatic report function;
- auto scope measurement – presents the waveform in the most appropriate format for the less experienced user.

Waveforms can generally be captured, recorded, reviewed and then stored for later analysis. Often they can then be added to a report or compared to another signal and printed out. Oscilloscopes are easy to connect and often have auto functions to aid less experienced users to sample waveforms. The sampling rate of a good specification digital oscilloscope is very quick enabling intermittent faults (glitches) to be captured and recorded. Sampling a signal allows you to check the signal pattern providing additional diagnostics.

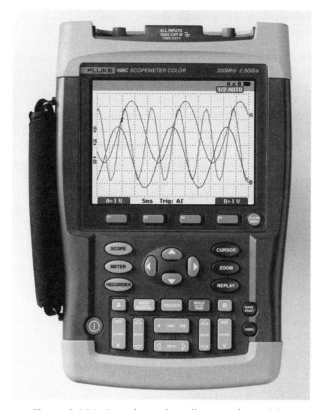

Figure 3.104 Two channel oscilloscope from Fluke
(courtesy of Fluke)

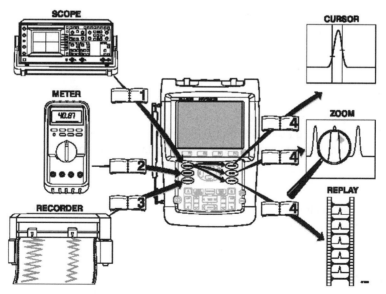

Figure 3.105 Functions of a modern oscilloscope
(courtesy of Fluke)

Information is available illustrating waveforms for a range of different sensors and actuators for comparative purposes. Waveforms generally have the following characteristics:

- Amplitude – voltage level.
- Frequency – the number of cycles of the waveform per second, dependent on the circuit or sensor's operating speed.
- Pulse width – the time current flows through the component (measured in ms).
- Duty cycle – a measurement of the on time compared to the off time of a cycle, measured in percentage. Duty cycle is often used to control the amount of current flowing in a circuit.
- Signal shape – a flat slow changing analogue signal, sawtooth signal, square-wave signal. Shape depends on the type and construction of the sensor or actuator.

Waveforms

DC voltage signal analogue waveform (amplitude manipulated)

An analogue DC waveform only tends to change in amplitude. It has no defined signal shape apart from being a flat line if there is no change in the measured variable, e.g. no change in temperature. Often the change is slow meaning a good method of recording such a variation is by recording the waveform using trend. Trend is used to record slow changing signals using slow sampling rates. For example, a sample rate of 5 ms for 6 hours can be used which means the oscilloscope will record the signal and plot a point on a graph every 5 ms for a total period of 6 hours. The sample can vary from sampling every 2 minutes for 48 hours. This may be used for data recording of a specific fault associated with the gradual loss of a signal or level of power in a circuit.

A DC analogue signal can also be measured using the Min and Max selection of a scope. This enables the measurement of the total variation in voltage to be known which can be compared to a known value for analysis.

Waveform

The plot in Figure 3.106 shows a reduced volt drop across an NTC sensor. As the temperature applied to the sensor by the A/C system increases the resistance reduces with a corresponding reduction in volt drop.

Signal checks:

1. Monitoring of the temperature of the air flowing past the sensor to carry out a comparison.
2. The waveform should correspond to an NTC or PTC graph.
3. The peak voltage should be referenced to the specification of the sensor.
4. Voltage transitions should be steady and reflect a change in temperature.

DC/AC

This is displayed as an alternating DC voltage. This is a DC voltage which alternates between two voltage points at and above the zero voltage line. For example, the voltage may alternate between 1.5 and 3.5 volts. The signal will not go into a negative voltage like an AC signal.

Step and pulse voltage

A step and pulse voltage are single occurrences which change from one state to another when the circuit is triggered. A step voltage occurs when a voltage steps from one level to another.

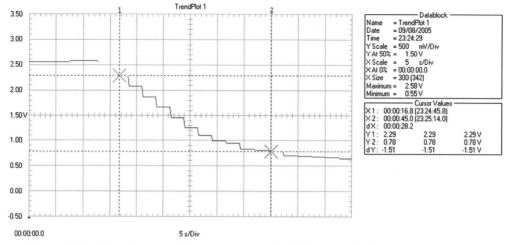

Figure 3.106 Slow changing DC analogue signal (left face vent outlet temperature sensor)
(courtesy of Fluke)

A typical example is a compressor clutch (Fig. 3.107). Diodes are connected to these circuits to prevent voltage spikes which are produced from the collapse of the magnetic field. The spikes could damage the driver circuit inside the control module. A clean transition from one state to another should be seen.

A pulsed voltage occurs when the voltage changes from one state to another and then back to its original state. This is a single pulse, characterised by an instant rise and fall time, and indicates where the voltage spike is being filtered. This is after the clutch is de-energised.

Pulse train

A pulse train can be categorised into a number of subgroups. For example:

- Frequency modulated (change in the frequency, cycles per second).
- Duty cycle (change in the percentage of the on to off time in one cycle).
- Pulse width modulated signal (the time in milliseconds the current is flowing through the component).
- Pulse shape, a change in the shape of the pulse over time.
- Amplitude, change in the voltage level of the signal.

A voltage signal can include a number of these traits.

Frequency modulated signal with fixed duty ratio, shape and amplitude

The *frequency* is determined by the number of pulses (oscillations per second). Accordingly, the frequency increases/decreases proportionally to the number of pulses per second. The frequency (formula symbol 'f') is measured in hertz (Hz). A frequency modulated signal will change in the number of 'cycles' that occur every second with a change in the sensor input of the control of the actuator. Hall effect speed sensors display this type of signal (Fig. 3.108). As the vehicle speed increases the signal is switched on and off quicker so the number of cycles increases. The shape and duty cycle of the signal is fixed by the very nature of the sensor construction. The signal should be clean with no under- or overshoot providing a constant amplitude throughout the pattern.

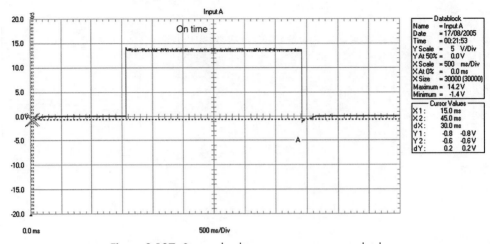

Figure 3.107 Stepped voltage on a compressor clutch
(courtesy of Fluke)

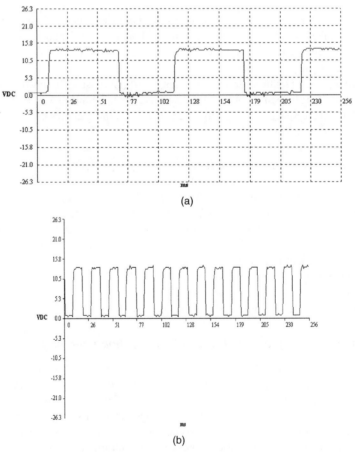

(a)

(b)

Figure 3.108 Hall effect speed signal with fixed amplitude and duty
cycle and variable frequency:
(a) 9.6 Hz, (b) 58 Hz

Signal checks:

1. Monitor a steady change in frequency directly proportional to a change in speed.
2. The peak voltage should be equal to the reference voltage (200–400 mV difference is acceptable). If different then carry out a power-to-power test.
3. Voltage transitions should be straight and vertical (unless switching an inductive component).
4. The lower horizontal lines should almost reach zero. A small volt drop is allowed approximately 200 mV, it should not exceed 400 mV. If high carry out earth-to-earth check.

AC voltage signal – amplitude, shape, frequency and pulse width are variable

An AC signal voltage alternates from zero to a maximum positive voltage and then back to zero and then to a maximum negative voltage (one cycle). These signals are generally produced by an inductive sensor (variable reluctance sensors) using the theory of movement and magnetism to produce an AC voltage. The sensor has no power supply only two shielded wires and a coil with a magnet inside. The trigger wheel is constructed from a permeable material, a low magnetic reluctance steel. As the trigger wheel rotates, small signal voltages are induced into the coil which can be measured by a module for the rate and change of speed of the rotor. If a tooth is missing on the rotor then this will cause one of the cycles to be missing due to no induced voltage.

The signal in Figure 3.109 shows the voltage constantly changing and repeating itself. The cycle is described as frequency. The number of cycles per second is presented in hertz. The signal should be smooth and progressive and the peak voltages should be identical with no change in speed.

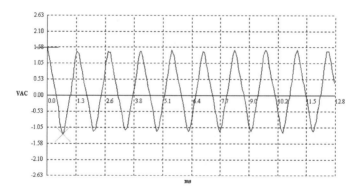

Figure 3.109 A/C PWM signal from an inductive speed sensor

As a change in speed occurs then a change in the shape, frequency, and amplitude will occur (Figure 3.110). The signal often displays a missing cycle. This is due to a missing tooth on the sensor's trigger wheel.

AC signals have the potential to induce electrical interference. This is why the cables are generally heavily shielded. This interference can be viewed on a scope.

Pulse width modulated signal (millisecond measurement)

PWM (Pulse Width Modulation) signals are square-wave or pulse train signals with a constant frequency, but variable on time. The *pulse width* is the *duration* of the active signal. This signal displays a number of characteristics but the technician is only interested in one. The time current flows in the circuit measured in time, often milliseconds. A solenoid can be viewed using this method. Figure 3.111 shows a saturated voltage, a solenoid (injector) with a full voltage applied to operate it. The solenoid is required to be open for a fixed time. This can be measured using PWM.

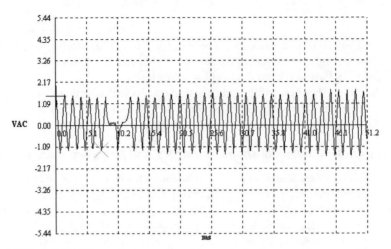

Figure 3.110 AC PWM waveform produced by an inductive speed sensor with a missing tooth on the rotor

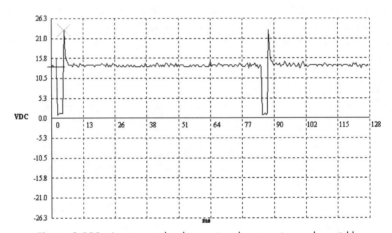

Figure 3.111 A saturated voltage signal measuring pulse width

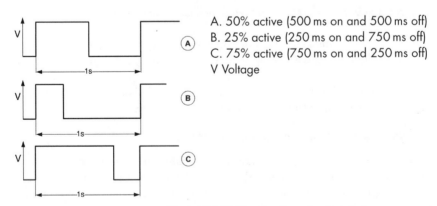

A. 50% active (500 ms on and 500 ms off)
B. 25% active (250 ms on and 750 ms off)
C. 75% active (750 ms on and 250 ms off)
V Voltage

Figure 3.112 Duty cycle ratio signal
(reproduced with the kind permission of Ford Motor Company Limited)

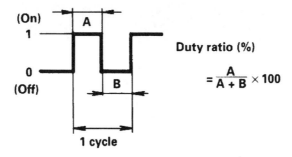

A: **Current flows (on)**
B: **Current does not flow (off)**

Figure 3.113 Formula for duty cycle ratio
(with the agreement of Toyota (GB) PLC)

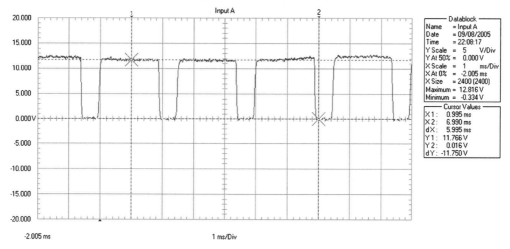

Figure 3.114 Speed demand signal, pulse width modulated signal with a
fixed frequency 400 Hz, duty cycle ratio 76%
(courtesy of Fluke)

Duty cycle ratio with fixed frequency

The *duty cycle* is the ratio between the *on* and *off times* of a PWM signal. The duty cycle is expressed as a percentage. Accordingly, a duty cycle of 25% means that the signal is active 25% of the time; over 1 second of pulse width modulation, for example, the signal is active for 250 ms and inactive for 750 ms. PWM signals can serve as output (e.g. for controlling a solenoid valve or blower motor) as well as input (digital sensor) (Fig. 3.112).

PWM signals can serve as outputs for controlling solenoid valves as well as inputs from sensors or control modules. Figure 3.114 provides a duty cycle ratio signal which is sent from the A/C module to the blower control module to indicate blower speed demand. The duty ratio variation indicates the required speed.

The duty cycle can be measured with the help of an oscilloscope or a serial tester.

Blower speed	Duty cycle ratio (%)	Frequency (Hz)
1	76.5	400
2	70	400
3	63.7	400
4	54.7	400
5	43.7	400
6	30	400
7	26	400

Figure 3.115 Blower control signal (demand signal)

Figure 3.116 PV 350 Fluke pressure transducer
(courtesy of Fluke)

Measuring pressure with transducers

Pressure transducers can be linked to scope meters for the direct measurement of pressure and vacuum. A highly accurate pressure transducer can measure from 0.5 to 350 psi (3.447 to 2413 kPa) and 0 to 29.9 inHg (0 to 76 cmHg). Most are compatible with any automotive fluid, such as R12, R134a, and possibly CO_2.

Testing multiplex signal communication between nodes (modules)

The CAN (Control Area Network) standard supports half duplex communication with only two wires to send and receive data forming the bus. The nodes have a CAN transceiver and a CAN controller for bus access. At both ends, the bus must be terminated with a resistor, typically 120 Ω. The CAN transmits signals on the CAN bus which consists of a CAN-high and CAN-low. These two wires carry anti-phase signals in opposite directions to minimise noise interruption that simultaneously interferes with the bus. The CAN bus line can have one of two logical states: 'recessive' and 'dominant'. Typically, the voltage level corresponding to recessive (logical '1') is

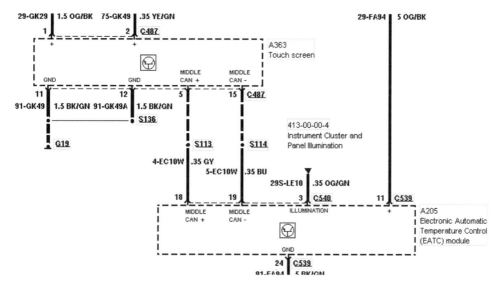

Figure 3.117 Connection between two nodes for A/C control

2.5 V and the levels corresponding to dominant (logical '0') are 3.5 V for CAN-high and 1.5 V for CAN-low. The voltage level on the CAN bus is recessive when the bus is idle.

When analysing the CAN bus signals, it is of interest to measure the peak-to-peak voltages and to verify that the CAN signals are disturbance free. Each input can be viewed on a dual scope (two measurements to be made simultaneously – Figure 3.93).

Multiplex wiring diagram example

Figure 3.117 shows a mid-speed CAN network. On the MS CAN bus, the transmission rate is 125 kbit/s. Cabling is between two nodes, the touch screen A363 module pins 15 and 5 and the A/C module A205 (EATC) pins 18 and 19. CAN-high signal is between pins 5 and 18 and CAN-low signal between pins 15 and 19.

Motor speed control

Blower motor example

Speed control is required for the interior fan of the ventilation system. Generally this is achieved by controlling the current within the circuit.

The current is controlled through a series of resistors which reduce the current flow. Figure 3.118 shows the motor M3 receives a switched feed to its positive terminal. The ground of the motor is then directed by switch N73 allowing current to flow through the resistor pack to ground or directly to ground. If the current flows to the resistor pack then it will reduce in value allowing the fan to operate at a reduced speed (some of the electrical energy will convert to heat and be wasted).

If the current goes straight to earth then the fan will operate at its maximum speed due to maximum current flow.

Figure 3.119 is a simplified circuit of the blower motor. The switches represent the multi switch N73 which has three speed positions. In position one (left switch) the motor has the maximum amount of current flowing and will operate at maximum speed. This can be seen on

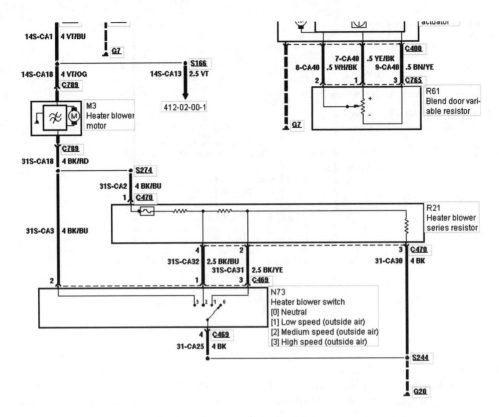

Figure 3.118 Diagram of heater blower motor speed controller
(reproduced with the kind permission of Ford Motor Company Limited)

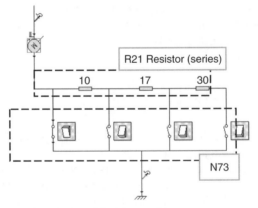

Figure 3.119 Simplified heater blower motor control
(courtesy of Crocodile Clips)

the waveform in Figure 3.120. For the first 2 seconds the maximum amount of current is flowing to the motor so it will be rotating at its maximum speed. To reduce the speed of the motor we reduce the current by including more resistors in the circuit.

Switches 2, 3 and 4 direct the current through more resistors which cause the motor to reduce in speed.

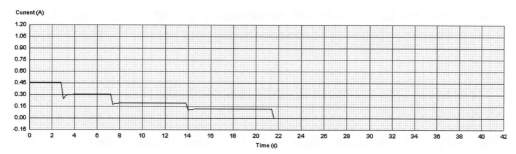

Figure 3.120 Waveform of the blower motor control (current/time)
(courtesy of Crocodile Clips)

The waveform shows the reduction in current over time which caused a reduction in rotational speed of the motor. The switching points can be seen on the graph by the sudden drop in current. In reality the blower circuit uses about 15 amps at full load. The above graph with its current values is used just as a simulation.

Electronic control of brushless blower motor

The above waveform is analogue and is a typical way of controlling the speed of a motor. It is very inefficient due to the wasted energy dissipated by the additional resistors. The resistor pack R21 has a large heatsink enabling it to do this job. Not all motors have to be controlled in this way. The ultimate aim is to reduce either the voltage or the current to the motor which can be achieved through switching the current on and off very quickly, often as quick as 500 times per second (500 Hz).

Pulse width modulation and duty cycle switching

By switching the motor on for one millisecond and then off for one millisecond and then on and so on, the motor accelerates during its 'on' phase and freewheels during its 'off' phase. The switching is very quick and the acceleration and deceleration are not noticeable until very low switching rates are achieved; like 1 cycle per second (1 Hz). When the motor is switched on it will be connected to a power supply of 12 volts (V). If the signal is on for 50% of the time and off for the other 50% of the time then the average voltage will be only 6 V. This waveform is said to have a duty cycle of 50%. If the on time is extended and the off correspondingly reduced, for example 75% on and 25% off, then the duty cycle will be 75% with average voltage being 9 V. A number of motors and solenoids are controlled in this manner on a motor vehicle. By reducing the voltage there is no waste heat created and a method of fine adjustment is achieved. There is neither any current being drawn during the off time except for the electronics to drive the circuit. This method of control is called Pulse Width Modulation (PWM).

3.5 Multiplex wiring systems

Many vehicles are already fitted with a large number of electronic control and regulation systems. The complexity of these systems means there is an increasing need for them to exchange data between each other. However, the increase in data means that it's no longer sensible to transfer data in the conventional way. In other words, if data were exchanged via individual lines, the sheer complexity of the additional sensors, wiring and connections would make this system scarcely manageable, not to mention the potential error sources that this would introduce.

Another advantage is that fewer sensors have to be installed, as their analogue signals are digitised in the dedicated control unit and made available to other control units via a 'data bus'.

Protocol

Before a data bus system is developed a communication protocol is defined. The protocol agrees the following:

1. Medium used to transmit information – copper wire, twisted pair of wires or fibre optic cable.
2. Speed of transmission – different systems on the vehicle require information quickly while other systems do not. Audio system (slow speed rate) compared to anti-locking braking (high speed rate). Also referred to as class of system – Class A low speed up to 10 Kbps, Class B medium speed up to 125 Kbps and Class C high speed up to 1 Mbps.
3. Method of addressing information (lengths of signals sent).
4. Signals – digital, analogue, voltage, current or frequency manipulated.
5. Error detection and treatment.

The protocol allows all control modules to communicate in the same language. The 'protocol' is provided in the form of software in each control unit.

Data bus

A data bus is the circuit and interface used to allow the information to be available. The modules are in technical terms referred to as 'nodes' in the data bus system and are configured to form a network. The nodes have integrated circuits in them and software to allow them to communicate using the protocol via the bus system. The system becomes more complex when different systems, speed rates and protocols are used on a vehicle. The data bus can also be used to transmit information or actual commands from one control unit to another, for example in order to operate its actuators (e.g. servomotor, relay). The DLC (Diagnostic Link Connector) is also connected to the data bus and the vehicle's systems can be checked with modern diagnostic systems (see section 3.10 for Tech 2 and WDS).

Network

In a bus system, several nodes (control units) are connected with each other via either copper cables or in some systems via fibre optic cables. This is called a network. The nodes communicate with each other through this network and exchange information and data in digital form.

Digital transmission

If you take a specific length of time and divide it into small time windows, it is possible to switch on (1) or off (0) a defined voltage (e.g. 5 volts) on an electric cable for the duration of such a time window, i.e. you can create two switching states (Fig. 3.121). A time window as mentioned above is called a bit (binary digit). As there are only two possible statements for a bit, 8 bits have been grouped together to form a block, which is known as a byte. A byte has 2^8 (2 to the power of 8) = 256 possible statements ($2 \times 2 \times 2 \times 2 \times 2 \times 2 \times 2 \times 2 = 256$). So, with 2 switching states and 8 bits (1 byte) there are 256 possible statements. If more are required, several bytes can be grouped together into one dataframe.

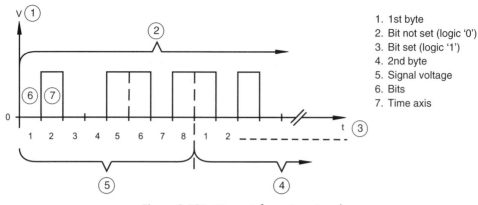

1. 1st byte
2. Bit not set (logic '0')
3. Bit set (logic '1')
4. 2nd byte
5. Signal voltage
6. Bits
7. Time axis

Figure 3.121 Binary information signal
(reproduced with the kind permission of Ford Motor Company Limited)

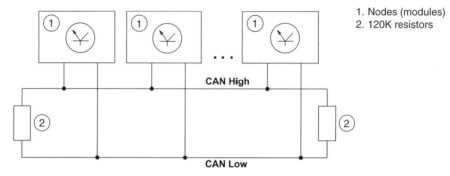

1. Nodes (modules)
2. 120K resistors

Figure 3.122 CAN bus (control area network)
(reproduced with the kind permission of Ford Motor Company Limited)

Networking principles

To enable data to be exchanged between the control units, the control units must be connected to each other electrically. Each control unit can send and receive data on one cable (Fig. 3.122); this is known as a serial exchange, i.e. the individual bits (binary digits) are sent one after the other, like marbles rolling through a tube, on the data bus cable and are read in this way too.

The typical transmission speed is around 50 kbit per second. To countercheck the information sent by a control unit and for the purposes of self-monitoring, a second cable is required; the data transmitted on the first line is also transmitted on this line at the same time, but is inverted. If the signal on one of the lines fails, this is detected by the self-diagnosis facility and stored in the error memory of the control unit as a communications error. If two control units are transmitting at the same time, the significance of the data blocks is stipulated based on the bit combination so that priority can be given to the more important information at the processing stage.

Types of bus system

Types of bus
Various bus systems are used in modern vehicles:

● ISO bus
● SCP bus

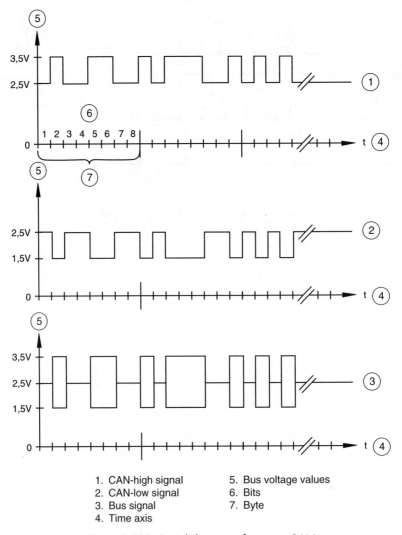

1. CAN-high signal   5. Bus voltage values
2. CAN-low signal   6. Bits
3. Bus signal   7. Byte
4. Time axis

Figure 3.123 Serial data transfer using CAN
(reproduced with the kind permission of Ford Motor Company Limited)

- ACP bus
- CAN bus

Control units that communicate via CAN, SCP and ISO buses can be checked via the DLC using serial diagnostic equipment; however, the ACP bus cannot be checked in this way.

ISO bus
The ISO (**I**nternational **O**rganisation for **S**tandardisation) data bus consists of a single communications line (K cable). This K cable is not used for communication between the control units but exclusively for diagnosing an individual control unit. On new vehicles, the ISO bus is being increasingly replaced by the CAN bus. However, the K cable is still present in most current

control units and is used to import and export parameters during vehicle production. If there is a fault on the cable caused by a power interruption, or a short circuit or short circuit to ground, the module and diagnostics tester will not be able to communicate with each other any more.

SCP bus

The SCP (**S**tandard **C**orporate **P**rotocol) databus consists of two twisted wires. It is used exclusively for communication between the PCM and the diagnostics tester. Even if only one of the two cables is faulty, the PCM and the diagnostics tester will not be able to communicate with each other any more. All information and data are placed in a frame containing bus control information and are transmitted as a packet (datablock) using the serial transmission method.

The complete packet consists of data in precisely defined sequences of bits. All nodes, i.e. control unit connecting points, have equal priority. Therefore, several control units can participate in performing a function. Both functional and physical addressing are possible. If there are several simultaneous messages, these are processed in turn according to their significance. At least one valid response must be returned for each message sent. If this does not happen, an error is stored in the error memory.

ACP bus

ACP (**A**udio **C**ontrol **P**rotocol) is similar to the SCP bus, has a simpler protocol and is exclusively for audio applications and telephone systems.

CAN bus

The CAN (**C**ontroller **A**rea **N**etwork) bus is also a serial data bus, but has a different protocol to the SCP bus and is therefore also faster. It operates at a transmission rate of up to 1 megabit/s (1 million) and comes close to real-time transmission. CAN was originally developed by Robert Bosch AG as a cost-effective network solution for use in vehicles. The CAN bus is an international standard and is described in ISO 11898 (high speed CAN) and ISO 11519 (low speed CAN).

The features of the CAN bus   CAN is a so-called multi-master bus system which means that all bus users (control and test equipment) can send and request data. In the CAN system, there is no addressing of the individual users in the traditional sense. Instead, 'identifiers' are assigned to the data packets to be sent. Each user can send its data on the data bus, the other users decide, using the identifiers, whether they read out and process this data further or not. The identifier also carries out prioritisation, i.e. it stipulates the priorities of the data packets, if several bus users want to send data simultaneously. One of the outstanding properties of the CAN bus is its high transmission reliability. Each user's CAN controller records transmission errors. These are recorded and evaluated statistically so that appropriate action can be taken. This can even result in the bus user which has produced the error being switched off from the bus. A data packet in its dataframe can consist of up to 8 bytes. More comprehensive transmission data is sent distributed across several dataframes. The maximum transmission rate is 1 Mbit/s, i.e. up to 1 million pulses per second. Because of the line resistance, this speed applies to networks with a databus length of up to 40 metres. For longer distances, the transmission rate has to be reduced: to 125 kbit/s for distances up to 500 metres and to 50 kbit/s for distances up to 1000 metres.

CAN bus variants   In vehicle engineering applications, three different CAN buses are used at present:

- A high-speed CAN bus (HS CAN).
- A mid-speed CAN bus (MS CAN).
- A low-speed CAN bus (LS CAN).

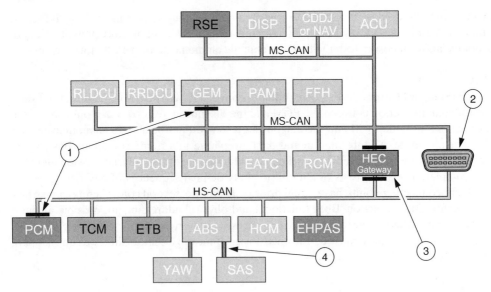

Figure 3.124 CAN data bus system with portals (gateways)

The bus connections consist of twisted cable pairs. On the HS CAN bus, the transmission rate is 500 kbit/s. On the MS CAN bus, the transmission rate is 125 kbit/s. On the LS CAN bus, the transmission rate is 50 kbit/s.

Portal (gateway)   On vehicles that operate with different network speeds (e.g. HS CAN and MS CAN), the information also has to be exchanged between the different bus systems and transmission speeds. As the two bus networks cannot be connected directly with each other, an interface is required via which the two networks can communicate with each other. This interface is also called a portal or a gateway. The gateway matches the different transmission speeds to each other and thereby enables the control units of the different networks to communicate.

Figure 3.124 shows the instrument cluster (HEC) module is the gateway between the MS CAN and the HS CAN.

1. 120 ohm matching resistor.
2. DLC.
3. Instrument cluster (HEC) as portal (gateway) between HS CAN and MS CAN.
4. Private bus between the ABS, yaw rate – lateral acceleration sensor (YAW) and steering angle position sensor (SAS).

The EATC (air-conditioning module with automatic temperature control) is on the MS CAN system. Not all sensors and actuators are 'busable'. This means they still operate in a conventional sense by manipulating an analogue voltage, current or through PWM, saturated voltage etc.

3.6 OBD and EOBD

OBD (On Board Diagnostics)

To combat its smog problem in the Los Angela's basin, the State of California started requiring emission control systems on 1966 model cars. The US federal government extended these controls nationwide in 1968. In 1970 the US Congress passed the Clean Air Act and established the Environmental Protection Agency (EPA). The Environmental Protection Agency has been charged with reducing 'mobile emissions' from cars and trucks and given the power to require manufacturers to build cars which meet increasingly stiff emissions standards. To meet these standards, manufacturers turned to electronically controlled fuel and ignition systems. Sensors were incorporated to measure engine performance and adjust the systems to provide minimum pollution. These sensors were also accessed to provide early diagnostic information. In 1988, the Society of Automotive Engineers (SAE) set a standard connector plug and a set of diagnostic test signals. The EPA adopted a number of the recommended standards. OBD-II is an expanded set of standards and practices developed by SAE and adopted by the EPA and CARB (California Air Resources Board) which were implemented in 1996. OBD-II provides a universal inspection and diagnosis method to be sure the car is performing to OEM standards. All cars built since 1, January 1996 have OBD-II systems.

EOBD (European On Board Diagnostics)

EOBD is an EU directive and is a part of the Euro Standard Stage 3. This means the compulsory introduction of a standard system which has the facility to record errors in a vehicle's engine management system that would affect the output of emissions.

EOBD was introduced on 1 January 2000 for vehicle type approval, which means that only new vehicles conforming to EOBD are licensed in Europe. In simplistic terms, the directive means that all European vehicles manufactured after this date will be fitted with *standard* 16-pin diagnostic connectors allowing access to emission data. Up to this date manufacturers used a wide range of diagnostic connectors. EOBD is based on the EPA OBD system. The vehicle's computer must communicate with one of the stated EOBD protocols. In the event of a failure the 'Malfunction Indicator Light' (MIL) on the instrument cluster will illuminate to advise the driver that they must take the vehicle to a repair outlet immediately. At the very least, these faults must be rectified before the vehicle will pass an MoT (Ministry of Transport) test.

OBD and EOBD systems were not only designed to reduce pollution being emitted from vehicles with faults that affect emissions. The systems were also designed to assist in diagnosing any OBD related faults which existed before the vehicle left the assembly line ensuring that the vehicle conformed to emission standards immediately. OBD also assists in providing information for the technician on sensor and actuator operation. This often allows technicians to evaluate system performance, providing valuable diagnostic information. If the system recognises a fault with a sensor or actuator, then the control module can use 'failsafe' measures to reduce the possibility of producing excessive pollutant from the vehicle. For example, if the engine coolant temperature sensor failed and provided a signal relating to 0°C for an extended period then the control module would ignore the signal and use a stored value of 80°C which would correspondingly reduce the amount of fuel injected into the engine thus reducing pollutants being emitted. The system will operate at a reduced performance but this is often acceptable enough to allow the vehicle to get back to base.

EOBD and OBD testers

An illuminated or flashing MIL will notify the driver prompting them to take the vehicle to the workshop for investigation. An EOBD compliant tool will be required to communicate with a system that complies with EOBD and OBD standards. It is important if working on a range of manufactured vehicles that your diagnostic tool has the facility to communicate using required protocols:

SAE protocol – J1850PWM, J1850VPW
ISO protocol – ISO 9141-2, ISO11898, 11519 (CAN) and ISO 14230-4KW2000

Note – *in the future (2008) diagnostic testers will communicate with control modules via the CAN bus (ISO 15765-4).*

A car manufacturer can choose any one of these protocols, so make sure that any tools are fully EOBD/OBD compatible. OBD/EOBD testers will automatically detect which communication protocol the vehicle uses. Not all EOBD/OBD testers give the same level of information. Some will only give a basic code reading facility while others will give access to full data stream, freeze frame and special test facilities.

Fault codes

The increase in the number of available codes can now give the technician a better understanding of where the problem exists. Vehicles with OBD II and EOBD may use up to seven or eight possible codes relating to an individual sensor or actuator fault. Fault codes must only be used as information or a guide to prompt further investigation into a system or component. A check of the associated wiring and components is always required before replacing the sensor or actuator. Remember it is the technician who makes the diagnosis, the machine only provides them with information and measurements. An appreciation between mechanical and electronic components must exist.

Fault codes are often referred to as Diagnostic Trouble Codes (DTC) and have been standardised. This means that all new EOBD and OBD fault codes have a 5 digit alphanumerical code.

A DTC is made up of 5 digits. The information below shows the composition of a DTC code 'P0100'.

The *first digit* of the code identifies the system that has the fault, for example:

- B for the body
- C for chassis
- P for powertrain
- U for network communication system

The *second digit* identifies the standard manufacturer's code 'P0100' or the standard ISO/SAE code 'P1100'.

The *third digit* identifies the diagnostic trouble code group the fault belongs to.

Diagnostic trouble code group
P01. Fuel and air metering
P02. Fuel and air metering
P03. Ignition system or misfire
P04. Auxiliary emission controls
P05. Vehicle speed and idle control system
P06. Computer output circuit, trip computer etc.
P07. Transmission
P08. Transmission
P09. Reserved for ISO/SAE
P00. Reserved for ISO/SAE

The *fourth* and *fifth* digits identify the fault, 0–99.

Example codes

Trouble code	Fault location
B1343	A/C sunlight sensor – short/open circuit
B1347	A/C foot well vent temperature sensor – short circuit
B1348	A/C foot well vent temperature sensor – short/open circuit
B1352	In-car temperature sensor – short circuit
B1353	In-car temperature sensor – short/open circuit
B1355	A/C/heater blower motor/in-car temperature sensor – supply voltage
B1360	A/C control module – keypad fault
B1493	A/C control module/combination control module/relay – signal fault
B1498	Heated rear window – short circuit
B1515	Common sensor earth – short circuit
B1605	A/C control module – defective
B1675	A/C/heater air direction motor – defective
B1676	A/C/heater recirculation motor – faulty
B1677	A/C/heater air mix flap motor – faulty
B2402	A/C/heater air direction motor – short circuit
B2403	A/C/heater air direction motor – open circuit
B2404	A/C/heater air direction motor – short circuit
B2405	A/C/heater air direction flap – loose
B2406	A/C/heater air direction flap – jammed
B2413	A/C/heater recirculation flap motor – open circuit
B2414	A/C/heater recirculation motor – short circuit
B2426	A/C/heater blower motor – control defective
B2427	A/C/heater blower motor – control defective
B2428	A/C/heater blower motor – control defective
B2492	A/C/heater air mix flap motor – short circuit
B2493	A/C/heater air mix flap motor – open circuit
B2494	A/C/heater air mix flap motor – short circuit
B2495	A/C/heater air mix flap – loose
B2496	A/C/heater air mix flap – jammed

There are two categories of DTC that apply to EOBD and OBD II.
Type A:

1. Emissions related.
2. Requests illumination of the MIL after one failed driving cycle.
3. Stores a freeze frame DTC after one failed driving cycle.

Type B:

1. Emissions related.
2. Sets a pending trouble code after one failed driving cycle.
3. Clears a pending trouble code after one successful driving cycle.
4. Turns on the MIL after two consecutive failed driving cycles.
5. Stores a freeze frame after two consecutive failed driving cycles.

Malfunction Indicator Light (MIL)

The MIL is located in the instrument cluster and takes the form of an engine symbol (international standard). Whenever the ignition is switched on the system carries out a self-test and upon successful completion will switch the MIL off. If the MIL does not go out a fault exists relating to emissions.

Fault detection and storage

A drive cycle begins when the engine is started and ends when the engine is switched off. During this trip all sensors and actuators are monitored if operated. A readiness trip is when the engine is started and ends when the monitoring of the systems has completed the tests. This can occur over a number of drive cycles (due to temperature ranges, distance covered, speed). A dealer test cycle (dealer drive cycle) is one readiness trip. This means that the vehicle is driven under a number of conditions (varying load, temperature etc.) to enable the monitoring of all EOBD systems.

MIL signals during these drive cycles are as follows:

1. Occasional flashes show momentary malfunctions. It stays on if the problem is of a more serious nature, affecting the emissions output or safety of the vehicle.
2. A constantly flashing MIL is a sign of a major problem which can cause serious damage if the engine is not stopped immediately.

In all cases a 'freeze frame' of all sensor readings at the time is recorded in the vehicle's central computer. Hard failure signals caused by serious problems will cause the MIL to stay on any time the car is running until the problem is repaired and the MIL reset. Intermittent failures cause the MIL to light momentarily and they often go out before the problem is located. The freeze frame of the car's condition captured in the computer at the time of the malfunction can be very valuable in diagnosing these intermittent problems. However, in some cases if the car completes three driving cycles without a reoccurrence of the problem, the freeze frame will be erased.

Data link connector (SAE J1962)

The EOBD interface connector allows technicians and traffic monitoring authorities to read data relating to EOBD and any faults that are present which affect the vehicle's emission output. The traffic authorities gain access to the system data using a generic scan tool. This information

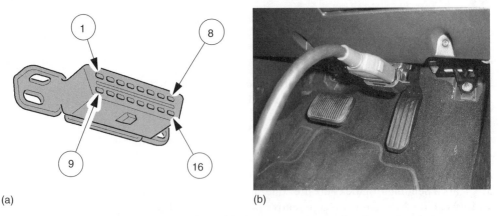

(a)   (b)

Figure 3.125 (a) Data link connector. (b) Cable from a serial tester fitted to a data link connector

may eventually be used to determine a penalty for a vehicle keeper who is driving a vehicle with an emission related fault causing the MIL to be illuminated over a prolonged period.

J1962 pin allocation

Pin	Definition	Purpose
1	Ignition control	Activation of the low-tension switch (relay etc.) for actuating the ignition circuit
2	Bus (+) SCP (J1850)	SCP (standard corporate protocol) communication
3	SCL (+)/STAR (out)/ MS-CAN (+)	SCL communication (self-test output)/mid-speed CAN communication
4	Chassis ground	Ground for power supply to data link connector
5	Signal ground	Signal return for programming
6	Class C link bus (+)	High speed data transmission bus (+)
7	K cable for ISO 9141	Communications cable for vehicles to ISO 9141
8	Tripping signal	Multiple module output
9	Battery power supply	Power supply through ignition switch
10	Bus (−) SCP (J1850)	SCP (standard corporate protocol) communication
11	SCL (−)/STAR (in)/ MS-CAN (−)	SCL communication (selftest output)/mid-speed CAN communication
12	Module programming	Programming of flash EEPROM
13	Module programming signal	Programming of flash EEPROM
14	Class C link bus (−)	High speed data transmission bus (−)
15	L cable to ISO 9141	Communications cable for vehicles to ISO 9141
16	B+	Battery positive power supply to data link connector

The serial link operates at a baud rate (bit rate) of 5–10 Kbaud on a single wire interface or as a two wire interface with a separate data line using a 'K' and 'L' line.

Dealer drive cycle (GM)

When a fault code in the memory of a module has been cleared some systems produce a readiness code that is stored in the KAM (Keep Alive Memory). The readiness code P1000 is only deleted after complete execution of a successful dealer drive cycle. A complete dealer driving cycle will perform diagnostics on all monitored systems and can usually be completed in less than 15 minutes.

Preparation – cold start below 50°C and at least 20% in the fuel tank.

Performing a GM OBD II driving cycle:

1. Cold start. In order to be classified as a cold start the engine coolant temperature must be below 50°C and within 6°C of the ambient air temperature at start-up. Do not leave the key on prior to the cold start or the heated oxygen sensor diagnostic may not run.
2. Idle. The engine must be run for 2½ minutes with the air-conditioner and rear defroster on. The more electrical load you can apply the better. This will test a range of sensor and actuators such as O$_2$ sensor, canister purge, misfire and if closed loop is achieved, fuel trim.
3. Accelerate smoothly. Turn off the air-conditioner and all the other loads and apply half throttle until 55 mph is reached. During this time the misfire, fuel trim, and purge flow diagnostics will be performed.

4. Hold steady speed. Hold a steady speed of 55 mph for 3 minutes. During this time the O_2 response, EGR, purge, misfire, and fuel trim diagnostics will be performed.
5. Decelerate. Let off the accelerator pedal. Do not change gear, touch the brake or clutch. It is important to let the vehicle coast gradually slowing down to 20 mph (32 km/h). During this time the EGR, purge and fuel trim diagnostics will be performed.
6. Accelerate. Accelerate at ¾ throttle until 55–60 mph. This will perform the same diagnostics as in step 3.
7. Hold steady speed. Hold a steady speed of 55 mph for 5 minutes. During this time, in addition to the diagnostics performed in step 4, the catalyst monitor diagnostics will be performed. If the catalyst is marginal or the battery has been disconnected, it may take five complete driving cycles to determine the state of the catalyst.
8. Decelerate. This will perform the same diagnostics as in step 5. Again, don't press the clutch or brakes or shift gears.

3.7 How to read wiring diagrams

Introduction

The automotive industry has been massively transformed by globalisation. Design work can be conducted in one country while manufacturing in another for vehicles to be shipped to nearly all. This can have problems when not all manufactures use the same standards to present information on the vehicles within the market. Something as simple as the results of measuring temperature may need converting °C to °F. The answer is standards. Standards have been an integral part of the automotive world since the earliest days of the automotive assembly line. Standardisation of parts allowed automakers to transform their businesses from a one-at-a-time proposition to a many-at-a-time operation.

In the US the Society of Automotive Engineers (SAE) is responsible for standards that apply to automobile manufacturing. For example, when selecting a bolt for a US domestic vehicle out of the bolt bin, chances are the standards and specifications concerning its thread pitch and hardness were originally defined by SAE. Thanks to standardisation, that bolt should thread into any nut made anywhere in the world, as long as it conforms to the same set of standards.

In Europe, the most widely recognised organisation responsible for establishing and publishing automotive standards is called Deutsches Institut für Normung e.V. Standards established by this organisation are often referred to as DIN standards.

DIN standards (Europe)

Manufacturers may not fully conform to DIN standards; they may only conform to a number of the standards and use what they feel work better for their service support data. For example, Figures 3.126 and 3.127 show some relationships to DIN standards on terminal designation but not all letter codes correspond to DIN standards, e.g. – C for connector DIN standard uses X. It is important whenever using wiring system manuals, CDs or intranet support material to ensure the charts and tables are checked against the current standard or codes.

Block diagrams (basic) and schematic (detailed) DIN 40 719/1

Description of an electrical system or circuit may begin with a circuit diagram. This is presented in the form of symbols to provide a quick overview of circuit and device functions. The circuit diagram illustrates the functional interrelationships and physical links that connect various devices. A block diagram (Fig. 3.126) is another simplified representation of a circuit, showing only

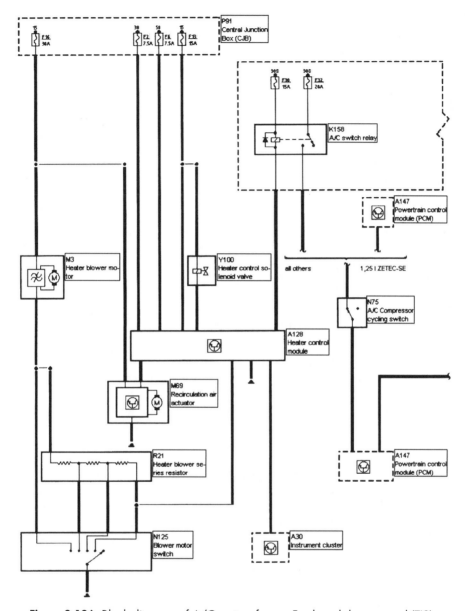

Figure 3.126 Block diagram of A/C system from a Ford workshop manual (TIS)
(reproduced with the kind permission of Ford Motor Company Limited)

the most significant elements. It is designed to furnish a broad overview of the function, structure, layout and operation of an electrical system. This format also serves as the initial reference for understanding more detailed schematic diagrams. Squares, rectangles, circles and symbols illustrate the components. Information about wire colours, terminal numbers, connectors etc. are omitted to keep the diagram as simple as possible.

The schematic diagram in Figure 3.127 shows a circuit and its elements in greater detail. By clearly depicting individual current paths, it also indicates how the electrical circuit operates. Most DIN schematic diagrams are current flow diagrams. They are arranged from top to bottom,

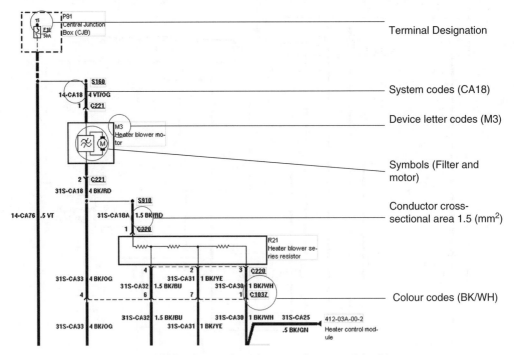

Figure 3.127 Example of part of a schematic diagram of the blower motor (reproduced with the kind permission of Ford Motor Company Limited)

so we can clearly see how the current flows through the circuit. Generally a terminal designation number 15 (ignition switch) and or 30 (power from battery) runs along the top and 31 runs along the bottom (earth). System codes, wire sizes and colours are also included. The symbols used to define the components also conform to DIN specifications. A key explaining these symbols will often be included with the schematic diagram. Even if you're fairly familiar with a circuit on a given car, a schematic diagram will help you find the correct location of a ground terminal, or help you identify a specific pin number in a connector. The following will help you understand the information within a schematic diagram. Where the components and wires are situated in relation to one another in the diagram usually bears no resemblance to how they are actually arranged on the vehicle but, generally, wiring diagram vehicle systems have loom layouts, component location charts and connector charts to aid in finding all the information required.

Terminal designations DIN standard 72 552
DIN standard 72 552 establishes the terminal numbering system that is used for any wiring schematic or component diagram that conforms to DIN specifications. The code tells the reader about the wire as a connection and not as a wire itself. Chart 1 in the appendix offers a full list of terminal numbers and definitions. All DIN standard wiring schematics have current flowing from the top to the bottom. This starts with a fuse (depicted by a symbol and a letter destination F meaning fuse). The fuse has the number 15 above it. 15 is the terminal destination number which tells you the electrical state of the wire is only live when the ignition switch is on start or run (chart 1). This DIN standard is used on any component connection or electrical connection. For example, the wire leading to the heater blower motor M3 starts with the number 14 which is voltage supplied in start or run and is overload protected. This simply means that voltage will be supplied when the ignition is switched to the start or run position and that the connection is fused. See chart in the appendix.

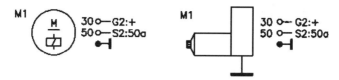

Figure 3.128 Symbol and destination and pictorial device and destination (reproduced with the kind permission of Ford Motor Company Limited)

Table 3.3 DIN basic colour codes

Colour	Symbol	Colour	Symbol
Blue	bl	Pink	rs
Brown	br	Red	rt
Yellow	ge	Black	sw
Grey	gr	Turquoise	tk
Green	gn	Violet	vi
Orange	or	White	ws

Table 3.4 British Standards colour coding

Colour	Symbol	Colour	Symbol
Brown	N	Pink	K
Blue	U	Light Green	LG
Red	R	Slate	S
Purple	P	Orange	O
Green	G	Yellow	Y
White	W	Black	B

Detached terminal diagrams DIN 40 719

These are used to show just one component and the connection details. This is common on aftermarket wiring diagrams like Autodata. Autodata sells books which include lots of wiring connections for a large range of manufacturers. This allows the information to be interpreted using individual components and is particularly useful with ECU connections. Technicians often test a range of input and output signals going to and from electronic control units. Because they are central to the operation of a system they are often useful as a testing point (if accessible). Figure 3.128 shows two detached device terminal diagrams. The left uses symbols and the right uses a pictorial representation.

Example:

30   See chart 1   G2: See Device letter codes   +   See chart 1
50   See chart 1   S2: See Device letter codes   50a   See chart 1
●—┤   Ground symbol (earth)

Table 3.5 A sample from the BSI colour codes with tracer wires

Main	Tracer	Purpose
Black		All earth connections
Black	Brown	Tachometer generator to tachometer
Black	Blue	Tachometer generator to tachometer
Black	Red	Electric or electronic speedometer to sensor
Black	Purple	Temperature switch to warning light
Black	Green	Relay to radiator fan motor
Black	Light green	Vacuum brake switch or brake differential pressure valve to warning light and/or buzzer
Black	White	Brake fluid level warning light to switch and handbrake switch, or radio to speakers
Black	Yellow	Electric speedometer
Black	Orange	Radiator fan motor to thermal switch

Symbols DIN 40 900

Symbols are the smallest component of a circuit diagram and are the simplest way to illustrate a whole or part of a device like a sensor or actuator. Standards are used so engineers can easily identify a component based on its symbol. For example, the M3 blower motor has the symbol of a filter and a motor. These placed together show that the assembly of the component includes a motor assembly and filter to reduce electron magnetic interference. Chart 3 in the appendices shows a range of symbols used.

Device letter codes Din 40 719 Part 2

With reference to Figure 3.127 F16 means F̲use number 16. F is a device letter code and is a standard letter used in all DIN schematics. This is the same for M̲ for Motor, R̲ for Resistor. Not all device codes resemble the letter designated to them, e.g. B for transducer. Examples of device codes can be found in the Appendices, chart 2.

Wire colour codes DIN 47 002

Colour codes for electrical wiring are defined in DIN 47 002. The main colour defines the purpose of the wire and the tracer defines its use.

BSI Standards (UK)

BS AU 7a: 1983 are the British Standards for colour coding (Table 3.4). The main colour defines the purpose of the wire and the tracer colour determines its use. Twelve colours are used for the main colour coding system.

Tracers are used to define the function of the wire. Table 3.5 shows a small sample. See the Appendices for more details.

Example of a manufacturer's wiring system – Ford FSC system

Since the introduction of the Mondeo the Ford Motor Company uses a system for circuit numbering and wire identification. The system is called Function System Connection (FSC).

The Function is the same as DIN terminal designation DIN 72 552. The base colour is directly related to the function of the wire (terminal designation).

Table 3.6 Sample of function codes from Ford

Number	Description	Basic colour
4	Datalink, bus positive	GY
5	Datalink, bus negative	BU
7	Voltage D.C. supplied at all times (sensor)	YE
8	Sensor signal	WH
9	Sensor signal return (reference/earth)	BN
14	Voltage supplied in Start and Run (overload protected)	VT
15	Voltage, ignition switch in Start or Run (not overload protected)	GN
29	Voltage supplied at all times (overload protected)	OG
30	Voltage supplied at all times (not overload protected)	RD
31	Earth	BK
32	Switch output supplies battery voltage, earth or open circuit to motor; positive voltage output is for front–down, open, mirror leftward, unlock, on	WH
33	Switch output supplies battery voltage, earth or open circuit to motor; positive voltage output is for front–up, close, mirror rightward, lock, off	YE
49	Pulsed power feed (overload protected)	BU
50	Voltage, ignition switch in Start (not overload protected)	GY
59	AC Power	GY
63	Variable D.C. voltage (overload protected)	WH
64	Variable D.C. voltage (not overload protected)	BU
74	Voltage in Run or Accessory (overload protected)	BU
75	Voltage: Ignition switch in Accessory or Run	YE
89	Voltage supplied at all times (overload protected electronic module)	OG
90	Voltage supplied at all times (not overload protected electronic module)	RD
91	Earth	BK
94	Voltage, ignition switch in Start or Run (overload protected electronic module)	VT
95	Voltage, ignition switch in Start or Run (not overload protected electronic module)	GN

(reproduced with the kind permission of Ford Motor Company Limited)

The next part are system codes (Table 3.7), denoted by two letters, the first is the system group and the second the actual system within that group.

Finally the connection is a number from 1 to 99 which defines which component the wire goes to. The information also includes a colour code and cable size. The base colour is related to the terminal designation of the wire, i.e. 29 colour OG – orange. Colour codes are related to the English name of the colour itself (DIN IEC 757).

Example schematic – heater control module

The heater control module is built into the A/C switch panel inside the vehicle (where the A/C and heating system are turned on/off). Using the manufacturer's information it should be possible to determine the function/terminal designation of most of the single wires, what systems they are related to, the cross-sectional area of the wire and the colour.

Table 3.7 System codes used to identify the connection and the system it belongs to

---

Distribution systems

DA:	Soldered connectors
DB:	Bridges
DC:	Circuit protection – battery junction box (BJB)
DD:	Circuit protection – central junction box (CJB)
DE:	Earth
DF:	Branch block
DG:	Circuit protection – auxiliary fuse box
DH:	Distribution – auxiliary fuse box
DJ:	Auxiliary distribution
DK:	Integrated components

Actuating systems

AA:	Door locking and luggage compartment lid (incl. infrared locking/global closing)
AD:	Power mirrors
AG:	Power sunroof
AH:	Power seats
AJ:	Power windows

Basic systems

BA:	Charging/power distribution (incl. ammeter/voltmeter)
BB:	Starter motor

Chassis systems

CC:	Suspension (adaptive damping)
CD:	Suspension (pneumatic ride height control)
CE:	Power steering (adjustable automatic power assistance)
CF:	ABS/traction control
CG:	Brakes (engine/exhaust)

Air conditioning systems

FA:	Air conditioning 1
FC:	Air conditioning – auxiliary heater

Information and warning systems

GA:	Measuring – level/pressure/temperature
GB:	Measuring – miscellaneous (incl. speedometer/clock)
GC:	Indicating – level/pressure/temperature
GE:	Auxiliary warning/bulb check module
GG:	Instrument cluster
GH:	Trip computer
GJ:	Horn
GK:	Navigation
GL:	Anti-theft alarm system
GM:	Audible warning systems (incl. warning buzzer/acoustic modules)
GN:	Parking aid

Heated systems

HA:	Auxiliary (incl. air/filter/cigarette lighter)
HB:	De-ice (incl. mirrors/nozzles/windows)
HC:	Heated seats

Passenger restraint systems

JA:	Air bags

Wash/wipe systems

KA:	Wash/wipe system

(reproduced with the kind permission of Ford Motor Company Limited)

In Europe, there is often a large difference between the quality and quantity of information supplied with aftermarket wiring manuals compared to those supplied by the manufacturer. The Ford TIS system has the ability of locating almost every component, connector on the wiring schematic as well as supplying a wealth of other information like technical updates,

Table 3.8 Ford colour coding system

Wire colours		
Abbreviation	English name	Colour
BK	BlacK	Black
BN	BrowN	Brown
BU	BlUe	Blue
GN	GreeN	Green
GY	GreY	Grey
LB	Light Blue	Light blue
LG	Light Green	Light green
OG	OranGe	Orange
PK	PinK	Pink
RD	ReD	Red
SR	SilveR	Silver
TN	TaN	Tan
VT	VioleT	Violet
WH	WHite	White
YE	YEllow	Yellow

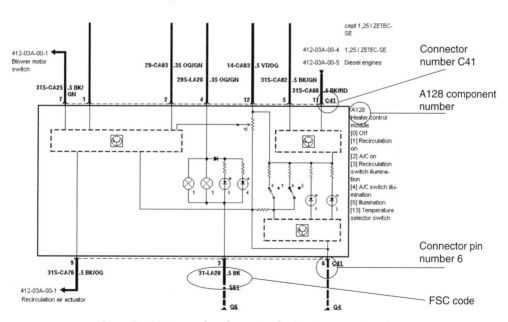

Figure 3.129 Example schematic of a heater control module
(reproduced with the kind permission of Ford Motor Company Limited)

Table 3.9 Two connection examples

Connector	Pin no.	Function	System	Connection	MM2	Colour
C41	4	29S	LA	28	.35	OG/GN
The connector to the heater control module		Voltage supplied at all times overload protected (fuse). 'S' means switched, must be switched on. So once the device has been switched on then it will remain live at all times	LA is the lighting system's switches and instruments. This means the wire is live when the lights are operated. This gives power to the lamps and bulbs inside the unit (numbered 5,5,3,4) allowing it to illuminate in the dark	Connected to item 28 in the lighting circuit, probably the headlamp switch	Low current carrying capacity	Orange with a green tracer. Orange due to its function number 29 and green tracer to help identify it
C41	5	31S	CA	82	.5	BK/GN
The connector to the heater control module		Ground/earth and 'S' meaning switched. This terminal is a switched earth. The module switches this connection to earth. This would operate a component which had a power supply but no earth	CA is the system air-conditioning. This means that the module operates a component 82 within the A/C system	Component 82 of A/C system	Low current carrying capacity	BK – Black, the colour which is associated with ground/earth connection. GN – Green, used to identify it

workshop manuals etc. The information is supplied by a CD-ROM system (recently updated via the intranet). European aftermarket technical information suppliers have not yet achieved the coverage required to be used as a single source for information. Often they can provide service information for a range of manufacturers and popular models. They may also provide technical support via email or telephone lines. Sometimes they offer specialist information on areas like engine management and air-conditioning via CD-ROM and textbooks. What they are very successful at is providing information that is easily understood and often presented in an easy to follow format.

Manual A/C systems

System components
To maintain a constant temperature within the vehicle using a simple manual system requires continual changes to fan speed and temperature selection by the occupant.

This will be due to the following:

1. Changes in vehicle speed affecting natural flow rates.
2. UV light (ultraviolet radiation) causing additional heating through glassed areas.
3. Number of occupants.
4. Interior and exterior temperature.
5. Humidity levels.

To allow for a system to automatically account for all these plus other variables a control system must be used. Because a manual control system uses the occupant to adjust the comfort zone, only a simple control system is required. The module is generally an ASIC type controller with no Electronically Programmable Read Only Memory (EPROM) containing data tables (used for comparing data) which allow for closed loop control of a single or dual comfort zone. The controller will not be attached to a data bus system, or allow for OBD fault codes or other serial type information. Often the A/C compressor will not be controlled by the controller built into the A/C controls on a manual system. The A/C compressor will be controlled by the powertrain control module (engine control module) based on information from pressure switches. The temperature and air distribution are manually controlled by switches and Bowden cables. Often the blower is a conventional DC blower motor with speed control via a resistor pack although some systems may use a brushless blower motor due to greater reliability and a reduction in noise.

The manual control system will soon be obsolete across the mass produced vehicle market. A/C systems are evolving and graphical interfaces, even LCD displays, require more advanced electronics which often include A/C selection. Electronic displays will replace conventional display systems and semi-automatic systems will be a part of the basic specification of a vehicle.

For more information see A/C modules and displays under section 3.2.

Semi-automatic systems

Semi-automatic systems always have an electronic control system using a module. Sensors detect the temperature in the interior of the vehicle. The module compares the measured temperature values (often integrated inside the module) with the temperature setting selected by the user. If this comparison reveals a difference in temperature, the module activates heating or air-conditioning according to needs and actuates the air temperature door control motor and the blower motor. The system is semi-automatic due to the manual control of air distribution. This means that the system does not include air distribution motors and air duct temperature sensors. Depending on the circumstances, it may be necessary to activate the air-conditioning

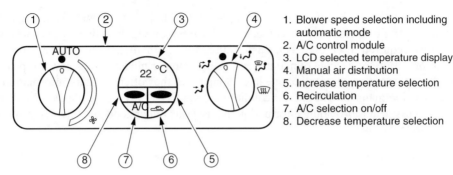

1. Blower speed selection including automatic mode
2. A/C control module
3. LCD selected temperature display
4. Manual air distribution
5. Increase temperature selection
6. Recirculation
7. A/C selection on/off
8. Decrease temperature selection

Figure 3.130 Semi-automatic control system
(reproduced with the kind permission of Ford Motor Company Limited)

automatically. This occurs in wet weather when the user selects 'DEFROST' mode. Warm air, previously dried by cooling, is best suited for drying internally misted windows. With semi-automatic temperature control, the air distribution (except in 'DEFROST' mode) must be set by the user. In addition, the user can switch off automatic operation and adjust the system manually.

For more information see A/C modules and displays under section 3.2.

Automatic climate control system

The main function of the automatic climate control system

As soon as a difference in the requested and actual temperature has been detected in the climate zone, the mixed air temperature and blower speed must be raised or lowered. The control module rotates the air mixing damper until the desired air mixture temperature is achieved and then directs air distribution in order to ensure that comfortable conditions are achieved as soon as possible and then maintained.

Electronic controlled temperature

The A/C control module supplies current to stepper motors (driver and passenger blend doors if dual zone) which rotate air blend doors. The air blend door mixes cold air that has passed through the evaporator with warm air that has passed through the evaporator and heat exchanger (heater matrix). The interior temperature is calculated for the front seat passenger and driver using a range of sensors. The calculated temperature is compared to the temperature set on the A/C module graphical display (if integrated). As soon as there is a difference, the mixed-air temperature must be increased or decreased. The control module turns the air blend door until the requested mixed-air temperature is achieved. When the lowest mixed-air temperature is requested, the control module closes the coolant flow to the heat exchanger to provide maximum cooling effect (only on some systems).

Electronic controlled air distribution

The control module supplies current to a stepper motor which rotates the air distribution door. The air distribution control is dependent on the requested mixed-air temperature. Demist/floor is selected at a high temperature value, floor/face vent or face vent is selected at low temperature value. For cold engine starts, demist is selected initially. The calculated position of the flap is dependent on how many pulses the control module has sent to the stepping motors. For example, a rotary air distribution door may use 0% face vents, 20% face/floor, 50% floor, 75% floor/windscreen (demist) and 100% windscreen (demist).

Electronically controlled blower speed

The A/C control module communicates with the fan control which supplies current to the fan motor. The fan current is at its lowest when the present mixed-air temperature matches the requested temperature. The current increases as soon as cold or heated air is needed. For cold starts, the value is determined by the coolant temperature and the outside temperature. The fan current is limited if the engine is not running.

Electronically controlled air recirculation

The A/C control module supplies current often to a direct current motor which rotates the air recirculation flap.

Recirculation is selected if the outside air temperature is high and the requested mixed-air temperature is low (that is, the cabin has to be cooled significantly). It is also selected if an air sensor is fitted and the ambient air contains pollution.

The automatic climate control system will often have a self-test facility which will be able to display codes via the graphics interface. The system will have multiplex communication and be able to provide data lists and fault codes via a serial connection.

For more information see A/C modules and displays under section 3.2.

3.8 Automotive A/C manual control system (case study 1)

This is the first of three case studies presenting a complete A/C system using OEM (Original Equipment Manufacturers) information. The case study is not written in textbook fashion. This is because the purpose of each case study is for the reader to appreciate the breadth and depth of information available and what information is required to enable the technician to successfully work on such systems. The author has attempted to try to present the information in a logical order. Photos of system components have been added to help visualise their position and explanations have been written to assist in the understanding of system operation. If the reader fails to understand a particular aspect of the case study then refer to a section which covers that topic.

Information from Ford is usually sourced electronically via an intranet. The Ford information system includes training and technical information, diagnostic routines, bulletins and wiring diagrams.

Ford manual control system (Fiesta)

Described using OEM information.

>   Manual temperature control
>   Single evaporator
>   Single zone

System information

The manual control system used by this manufacturer has a manual blower control selection, manual air distribution, manual temperature control, and switch operated A/C and recirculation. The only information the module must process is based on A/C switch input, recirculation input, and temperature variation for the solenoid operated water control. The blend door and air distribution are all Bowden cable operated except for the recirculation door which is DC motor operated. The electronic circuitry required for the operation of such a system is very simple. The module has no memory functions (EPROM) cannot be programmed and is not a part of a multiplex network. This means that the circuitry is designed as an ASIC (Application Specific Integrated Circuit). The output from the unit can be monitored by more electronically advanced modules like the Powertrain Control Module (PCM).

Filtering

A pollen filter is fitted as standard. This cleans the incoming fresh air of pollen and dirt particles. It must be renewed after every 40 000 km under normal operation conditions, or after 20 000 km under difficult conditions.

Air distribution

The blower motor has four speeds, controlled by the blower motor operating switch through a resistor pack. The position of the air distribution flap is adjusted by a splined shaft. The recirculation air flap is operated by a servo motor. If the air-conditioning system is switched on but the blower motor switched off, the air-conditioning system will not operate. The air-conditioning system will only operate if the blower motor is on. The temperature control and air distribution flaps are controlled by two operating cables. Air is supplied to the windscreen and side windows regardless of the air distribution setting. If the air distribution rotary control is turned to the 'Defrost' or 'Foot well/Defrost' position, the air-conditioning is switched on regardless of the position of the air-conditioning switch or blower switch. As a result, the air in the vehicle used to defrost the windows is dehumidified before being distributed.

Face level

The main flow of air enters the passenger compartment at face level.

Face level/floor level

The main flow of air enters the passenger compartment in equal parts at face and floor level, and a small amount flows to the demister nozzles of the side windows.

Floor level

The main flow of air enters the passenger compartment at floor level, and a smaller amount flows to the demister nozzles of the side windows.

Demist/floor level

The main flow of air enters the passenger compartment in equal parts at the windscreen and at floor level, and a smaller amount flows to the demister nozzles of the side windows.

Demist

The main flow of air enters the passenger compartment at the windscreen and a smaller amount flows to the demister nozzles of the side windows.

A/C system

The system uses an FOV valve which is explained in detail in Chapter 1, section 1.8. Later systems can have a variable orifice valve fitted.

System layout and components

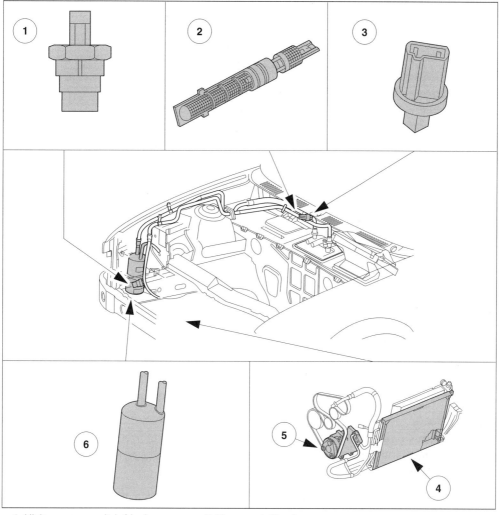

1. High-pressure switch (dual pressure switch)
2. Fixed orifice tube
3. Low pressure switch (cycling switch)
4. Condenser
5. Compressor
6. Suction accumulator/drier

Figure 3.131 A/C component layout
(reproduced with the kind permission of Ford Motor Company Limited)

A/C system operation
The temperature is set manually through the heater control module. The heater control module operates the heater control valve cyclically (up to 18 cycles per minute) according to the occupants' temperature selection which is sensed by a potentiometer inside the heater control module. This controls a heater control solenoid which regulates the flow of coolant to the heat exchanger. The air-conditioning switch activates the air-conditioning relay through the heater control module. This relay applies voltage to the low pressure switch. The low pressure switch controls the operating cycles of the compressor according to the pressure on the low pressure

side of the refrigerant circuit. If the low pressure switch is closed then the PCM will receive a voltage signal. When the air-conditioning is activated, the compressor clutch and cooling fan (stage 1) are energised.

The A/C system will be deactivated in the event of the following:

1. Engine overheating sensed by the coolant temperature sensor >120°C.
2. Engine under load, TP sensor output >3.4 volts. So the full engine power is available for vehicle acceleration.
3. Compressor cycling switch is below lower threshold approximately 1.6 bar.
4. High pressure switch is open, indicating excessive pressure in the system >30 bar.
5. Internal blower fan not operating.
6. Compressor thermal protection device becomes open circuit.

When the air-conditioning is switched off, the compressor clutch is immediately de-energised. The cooling fan continues working for a further 40 seconds. The maximum cut-off time for the compressor clutch with a wide open throttle is 12 seconds. When the air-conditioning is switched on, the cooling fan (stage 1) is switched on at once. Then, after a delay of about 2 seconds, the compressor clutch is switched on. If the cooling fan is already switched on, this time is reduced to 1 second. Switching the cooling fan on in advance ensures that the condenser is cooled before hot refrigerant gas flows through it. For the first stage the voltage of the cooling fan is restricted by a ballast resistor. This circuit is closed by means of a relay. For the second stage the full voltage is provided for the cooling fan through a second relay. Both relays are controlled by the PCM. The cut-in conditions for the relay for the first stage are as follows:

- Coolant temperature exceeding 95°C once and then remaining over 92°C.
- Air-conditioning switched on.
- Relay for the second stage pulled up.

The cut-in conditions for the relay for the second stage are as follows:

- High pressure switch (P1) closed for at least 5 seconds.
- Coolant temperature exceeding 100°C and then remaining over 97°C.
- Coolant temperature over 95°C for more than 15 seconds and engine speed over 4000 rpm at full load (WOT).

The vehicle battery has two terminals. A positive terminal and a negative (earth/ground) terminal. All the electrical power required to run electrical systems will come from this source. The power supply the battery provides would eventually diminish unless it was topped up. The alternator carries out this task and maintains the battery's power supply while the engine is running. There is a network of cables that runs from the battery power supply to various electrical systems. Most electrical systems are protected by a fuse. These are located in fuse boxes.

Most manufacturers use a Battery Junction Box (BJB fitted as close to the battery as possible) and Central Junction Box (CJB generally inside the vehicle) to house fuses, relays, diodes and sometimes diagnostic connectors. The BJB will contain fuses and relays mainly related to systems under the bonnet (hood) of the vehicle. The CJB will support the rest of the power supply to all the other systems.

Battery (01) supply has a DIN designation 30 BA610RD (Fig. 3.132). There is a permanent feed (not switched), charging system designation (BA), large diameter (10 mm^2), high current carrying capacity wire unprotected by a fuse going to the generator (alternator). This is the main power supply of the starter motor and the main feed to the battery from the alternator to keep the power supply above 12 V.

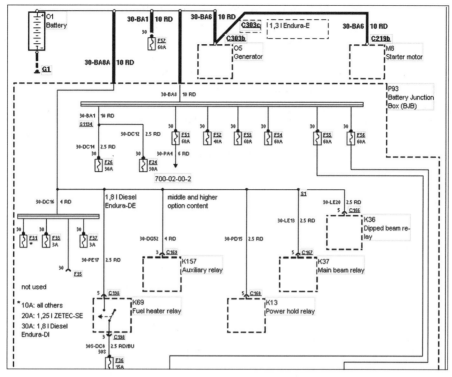

Figure 3.132 Ford power distribution – battery (01) and BJB (P93 Battery Junction Box) (reproduced with the kind permission of Ford Motor Company Limited)

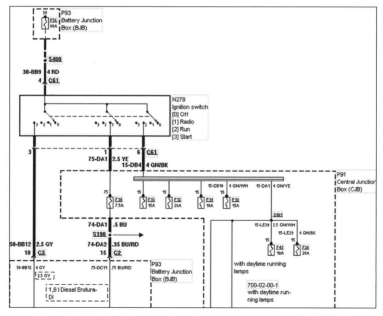

Figure 3.133 Ford power distribution – ignition switch (N278) and CJB (P91 Central Junction Box) (reproduced with the kind permission of Ford Motor Company Limited)

This power supply from the battery also provides power to the BJB bus bar, a bar inside the BJB which is fitted to all the fuses and acts as a supply line. In Figure 3.131 the FSC codes are all 30 meaning permanent feed. This means that the fuses will always remain live at this point. It does not mean that the circuit will be fully operational all the time. Remember the circuit will only be live when current can flow through it. Often the BJB supplies a feed to the CJB (Fig. 3.133). This is why there are often high current rated fuses in the BJB because it feeds a range of other circuits elsewhere in the system.

This also means that if one of these fuses blows (becomes open circuit) a large number of systems will not operate. Power distribution is like a pyramid system, at the very top is the battery positive terminal and at the bottom all the circuits that receive their power from a fuse, control unit/module or the ignition switch.

Manual A/C schematics
Block diagram

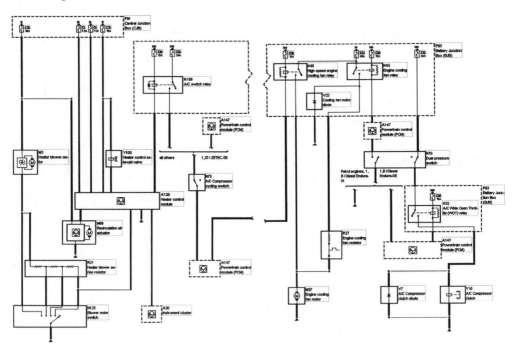

Figure 3.134 A/C system block diagram
(reproduced with the kind permission of Ford Motor Company Limited)

Components
A128 heater control module
The heater control module has the heating, ventilation and air-conditioning switches built into the unit (see Figure 3.88). It is this unit which communicates the information on heating temperature, air recirculation and A/C demand to other components. The heater temperature is set using the temperature control switch built into the heater control module called a potentiometer. The potentiometer operates a heater control solenoid valve (Y100), which controls the flow of hot coolant through the heat exchanger. Depending on the setting of this switch, the heater control module monitors the position of Y100 and varies the voltage applied to it to control the coolant flow through the heat exchanger.

Figure 3.135 Ford A/C compressor cycling switch (integrated into the accumulator/low pressure service port)

A147 powertrain control module
The powertrain control module controls all of the electrical operations of the engine, for example fuelling, ignition and diagnostics. It is also used to control the A/C. The block diagram in Figure 3.134 shows that a wire goes from K32 to the engine control module A147. K47 is used to supply power to the compressor clutch and can only be operated by A147 when providing an earth path to the relay.

N75 A/C compressor cycling switch (low pressure and cycling switch)
As described earlier in this chapter this switch acts as a safety device to prevent icing of the evaporator and operation if no refrigerant is present in the system by becoming open circuit (open so no current can flow).

Air-conditioning system – low pressure switch/cycling switch
The low pressure/cycling switch is fitted on the accumulator housing which is situated between the evaporator and the compressor. When the air-conditioning is switched on, the clutch only receives 12 V if the low pressure switch is closed. When the pressure falls below 1.6 bar, the low pressure switch is opened and the supply of current to the compressor clutch is interrupted. The low pressure switch functions as a de-icing switch and controls the compressor clutch on/off cycle. The pressures in the low pressure line and evaporator are practically equal. As the pressure in the low pressure line drops, the temperature in the evaporator approaches freezing point. There is a risk that the evaporator will ice up and lose its cooling power. The low pressure switch opens the compressor clutch when the pressure drops to a certain level and will not close it again until the pressure has risen to a suitable level.

N76 dual pressure switch
Switch one is normally open (high speed fan switch). The high pressure switch is fitted in the high pressure line between the condenser and orifice valve. The function of the high pressure switch is to protect the high pressure section of the refrigerant circuit. When pressure exceeds the maximum value of 31.4 bar because, for instance, the air flow through the condenser is impeded or the high pressure line is blocked, the high pressure switch turns the compressor off. The compressor will not be switched on again until the pressure has dropped back to 17.2 bar.

Switch two is normally closed (high pressure switch). The fan speed must increase when the pressure is more than 20.7 bar to enable additional cooling of the condensor. A147, the

Figure 3.136 Ford dual pressure switch and Schrader valve attachment

powertrain control module (main engine control module), will not switch the fan to a lower speed until the pressure has dropped back to 17.2 bar.

K158 A/C switch relay
Located in the BJB this relay is operated by the heater control module once A/C has been activated by the occupants. Upon activation a voltage will go to the low pressure switch and then to the A147 powertrain control module where under the right conditions the module will activate K32 to send power to the A/C compressor clutch.

K45 engine cooling fan relay
Operated by A147 (providing an earth path to the relay) this allows the fan to operate at a low speed due to reduced current flow via a resistor R27.

K46 high speed engine cooling fan relay
Operated by A147 (providing an earth path to the relay) this allows the fan to operate at maximum speed to a direct power supply.

K32 A/C wide open throttle relay
Operated by A147 and provides power to the compressor clutch Y16.

N125 blower motor switch
A four speed, five position switch which directs current through a resistor pack R21.

R21 heater blower series resistor
A series resistor pack, providing a range of resistances to reduce the current flowing through the motor M3 enabling speed control.

R27 engine cooling fan resistor
A resistor which enables reduced current flow to the engine cooling fan motor M37.

P91 Central Junction Box (CJB)
Situated inside of the vehicle and contains relays and fuses. The wiring schematics provide all the information on the fuse and relay requirements.

Figure 3.137 Ford resistor pack and diode

Figure 3.138 CJB, relays are behind the fuse box

Figure 3.139 Ford BJB contains A/C switch and WOT relay

P93 Battery Junction Box (BJB)
Situated near the battery and contains fuses and relays.

Y100 heater control solenoid valve
A solenoid operated valve controlling the flow of heated coolant (hot water mixed with antifreeze) to the heat exchanger inside the vehicle.

Figure 3.140 Ford compressor and thermo switch operates at 115°C

Figure 3.141 Ford M3 blower motor

Y16 A/C compressor clutch with built-in thermo-time switch
The compressor clutch is an electromagnetic unit with an internal resistance of $3\,\Omega$. A thermo switch is integrated in the compressor; at a temperature of 115°C in the compressor outlet (high pressure side), this switch cuts off the power supply to the magnetic clutch in the event of the unit overheating.

V7 A/C compressor clutch diode
A diode is used to protect the electrical system from any back-emf (electromotive force) or voltage spikes caused by the clutch being operated.

M69 recirculation air actuator
The recirculation air flap is operated by an electronic stepper motor (see section 3.2). The electric stepper motor has two cables attached to a lever which when operated opens and closes the recirculation doors (see Figure 1.29).

M3 blower motor
The blower motor has four speeds, controlled by the blower motor switch (N125) through a series resistor pack (R21). If the air-conditioning system is switched on but the blower motor switch is off, the air-conditioning system will not operate unless the rotary control switch is turned to the 'Defrost' or 'Footwell/Defrost' position, the air-conditioning is switched on regardless of the position of the air-conditioning switch or blower switch. As a result, the air in the vehicle used to

defrost the windows is dehumidified before being distributed. In demist/defrost mode the heater control module provides an earth path to the blower motor switch enabling the blower motor to operate.

Electrical system

Explanation of manual control system wiring schematic

When vehicle electricians analyse a vehicle wiring schematic they generally follow a set pattern:

1. Power supply signals.
2. Earth/ground signals.
3. Input – sensors.
4. Output – actuators.

When working on a system which incorporates an ECU (Electronic Control Unit, A128, A147) electricians will often take all electrical readings from the ECU connector. Some systems have built-in diagnostics and will display a fault via a code; other systems require a diagnostic tester to be plugged into a standard diagnostic connector on the vehicle to retrieve information based on the system's performance. Even when this has been achieved wiring schematics are still required to be examined. What happens if the diagnostic plug is faulty? Often technicians are not taught to read wiring schematics and rely too heavily on diagnostic test equipment. Research and development engineers still set up rigs in test cells and have to wire systems up manually. Understanding wiring diagrams promotes a greater depth of system operation and develops excellent diagnostic skills.

Heater control module circuit
Voltage supply

Voltage is always present at the heater control module (A128) pin 2 C41 (connector 41). With the ignition switched on, the heater control module (A128), the heater control solenoid valve (Y100), the recirculation air actuator (M69) and the heater blower motor (M3) receive voltage. The heater control motor is controlled by switching the earth path hence 31S at pin 5. When the air-conditioning is switched on using the switch at the heater control module (A128), the heater control module (A128) provides a ground signal via pin 11 of A128 (earth, 31S) for the A/C switch relay pin 1 (K158). The A/C will not be switched on unless the blower motor is on or the air distribution is switched to defrost. Pin 4 of A128 is live when the sidelights are switched on to illuminate the display. The heating temperature is controlled by the temperature selector switch (13) at the heater control module (A128). The heater control solenoid valve (Y100) is timed by the heater control module (A128 – see Figure 3.129 on page 201).

In this way the supply of heating water to the heat exchanger is controlled. Changes from outside air to recirculated air operation are made using the switch (1) at the heater control module (A128). The heater control module (A128) then sends a switched ground signal to the recirculation air actuator (M69) via pin 9 of the module (A128).

Blower motor circuit

The blower motor and recirculation air actuator receive power from F16 once the ignition switch has been operated to start or run (DIN code 15). The speed of the heater blower motor (M3) is controlled via the heater blower switch (N14). If the blower motor is off and the air-conditioning is operating in demist conditions then the heater blower motor (M3) receives a ground signal from the heater control module (A128) pin 7 – see Figure 3.127 on page 196.

Testing blower motors

All tests should include power and earth connections/fuses etc. Testing the circuit operation is best suited to current measurement. This is also the most dangerous. This could be achieved by removing fuse F16 and placing an ammeter capable of drawing 25–30 amps or using an amp clamp around the wires. This will give the health of the electric motor and resistors. Alternatively disconnect switch N125 and provide an earth path with a fusible link. Be very careful the blower motor draws a high current which can create heat or damage testers. Often shunts are used to reduce the possibility of damage and provide the ability to measure high current flow. Testing using the wrong cable thickness could act as a hot wire! Always look at the cable thickness and fuse rating if in doubt: M3 pin 2, wire code 31S – CA18 4BK/RD which is a $4\,mm^2$ cross-sectional area of high current carrying capacity. Any switched earth/grounds 31S can often be tested using an LED or power probe. If the circuit is constantly being switched on and off the LED will flash green for earth and red for live.

A/C signal to the PCM (Fig. 3.142)

The relay (K158) then closes, thereby sending a signal to pin 10 of the Powertrain Control Module (PCM) (A147) via the A/C compressor cycling switch (N75). If there is no refrigerant in the system (below 1.6 bar) then the switch will be open circuit and the signal will not reach A147, thus no A/C system operation. The PCM (A147) controls the A/C Wide Open Throttle (WOT) relay (K32) via pin 47. By providing a ground (earth) to the relay, power can flow from fuse 32 (F32) to dual pressure switch (N76) and providing system pressure is not excessive the power will go to the WOT relay (K32). The A/C compressor clutch (Y16) and the A/C compressor are switched on. The A/C compressor cycling switch (A75) interrupts the voltage supply to pin 41 of the PCM (A147) when the system pressure decreases. Consequently, the compressor is turned off by the PCM (A147) via the A/C WOT relay (K32) in order to avoid ice formation in the evaporator. When the pressure increases, the A/C compressor is turned on again.

High speed condenser fan (Fig. 3.143)

At the same time the second contact of the dual pressure switch (N76) closes and signals the Powertrain Control Module (PCM) (A147) to run the engine cooling fan motor (M37) at high speed. The PCM (A147) then sends a signal to the high speed engine cooling fan relay (K46) which bridges the engine cooling fan resistor (R27).

System diagnostics

Assuming the clutch is not engaging, refrigerant pressure exists and the technician has selected to carry out some electrical tests.

Initial diagnostics

Initially, treat the A/C system as operating between two control systems, one being the heater control module and the other the Powertrain Control Module (PCM). The heater control module sends an A/C demand signal via the low pressure switch to the PCM. This is a good starting point if the A/C cycling switch is easy to access. The voltage level should be battery voltage. A connector view is available from the manufacturer if unsure which wire to check. If the A/C cycling switch is difficult to access then remove the A/C relay K150 and place either a power probe, LED tester, or multimeter (continuity or current test) on the relay coil pin which is provided with a ground signal by the heater control module. If using an LED/power probe the green LED will light up. Switch the A/C on and off to check the operation of the heater control module. Remember the A/C switch and blower fan must be running for the ground signal to be sent. If the signal is present then system diagnostics will focus on the PCM receiving the signal and the A/C coil being operated. If the signal is not present then diagnostics will work backward to verify why the signal is not being sent.

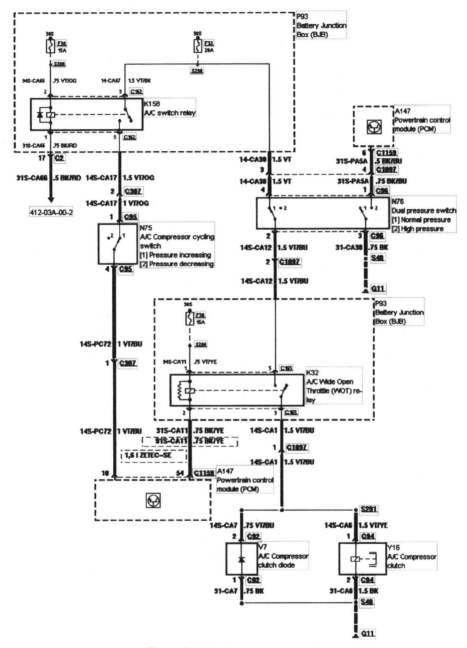

Figure 3.142 Compressor control
(reproduced with the kind permission of Ford Motor Company Limited)

If the signal is being sent, carry out a relay test on K32. Test to see if the PCM is providing a ground signal to the relay to activate the A/C system. If the ground is not provided then the fault exists where the PCM does not want the A/C system to run. This is due to the following reasons:

- It has not received the signal from the A/C relay.
- It has received an input from a sensor providing a signal which is out of operational parameters, i.e. engine coolant over 120°C or below 5°C, throttle position sensor over 3.8 V.

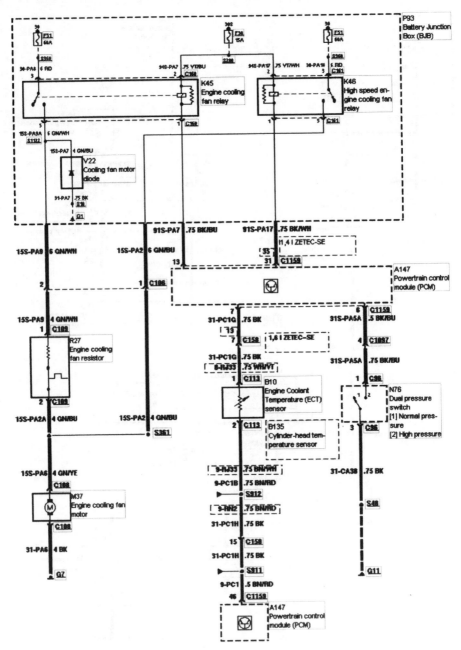

Figure 3.143 Condenser fan circuit
(reproduced with the kind permission of Ford Motor Company Limited)

In this event the PCM will not provide a ground signal to the WOT relay.

A/C relays are always a good starting point for initial electrical diagnostics.

WOT relay is not receiving a ground signal

Check that the A/C signal is being received by the PCM. This can be done at the module connector pin 10 C1159. A connector view can be obtained. A break-out box is sometimes useful and reduces the possibility of any damage to the connector block. A multimeter or oscilloscope can be used to verify the correct voltage level. First measure the voltage at pin 10 by placing the meter probe (red) on pin 10 and the ground probe (black) on ground (for zero reference). Then carry out a power-to-power check. This is done by placing the red probe on the battery positive terminal and the ground probe on pin 10. This will check the potential difference between the two signals. A difference no greater than 0.2 V should be present. If greater than this amount then a resistance exists in the system. If no voltage is present is the first instance then an open circuit exists possibly due to a faulty A/C switch.

If the signal is being received then a full 'PCM pin' test is required starting with all PCM power supplies and ground signal. 'Power-to-power' and 'ground-to-ground' tests must also be carried out to check pds across supplies. On successful completion the coolant, throttle and vehicle speed sensors should be checked.

During these tests you can provide your own feeds and earths using power probes to test the system's response. Never trick the A/C system to switch on if you have no evidence that the system has sufficient refrigerant and oil. Always seek advice when unsure. All tests should be carried out at the connector pins using the correct connector terminal fittings (connector pin kits or break-out box kits).

Serial testing

The heater control module is not accessible via serial testing. The PCM will provide DTCs and data lists but the A/C information will be limited to the cycling switch, throttle position sensor, vehicle speed sensor and engine coolant temperature sensor. The data list will provide information on whether the PCM is functioning correctly and if it is receiving power.

The serial test is certainly a quick and easy test to carry out and is worth conducting before any detailed electrical diagnostic testing is done.

Erasing trouble codes

The manual control system uses the PCM to operate the A/C clutch. The PCM receives a signal from the heater control module via the A/C relay and cycling switch. The only relevant diagnostic trouble codes available are A/C cycling switch signal, engine related vehicle speed sensor, engine coolant temperature sensor, and throttle position sensor. Diagnostic trouble codes are removed using a diagnostic scan tool. Carry out a dealer drive cycle upon retrieval and removal of codes.

Self-diagnostics

No self-diagnostics are available on this model and year.

System control

Three pressure switches are used to control A/C operation. A compressor cycling switch is used due to an FOV being fitted. A dual pressure switch is used to protect the system from overpressure and two operate the secondary condenser fan when the system is under load.

Low pressure switch closed	1.6 bar
High pressure switch opens (to protect system)	31.4 bar
High pressure switch operates secondary fans	20.7 bar

Refrigerant pressures
Preparatory conditions:

1. Pollen filter in good condition.
2. Engine idling and above 85°C.
3. A/C running (5 minutes).
4. Blower speed maximum.
5. Recirculation on.
6. A/C temperature to maximum cold selection.
7. All vents open.

Ambient temperature (°C)	A/C high pressure side (bar)	A/C low pressure side (bar)
15	5–12	1.6–3.1
20	6.5–14	1.6–3.1
26	9–15.5	1.6–3.3
32	11–18	1.8–3.5
37	13.8–20	2.4

Delivery temperature
Preparatory conditions:

1. All windows and doors closed.
2. All ventilation outlets fully open.
3. Engine idling.
4. Select cold on the A/C temperature control knob.
5. A/C/heater blower motor set to maximum speed.
6. Recirculation mode selected.
7. Air distribution knob set to face (centre vent).

Checking:

1. Run engine at 1500–2000 rpm.
2. Position temperature probe 100 mm into fascia ventilation centre outlet.
3. Measure temperature after 5 minutes.

Ambient temperature	Delivery temperature
20°C	6–12°C

Technical data

Description	Fill capacity/adjustment
Air-conditioning system capacity	200 ml
Ford product code WSH-M17B19-A	
Viscosity	ISO 46
Refrigerant (R134a)	740 grams ± 15 grams
Air-conditioning compressor if the quantity of oil removed from the compressor is less than 90 ml	Add 90 ml
Air-conditioning compressor if the quantity of oil removed from the compressor is between 90 ml and 148 ml	Fill with the same quantity plus 30 ml
Air-conditioning condenser	Add 30 ml
Air-conditioning evaporator	Add 90 ml
Dehydrator/accumulator	Pour the oil from the air-conditioning dehydrator/accumulator, which is being renewed, into a measuring cylinder. Refill the new dehydrator/accumulator with the same quantity of oil plus 90 ml
Every time refrigerant is extracted	Add 20 ml (measure whenever possible)
When all lines and components are renewed	Add 60 ml
Gap between the drive plate and the pulley (shim adjustment)	0.35 mm–0.85 mm
Tracer dye	Already in system (1999 onwards)
Fixed orifice tube colour	Green
Pressure	
Low pressure switch closed	1.6 bar
High pressure switch opens (to protect system)	31.4 bar
High pressure switch operates secondary fans	20.7 bar

Since 07/99 vehicles have a fluorescent tracer dye tablet inserted into the A/C system. If tracer dye is present, there is a green cross on the suction accumulator.

System repairs
The dash panel must be removed to gain access to the evaporator on this model.

3.9 Automotive A/C auto temp control system (case study 2)

This is the second of three case studies presenting a complete A/C system using aftermarket and OEM (Original Equipment Manufacturers) information. The key to this particular case study is the appreciation of the amount of interpretation which exists with information presented by Autodata. All diagrams supplied by this company are clearly referenced enabling the reader to distinguish between aftermarket and OEM. The diagrams provide some excellent examples of how information can be alternatively presented to assist the technician in visualising where system components are fitted. The wiring schematics are very easy to follow and adhere to most of the European requirements with respect to current flow and coding. Information from Autodata is available in a range of formats. This includes technical helplines, books and CD-ROMs. The information from Autodata used within this case study has been accessed from an Autodata A/C CD-ROM.

The case study is not written in textbook fashion. This is because the purpose of each case study is for the reader to appreciate the breadth and depth of information available and what information is required to enable the technician to successfully work on such systems. The author has attempted to try to present the information in a logical order. Explanations have been written to assist in the understanding of system operation. If the reader fails to understand a particular aspect of the case study then refer to a section which covers that topic.

Described using after market information courtesy of:

SAAB 900 ACC system (automatic climate control)
Automatic temperature control
Automatic air distribution
Single evaporator – single zone
Expansion valve

Air distribution

The vehicle has a pollen filter that requires replacement at 18 000 miles or 12 months, whichever occurs first, then every 12 000 miles or 12 months, whichever occurs first.

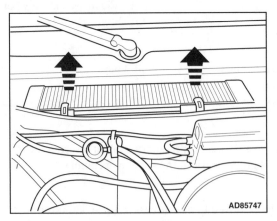

Figure 3.144 Pollen filter location
(kind permission of Autodata Ltd)

A/C system information

The main purpose of the ACC control module is to control the air-conditioning system so that a comfortable cabin temperature is achieved as soon as possible after starting and then maintained throughout the journey. The ACC control module consists of a display panel with pushbuttons

(S292) and an integral control module (A63). The ACC system communicates with other modules to obtain information on other sensor inputs related to A/C operation. Communication is via a data bus link or PWM signal.

ACC (A63) inputs and outputs:

1. ACC unit consisting of panel with integral control module (A63 and S292).
2. Stepper motor for air distribution (M112).
3. Stepper motor air blend damper (M114).
4. Cabin temperature sensor (B37).
5. Solar sensor (B102).
6. DC permanent magnet stepper motor for recirculation damper (M59).
7. Blended-air temperature sensor (footwell vent B160).
8. Fan control unit (feedback and control signal).
9. Engine coolant and heater regulator valve (Y36).
10. Instrument illumination rheostat (R4).

ECM (Engine Control Module) (A35) inputs and outputs:

1. Compressor clutch (Y11).
2. Vehicle speed sensor.
3. Evaporator temperature switch (S51).
4. Triple pressure switch (S341).

SID (SAAB Information Display). A digital multifunction display (A161) with multiplex facility:

1. Outside temperature (B61).

ICE (Integrated Central Electronics). A combination control module (A94):

1. Engine coolant temperature.
2. Condenser fan stages.

A/C amplifier (A175):

1. Heater blower motor.

System layout and components

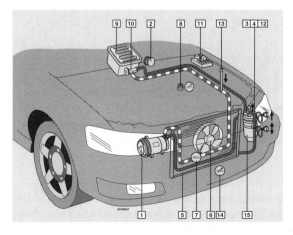

No.	Components
1	A/C compressor clutch
2	A/C/heater blower motor
3	A/C refrigerant high pressure switch
4	A/C refrigerant triple pressure switch
5	Condenser
6	Engine coolant blower motor
7	Engine coolant blower motor resistor
8	Engine coolant temperature (ECT) sensor
9	Evaporator – behind fascia
10	Expansion valve
11	Fuse box/relay plate
12	High pressure service connector
13	Low pressure service connector
14	Outside air temperature sensor – under front bumper
15	Receiver-drier

Figure 3.145 A/C components and accompanying table
(kind permission of Autodata Ltd)

No.	Components
1	A/C amplifier
2	A/C control module – behind heater controls
3	A/C evaporator temperature switch behind fascia
4	A/C footwell vent temperature sensor
5	A/C/heater air direction motor – behind heater controls
6	A/C/heater air mix flap motor
7	A/C recirculation motor – behind heater controls
8	A/C sunlight sensor
9	Combination control module/relay
10	Data Link Connector (DLC)
11	Engine Control Module (ECM)
12	Engine coolant heater regulator valve – behind fascia, centre
13	Engine coolant heater valve control switch – behind fascia, centre
14	Evaporator
15	Fuse box/relay plate – fascia I
16	Fuse box/relay plate – fascia II
17	In-car temperature sensor – fascia centre

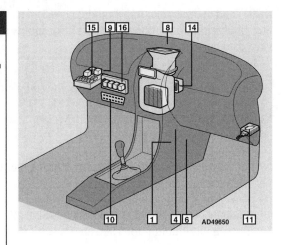

Figure 3.146 A/C electrical/electronic controls and accompanying table
(kind permission of Autodata Ltd)

Cabin temperature sensor

The cabin temperature sensor is mounted in the dashboard below the ACC control module. It has an integral suction fan which sucks the cabin air past an NTC resistor. Cabin temperature is the ACC control module's most important parameter. It is compared with the selected cabin temperature to determine whether the temperature of the blended air should be raised or lowered. Cabin temperature is corrected so that it corresponds to the physical perception of the selected temperature, regardless of the outside temperature. When the difference between selected temperature and corrected cabin temperature increases, the speed of the ventilation fan will also increase. When the ignition is switched off the suction fan will continue to run for about 4 minutes. This reduces the risk of incorrect temperature settings if the car is restarted within a fairly short time.

Suction fan motor and temperature sensor

Suction fan motor power supply 12 V 50 mA	Pin no. 4 (from ACC control module)
Suction fan motor ground	Pin no. 1 (from fan control)
Temperature sensor (NTC) 5 V power supply	Pin no. 2 (from ACC control module)
Temperature sensor (NTC) ground	Pin no. 3 (from ACC control module)

Evaporator switch

An anti-frost thermostat on the evaporator prevents freezing and ice building on the evaporator. There is a sensor lying against the evaporator low pressure pipe. When the temperature drops below +2°C the voltage to the compressor is broken. When the temperature exceeds +5°C the thermostat closes again and the compressor is engaged.

Outside temperature sensor

The outside temperature is obtained from the SID via a data bus connection. The sensor is fitted under the front bumper. The value of the outside temperature is used by the ACC control module to correct the value of the cabin temperature and to control the speed of the fan. The cabin temperature is corrected so that it corresponds to the physical perception of the selected temperature. This means that the measured cabin temperature is higher than the low outside temperatures. This is true even at high outdoor temperatures, but the difference is less. The condition is dependent on whether heat is passing out of the cabin or into it. At outside temperatures below +5°C and over +20°C the fan speed increases to achieve a more consistent cabin temperature.

Sunlight sensor

The solar sensor is placed on top of the dashboard. It measures mainly infrared radiation (heat radiation). In the event of an increase in solar radiation and an outside temperature exceeding +15°C the ACC control module increases the ventilation fan speed since more cold air must be supplied. The fan speed is immediately changed in the event of a change in the solar radiation. The fact that the solar sensor measures mainly heat radiation means that it cannot be tested with fluorescent light, only sunlight or light from a bulb must be used.

Rheostat

Display panel lighting value is obtained from the SID. This value regulates the lighting intensity of the ACC panel's display. The same value also regulates the mileometer display in the main instrument display panel and the SID's own display. In darkness the value is determined by the rheostat and in daylight by the brightness of the light in the cabin so that good readability is always obtained.

Pressure switches (trinary switch)

Three switches are built into the receiver drier assembly: a low pressure, high pressure and condenser fan switch. The receiver drier is positioned on the high pressure side of the A/C system.

		Low pressure switch	Condenser fan switch	High pressure switch
Switch-off pressure	bar	2.0 ± 0.25	12.5 ± 1.5	30.0 ± 2.0
Switch-on pressure	bar	2.15 ± 0.35	16.5 ± 1.2	24.0 ± 2.0

Engine coolant temperature sensor

The temperature of the engine is monitored by the ICE module. This signal is monitored to ensure that if the temperature of the cooling system exceeds 126°C then the A/C compressor will not be energised due to the absence of a signal from the ICE module to the ECM module. The coolant temperature is also communicated to the instrument panel. The information is communicated via a PWM signal with a fixed frequency of 122 Hz which varies in duty cycle ratio (Figs 3.147 and 3.148). The variation in duty ratio from 10% to 90% corresponds with the coolant temperature.

Signals between ICE and panel display

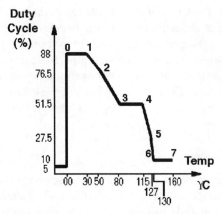

Figure 3.147 Duty cycle ratio and corresponding temperature programming points (courtesy of SAAB)

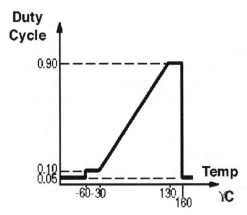

Figure 3.148 Duty ratio and temperature relationship (courtesy of SAAB)

Fan speed

The fan is powered directly with a +54 V supply from fuse 12. On the ground side the fan is connected to a fan control which receives a control voltage of 0–5 V from the ACC control module. The fan control also receives a special voltage supply from the ACC control module. The ACC control module receives feedback from the ground side of the fan motor, which gives it information on the actual voltage across the fan motor. The feedback voltage increases in relation to ground as fan speed decreases. Fan speed is affected in the auto mode of operation as follows:

1. As the difference between selected temperature and cabin temperature increases, fan speed also increases.
2. Outside temperature below +5°C or exceeding +20°C gives an increase in fan speed.
3. An increase in the intensity of the sun at outside temperatures exceeding +15°C gives an increase in fan speed.

The desired fan speed can also be selected manually on the ACC panel. It can be set in ten steps of about 2 A each. If the button is depressed for more than one second, fan speed will increase or decrease automatically. Fan speed is shown on the ACC panel display.

Recirculation damper

The recirculation damper is controlled by a DC permanent magnet stepper motor. The damper has only two positions. When the motor has turned the damper to one of the end positions, the current through the motor winding is limited by two PTC resistors, built into the motor.

Air blend motor

The blend door is operated by a hybrid stepper motor. The stepper motor has two windings. A voltage is applied to the windings in special order with short pulses. This causes the motor to move in short steps, hence its name. The direction of rotation can be changed. When the motor is stationary, current is applied continuously to both windings. A stepper motor requires no feedback to the ACC control module. By sending a definite number of pulses, the ACC control module always knows how much the damper moves. A condition for this is that the control module calibrates itself by rotating the damper to an end position so that the exact position of the damper is known. The air blending damper is set by the ACC control module with the aid of the blended air temperature sensor so that a suitable air temperature will be obtained. The cabin temperature, selected temperature and outside temperature are decisive for the blended air temperature. If the air blending damper is set at the position for maximum cold and this is still insufficient to maintain the selected temperature, recirculation will be selected. If the selected temperature is 'HI' or 'LO' the air blending damper will be set to the maximum heat or maximum cold position.

Air distribution

The air distribution stepper motor operates with the same principle as the air blend motor. The air distribution motor varies its rotation to direct air to the following distribution vents:

Distribution	Door angle manual mode	Door angle auto mode
Defrost	270°	250°–270°(only cold starts)
Defrost/floor	207°	207°–250°
Floor	180°	Not used
Panel/floor	150°	Not used
Panel	90°	93°–96°

If the requisite blended air temperature is high, air distribution will be set to the defrost/floor position. If the requisite blended air temperature is low, air distribution will be set closer to the defrost position. In the case of requisite blended air temperatures that are extremely low, air distribution will be set to the panel position. As will be realised, air distribution is a function of the requisite blended air temperature. The desired air distribution can also be selected manually on the ACC panel.

System operation – compressor engagement (Fig. 3.152)

A +54 voltage is fed via fuse 3 to the ventilation fan switch. If the switch is in one of the positions 1 to 4 and the A/C button is ON, the voltage is fed to the ICE (Integrated Central

Electronics) module. The ICE module monitors the engine temperature. If the engine temperature is below +126°C a signal is sent to the evaporator temperature switch. If the temperature in the evaporator is approximately +5°C or higher, the voltage is fed to the engine management system. The engine management system increases the idling speed and grounds the cable to the three stage pressure switch. If the pressure in the A/C system is higher than 2 bar but lower than 30 bar, the pressure switch grounds the cable to the A/C relay. From fuse 5 (15 A), current is fed via the A/C relay to the thermal fuse in the compressor overheating protection. If the temperature in the compressor is lower than +140°C, the current is fed to the solenoid clutch in the compressor.

The engine management system breaks the circuit to the A/C relay in the event of a powerful acceleration which is received from the throttle position sensor. The evaporator temperature sensor determines if the A/C compressor is running based on the evaporator's surface temperature.

Auto mode
Select the auto mode of operation. All functions will be controlled automatically. AUTO and the selected temperature appear in the display. Press the Auto button again and all automatically selected settings will be displayed. The electrically heated rear window function can be disabled or activated without leaving the auto mode of operation.

Temperature up or down
Selection of inside temperature is in steps of 1°C between +15°C and +27°C, alternatively steps of 2°F between +58°F and +82°F. If a temperature above +27°C (+82°F) is selected, 'HIGH' is displayed as the selected temperature. Correspondingly 'LOW' is displayed if the temperature selected is below +15°C (+58°F). When both buttons are pressed at the same time for more than 2 seconds the temperature display changes between Celsius and Fahrenheit.

If the selected temperature is 'HIGH' in Auto mode, the system sets itself as follows:

Air distribution damper	Defrost/floor
Blended air damper	Max. heat
Fan	Max. fan current
Recirculation damper	Fresh air

AUTO is extinguished and fan speed and air distribution are shown.

If the selected temperature is 'LOW' in Auto mode, the system will be set as follows:

Air distribution damper	Panel and rear centre vent
Blended air damper	Max. cold
Fan	Max. fan current
Recirculation damper	Recirculation

AUTO is extinguished and fan speed, air distribution and recirculation are displayed.

Defrost

Air distribution damper	Defrost position, 270°
Fan current	15 A
A/C	On
Electrically heated rear window	On
Recirculation	Off

Automatic temperature control still in effect, recirculation cannot be selected. Symbols for the button and activated functions are shown on the display.

Defrost/floor

The air is distributed to defrost as well as the floor and rear side windows, door angle 207°. The button symbols are shown on the display.

Panel

The air is distributed to the panel as well as the rear centre vent, door angle 90°. The button symbol is shown on the display.

Panel/floor

The air is distributed to the panel and rear centre vent as well as to floor and rear side windows, door angle 150°. The button symbols are shown on the display.

Floor

The air is distributed to the floor and rear side windows, door angle 180°. The button symbol is shown on the display.

Recirculation

Select recirculation or fresh air. If recirculation is selected the button symbol will be shown on the display.

ECON

A/C and recirculation are turned off. In other respects control takes place as in the AUTO mode. ECON is shown on the display.

Electrical information

Autodata provide information on fuse box and relay layout. This is useful when checking and testing fuses and relays.

Fuse box/relay plate

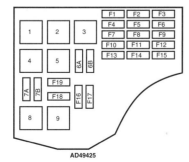

Fuse (amps)	Circuit
F3 (40A)	Engine coolant blower motor relay I
F5 (15A)	A/C compressor clutch cut-off relay
F16 (30A)	Engine coolant blower motor relay II

Location	Circuit
4	Engine coolant blower motor relay I
8	Engine coolant blower motor relay II
9	A/C compressor clutch cut-off relay

Figure 3.149 Battery junction box
(kind permission of Autodata Ltd)

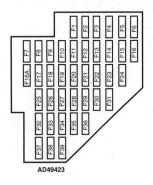

Fuse (amps)	Circuit
F1 (30A)	Heated rear window relay
F3 (30A)	A/C/heater blower motor
F19 (15A)	A/C compressor clutch cut-off relay
F21 (10A)	A/C control module, engine coolant heater valve control switch, engine coolant heater regulator valve
F27 (15A)	A/C control module

AD49423

Figure 3.150 Central junction box
(kind permission of Autodata Ltd)

Wire colour coding key (DIN standard)
The table below provides information on the wiring colour used on this vehicle.

bl = blue	br = brown	el = cream	ge = yellow
gn = green	gr = grey	nf = neutral	og = orange
rs = pink	rt = red	sw = black	vi = violet
ws = white	hbl = light blue	hgn = light green	rbr = maroon

Harness plug identification

Harness plug identification is very important when trying to identify wiring. Technicians often back probe using adaptors to sample live data from such connectors. Pins should be checked to ensure that no corrosion exists and the pins are straight.

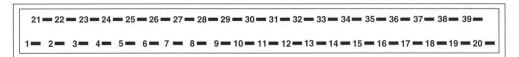

Figure 3.151 A63 A/C harness multiplug (male) 39 pin
(kind permission of Autodata Ltd)

Circuit diagram key (DIN standard)
The circuit diagram key helps identify the components within the wiring schematic.

Code	Component description
A5	Instrument panel
A16	ABS control module

(continued)

Code	Component description
A35	Engine Control Module (ECM)
A63	A/C control module
A94	Combination control module/relay
A161	Digital multifunction display
A175	A/C amplifier
B24	Engine Coolant Temperature (ECT) sensor
B37	In-car temperature sensor
B61	Outside air temperature sensor
B102	A/C sunlight sensor
B160	A/C footwell vent temperature sensor
K12	Engine coolant blower motor relay
K13	Heated rear window relay
K263	A/C compressor clutch cut-off relay
M6	Engine coolant blower motor
M7	A/C/heater blower motor
M59	A/C recirculation motor
M112	A/C/heater air direction motor
M114	A/C/heater air mix flap motor
R4	Instrument illumination rheostat
R46	Engine coolant blower motor resistor
S51	A/C evaporator temperature switch
S152	A/C refrigerant high pressure switch
S292	A/C/heater function control panel
S341	A/C refrigerant triple pressure switch
S346	Engine coolant heater valve control switch
X1	Data Link Connector (DLC)
Y11	A/C compressor clutch
Y36	Engine coolant heater regulator valve
15	Ignition switch – ignition ON
30	Battery+
31	Battery−

Explanation on wiring diagram (Figure 3.152)
An ignition switch voltage is fed via fuse 3 to the ventilation fan switch pin 37 which is built into the A/C control panel A63/S292. If the blower switch is in one of the positions 1 to 4 and the A/C button is ON, then battery voltage is fed to the ICE module A94 via pin 38 of the A/C module. ICE module checks the engine coolant temperature across pins 5 and 58 and if the volt drop corresponds to a temperature which is lower than +126°C then battery voltage is sent to the evaporator temperature switch. If the temperature in the evaporator is above 5°C, the voltage is fed to the engine management system A35. The engine management system increases the idling speed and grounds the cable to the three stage pressure switch. If the pressure in the A/C system is higher than 2 bar but lower than 30 bar, the pressure switch grounds the cable to the A/C relay. From fuse 5 (15 A), current is fed via the A/C relay to the thermal fuse in the compressor overheating protection. If the temperature in the compressor is lower than +140°C, the current is fed to the solenoid clutch in the compressor and the system engages.

The engine management system breaks the circuit to the A/C relay in the event of a powerful acceleration which is received via a voltage output from the throttle position sensor. The evaporator temperature sensor determines if the A/C compressor is running based on the evaporator's surface temperature. The trinary switch will also break the voltage supply to the ECM in the event of a loss of refrigerant or a blockage causing high pressure in the system.

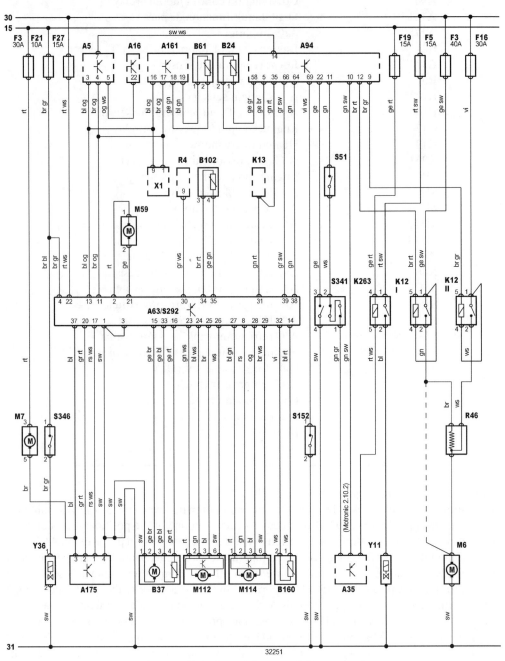

Figure 3.152 Circuit diagram DIN standard
(kind permission of Autodata Ltd)

Diagnostic information

The table below provides details of the pin and wire configuration from the module to other components within the A/C system. If a pin is not included in the table then it is not used on the module. The test conditions are the operations which are required to obtain the reading.

Pin	Component/ function	In/out	Test condition	Test reading	Across X–Y	Signal
1	Ground	In		<0.4 V	1 – B neg	
2	Recirculation motor	Out	Activate 'ON'	Battery voltage	2–21	Current is flowing
			Activate 'OFF'	0 V		Current not flowing
3	Parking heater	In	Parking heater off	Batt+	22–3	Battery voltage
			Parking heater on			
4	+54 supply	In		Batt+	3–1	Power to power test no more than 500 mV
				<0.5 V	Batt+ − pin 54	
			Ignition OFF	Batt+		
8	Air blending flap motor	Out	Stationary flap motor	Pulse battery voltage	8–1	Waveform analysis using oscilloscope – section 3.4
11	SI-BUS(+)	In/Out		approx. 2.5 V	11–1	Waveform analysis using oscilloscope – section 3.4
13	SI-BUS(−)	In/Out		approx. 2.5 V	13–1	Waveform analysis using oscilloscope – section 3.4
14	Blended air sensor, ground	Out		<0.1 V	14–1	Ground- to-ground signal
15	Cabin temperature sensor fan, ground	Out	Ignition on	<0.1 V	15–1	Ground- to-ground signal
16	Power supply, cabin temperature fan	Out		Batt+	16–1	Battery voltage
17	Power supply, fan control	Out		Batt+	17–1	Battery voltage
20	Control voltage, fan	Out	No fan	0 V	20–1	
			Full fan	5 V		

(continued)

Pin	Component/ function	In/out	Test condition	Test reading	Across X–Y	Signal
21	Recirculation motor	Out	Activate 'ON'	11–12 V	2–21	Current is flowing
			Activate 'OFF'	0 V		Current not flowing
22	30 supply	In		<0.5 V	B+ 22	
23	Air distribution flap motor	Out	Stationary flap motor	Pulse battery voltage	23–1	Waveform analysis using oscilloscope – section 3.4
24	Air distribution flap motor	Out	Stationary flap motor	Pulse battery voltage	24–1	Waveform analysis using oscilloscope – section 3.4
25	Air distribution flap motor	Out	Stationary flap motor	Pulse battery voltage	25–1	Waveform analysis using oscilloscope – section 3.4
26	Air distribution flap motor	Out	Stationary flap motor	Pulse battery voltage	26–1	Waveform analysis using oscilloscope – section 3.4
27	Air blending flap motor	Out	Stationary flap motor	Pulse battery voltage	27–1	Waveform analysis using oscilloscope – section 3.4
28	Air blending flap motor	Out	Stationary flap motor	Pulse battery voltage	28–1	Waveform analysis using oscilloscope – section 3.4
29	Air blending flap motor	Out	Stationary flap motor	Pulse battery voltage	29–1	Waveform analysis using oscilloscope – section 3.4
30	Rheostat	In	Max. rheostat Min	Batt+ 2 V	30–1	
31	Electrically heated rear window	In	Heated rear window on	0 V	31–1	
			Heated rear window off	Batt+		
32	Blended-air sensor	In	At approx. +23°C	approx. 2.5 V	32–1	See temperature technical data

Pin	Component/ function	In/out	Test condition	Test reading	Across X–Y	Signal
33	Cabin temperature sensor	In	At approx. +20°C	approx. 2.7 V	33–1	See temperature technical data
34	Solar sensor	In	Illuminate with light bulb	0–0.5 V	34–1	
35	Solar sensor ground	Out		<0.1 V	35–1	Ground-to-ground test
37	Test voltage, fan	In	Full fan No fan	0 V Batt+	37–1	
38/39	A/C (to ICE)	Out	AUTO MODE ECON MODE	Batt+ 0 V	38–1	

Self-test

Calibration must be carried out if the battery has been disconnected or discharged, or if the ACC panel has been replaced. To start calibration, press the 'AUTO' and 'OFF' buttons simultaneously. This also starts a self-test. Calibration and the self-test are carried out in parallel and take less than 30 seconds. All stored faults are cleared at the start and while calibration and self-testing are in progress the number of faults found will be shown on the ACC panel's display. Calibration must also be carried out if any of the stepper motors have been replaced. Calibration can also be carried out by means of a serial tester.

Erasing trouble codes

Suitable diagnostic equipment is required to erase data from A/C control module fault memory.

Table 3.10 Fault code identification (kind permission of Autodata Ltd)

Trouble code	Fault location
B1343	A/C sunlight sensor – short/open circuit
B1347	A/C foot well vent temperature sensor – short circuit
B1348	A/C foot well vent temperature sensor – short/open circuit
B1352	In-car temperature sensor – short circuit
B1353	In-car temperature sensor – short/open circuit
B1355	A/C/heater blower motor/in-car temperature sensor – supply voltage
B1360	A/C control module – keypad fault
B1493	A/C control module/combination control module/relay – signal fault
B1498	Heated rear window – short circuit
B1515	Common sensor earth – short circuit
B1605	A/C control module – defective
B1675	A/C/heater air direction motor – defective

(*continued*)

Trouble code	Fault location
B1676	A/C/heater recirculation motor – faulty
B1677	A/C/heater air mix flap motor – faulty
B2402	A/C/heater air direction motor – short circuit
B2403	A/C/heater air direction motor – open circuit
B2404	A/C/heater air direction motor – short circuit
B2405	A/C/heater air direction flap – loose
B2406	A/C/heater air direction flap – jammed
B2413	A/C/heater recirculation flap motor – open circuit
B2414	A/C/heater recirculation motor – short circuit
B2426	A/C/heater blower motor – control defective
B2427	A/C/heater blower motor – control defective
B2428	A/C/heater blower motor – control defective
B2492	A/C/heater air mix flap motor – short circuit
B2493	A/C/heater air mix flap motor – open circuit
B2494	A/C/heater air mix flap motor – short circuit
B2495	A/C/heater air mix flap – loose
B2496	A/C/heater air mix flap – jammed

System repairs

Access to evaporator housing from vehicle interior, removal of fascia panel required.

Access to A/C/heater blower motor from engine bay, removal of fascia panel not required.

Refrigerant charging

Ensure refrigerant circuit is evacuated for a minimum of 30 minutes prior to charging. No further instructions specified. Refer to refrigerant charging equipment operating instructions.

Refrigerant pressures

To measure the A/C system pressure the following conditions must exist.

Preparatory conditions:

1. All windows and doors closed.
2. All ventilation outlets fully open.
3. Engine idling.
4. Select 'LO' on the A/C control module.
5. A/C/heater blower motor set to maximum speed.
6. Recirculation mode selected on the A/C control module.
7. A/C control module set to vent.
8. Run air-conditioning for 5 minutes prior to testing.

Checking:

1. Run engine at 1500–2000 rpm.

Ambient temperature	High pressure	Low pressure
20°C	12–16.5 bar	1–3 bar

Delivery temperature

To measure the A/C system delivery temperature from the air vents the following conditions must exist.

Preparatory conditions:

1. All windows and doors closed.
2. All ventilation outlets fully open.
3. Engine idling.
4. Select 'LO' on the AC control module.
5. A/C/heater blower motor set to maximum speed.
6. Recirculation mode selected on the A/C control module.
7. A/C control module set to vent.

Checking:

1. Run engine at 1500–2000 rpm.
2. Position temperature probe 100 mm into fascia ventilation centre outlet.
3. Measure temperature after 5 minutes.

Ambient temperature	Delivery temperature
20°C	6–12°C

Technical data (by kind permission of Autodata Ltd)

Refrigerant	
Type	R134a
Quantity	725 grams
Refrigerant oil	
Type	SP10
Viscosity	ISO 46
Quantities	
Compressor Sanden TRS 105	70 ml
Condenser	40 ml
Evaporator	40 ml
Expansion valve	20 ml
Hose/lines	40 ml
System	150 ml
Compressor clutch	
Adjustment type	Shim
Clearance	0.3–0.6 mm
Resistance	3.0 Ω
A/C footwell vent temperature sensor	
Temperature	*Resistance*
0°C	28 kΩ
10°C	18 kΩ

(*continued*)

20°C	12 kΩ
30°C	8.3 kΩ
40°C	5.8 kΩ
50°C	4.1 kΩ
60°C	3 kΩ
70°C	2.2 kΩ
80°C	1.6 kΩ
90°C	1.2 kΩ

A/C sunlight sensor

Condition	Voltage
Shade	0 V
Sunlit	0.5 V

Engine coolant temperature sensor

Temperature	Resistance
20°C	2 kΩ
30°C	1.6 kΩ
50°C	800 Ω
85°C	300 Ω
110°C	140 Ω

In-car temperature sensor

Temperature	Resistance	Voltage
0°C	28 kΩ	3.7 V
10°C	18 kΩ	3.2 V
20°C	12 kΩ	2.7 V
30°C	8.3 kΩ	2.3 V
40°C	5.8 kΩ	1.8 V
50°C	4.1 kΩ	1.5 V
60°C	3.0 kΩ	1.2 V
70°C	2.2 kΩ	0.9 V
80°C	1.6 kΩ	0.7 V
90°C	1.2 kΩ	0.5 V

Outside air temperature sensor

Temperature	Resistance
0°C	5.8–6.2 kΩ
10°C	3.8–4.1 kΩ
20°C	2.5–2.8 kΩ
30°C	1.7–1.9 kΩ

3.10 Automotive climate control system (case study 3)

This is the last of three case studies presenting a complete A/C system using OEM (Original Equipment Manufacturers) information. The case study is not written in textbook fashion. This is because the purpose of each case study is for the reader to appreciate the breadth and depth of information available and what information is required to enable the technician to successfully work on such systems. The author has attempted to try to present the information in a logical order. Explanations have been written to assist in the understanding of system operation. If the reader fails to understand a particular aspect of the case study then refer to a section which covers that topic. Information from Vauxhall is usually sourced electronically on

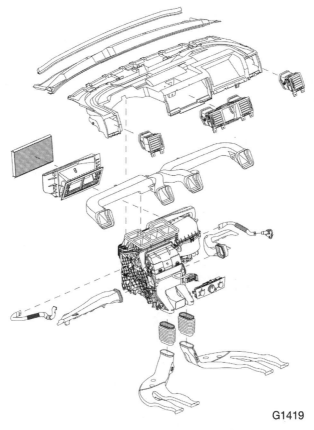

G1419

Figure 3.153 Heating and ventilation system (courtesy of Vauxhall Motor Corporation)

CD-ROMs. The Vauxhall information system is called TIS 2000 which includes technical information, diagnostic routines, bulletins, wiring diagrams, key programming and the facility to link with diagnostic equipment like TECH 2.

Described using OEM information:

Vauxhall
Electronic Climate Control (ECC)
Automatic temperature control
Single evaporator – dual zone
Expansion valve control

System information

Available for the Vectra-C is a newly developed Electronic Climate Control (ECC) system.

The Vectra-C features a new user interface concept for its ECC system. The system is operated via a control panel through a menu structure that is displayed on the centrally placed Graphic Info Display (GID) or Colour Info Display (CID).

Furthermore the system includes features such as:

● air quality sensor;
● dual sun sensor;

- rest-climatisation (reduces hot spots and measures UV intensity);
- remote controlled parking heater (future).

Heating and ventilation

The control panel for the heating and ventilation system incorporates a control module. This control module controls:

- the position of the flaps;
  - air temperature/mix flap (stepper motor);
  - air distribution flap (stepper motor);
- the blower speed (four different speeds: supply via shunt resistors or directly by a relay);
- rear window heating.

Without air-conditioning or ECC, the Vectra-C does not feature a recirculation valve.

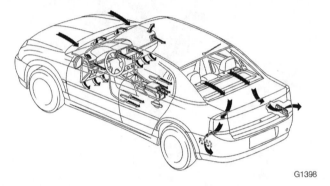

G1398

Figure 3.154 Air flow through system
(courtesy of Vauxhall Motor Corporation)

Air filtration

The pollen filter is located under the water deflector and is available as a filter with activated carbon layer.

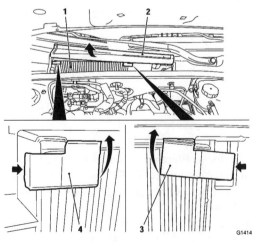

1. Pollen filter
2. Water deflector
3. Left filter locating clip
4. Right filter locating clip

G1414

Figure 3.155 Location of the pollen filter
(courtesy of Vauxhall Motor Corporation)

Control panel heating and ventilation

There is no feedback on the actual position of the mix and distribution flap. Using so-called overstepping ensures the actual position. If a flap is moved to the fully opened or closed position, under certain conditions, the nominal number of steps is increased by 5%, to make sure the flap reaches this position. Also, when switching OFF the ignition, the flaps are set in the parking position (fully opened or closed) with 5% overstepping. This means that synchronisation of the flaps is no longer needed (or even possible).

Controls for electronic climate control

The communication between the A/C control panel and display takes place via the mid-speed CAN bus. The controls for the ECC feature four switches and three rotary knobs for selection of the ECC functions. The defrosting, rear screen heating, automatic and recirculating air functions can be activated via the four switches. Each of these functions has a status LED integrated in the switch, which indicates which function is currently active. The temperature is set separately for the driver and passenger via the left and right rotary knobs. The central knob serves to manually adjust blower speed and select the ECC functions from the menu. It is displayed on the GID (Graphics Interface Display) or CID (Colour Interface Display). The knob can be turned to the right or left to scroll down the menu, and the menu options are selected by pressing the centre knob.

Multiplex communication

The ECC is the central control unit for controlling the interior vehicle temperature. For this purpose, it is provided with temperature, air quality and sun sensors in order to measure the current environmental conditions as well as actuators to actuate the air flaps, the fan and the AC compressor (via the CAN bus system). ECC also actuates the rear screen heating and the auxiliary heating. In order to prevent overloading the engine (in particular during idling), the ECC exchanges the ECC operating conditions and request signals with the engine control unit via the CAN bus. These messages are transmitted to and from the HSCAN and MSCAN via the CIM (Column Interface Module) (HSCAN to LSCAN) and the GID/CID interface (LSCAN to MSCAN).

Bus interface to mid-speed CAN (MSCAN)

The ECC has two connections to the mid-speed CAN bus. They serve to loop the mid-speed CAN bus through the control unit. The control unit electronics are then internally connected to the looped-through mid-speed CAN bus. If the ECC control unit is not connected, the mid-speed CAN bus is interrupted.

Display actuation function

Among other tasks, the GID/CID is responsible for the display of all ECC menus. Furthermore, information that needs to be transmitted from the ECC to other control units in the CAN network is transmitted to the low speed CAN bus by the GID/CID (interface function). If this information is destined for control units connected to the high speed CAN bus, the CIM effects the transmission from the low speed CAN bus to the high speed CAN bus.

Dimmed illumination function

The BCM (Body Control Module) transmits the dimming value for the interior illumination via the low speed CAN bus. After receiving this message, the ECC adjusts the intensity of illumination of the control panel according to the dimming value.

K0831

Figure 3.156 Control panel ECC system
(courtesy of Vauxhall Motor Corporation)

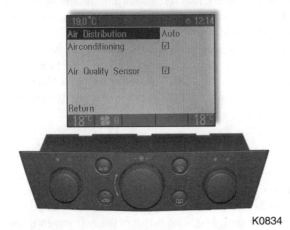

K0834

Figure 3.157 ECC control panel and display
(courtesy of Vauxhall Motor Corporation)

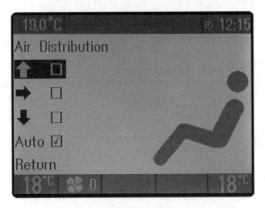

K0835

Figure 3.158 Air Distribution menu
(courtesy of Vauxhall Motor Corporation)

From the main menu it is possible to select:

- air distribution;
- switching the air-conditioning ON/OFF (similar to ECC on the previous systems);
- switching the auto recirculation (by means of the air quality sensor) ON/OFF.

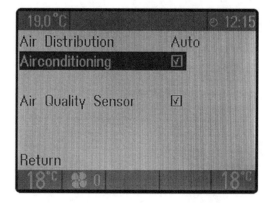

Figure 3.159 Switching the air-conditioning ON or OFF
(courtesy of Vauxhall Motor Corporation)

When selecting 'Air Distribution' and confirming by pressing the centre rotary control, the Air Distribution menu is displayed.

By selecting the three possible air outlets with the centre rotary control, all possible combinations can be made manually. To return to 'Auto Air Distribution' select 'Auto' from this menu or press the Auto button on the ECC control panel.

System activation and deactivation
To switch the compressor ON/OFF, select or deselect the option 'Air-conditioning'. As with the Eco button, the heating and ventilation remains in Auto mode but without the help of air-conditioning.

ECC features an air quality sensor that enables auto recirculation. This feature can be activated by pressing the recirculation button twice. On Vectra-C auto recirculation is enabled as default. Auto recirculation can be disabled with the option 'Air Quality Sensor' from the ECC main menu. Recirculation can also be switched ON manually, by using the switch on the ECC control panel. The manual recirculation has priority over the Auto Recirculation function. During manual recirculation, the status of the Auto Recirculation remains active. When the manual recirculation is switched OFF, the Auto Recirculation remains enabled. Again, this is indicated in the ECC main menu.

ECC components (Figs 3.160 and 3.161)
The ECC system consists of the following components:

- ECC control module/control panel.
- Display (GID or CID).
- Evaporator sensor.
- Output air temperature sensor, outlet footwell left/right and outlet passenger compartment left/right.
- Dual sun sensor with integrated in-car temperature sensor.
- Air quality sensor.
- Air mix flap stepper motor, left and right.
- Passenger compartment outlet flap stepper motor.
- Footwell outlet flap stepper motor.

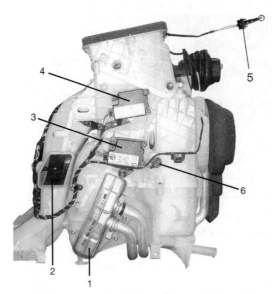

1. Heater core
2. Sensor – outlet temperature footwell, right
3. Stepper motor – mixed air flap, right
4. Stepper motor – footwell flap
5. Sensor – outlet temperature passenger compartment, right
6. Sensor temperature evaporator

Figure 3.160 ECC heater assembly
(courtesy of Vauxhall Motor Corporation)

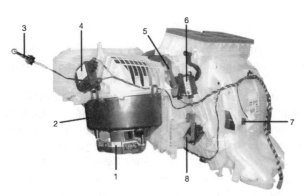

1. Blower voltage regulator
2. Blower motor, passenger compartment
3. Sensor – outlet temperature passenger compartment, left
4. Control motor – recirculation flap
5. Stepper motor – defrost flap
6. Stepper motor – passenger compartment flap
7. Sensor – outlet temperature footwell, left
8. Stepper motor – mixed air flap, left

Figure 3.161 ECC heater assembly

- Defrost outlet flap stepper motor.
- Recirculation flap DC motor.
- Blower with blower control unit.
- Electrical coolant recirculation pump (only when equipped with a parking heater).
- Compressor clutch control coil.
- Parking heater (future option).
- Parking heater remote control (future option).

Pulsation damper
Pulsation dampers are installed to prevent noise, which is generally caused by the pressure pulses of the compressor.

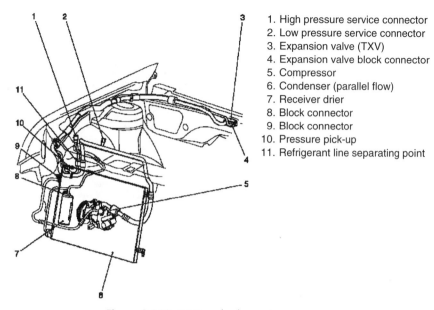

1. High pressure service connector
2. Low pressure service connector
3. Expansion valve (TXV)
4. Expansion valve block connector
5. Compressor
6. Condenser (parallel flow)
7. Receiver drier
8. Block connector
9. Block connector
10. Pressure pick-up
11. Refrigerant line separating point

Figure 3.162 ECC under bonnet components

Pressure sensor

To ensure that the air-conditioning system operates safely at all times, the refrigerant circuit is monitored on the high pressure side. The pressure sensor responds and switches off the compressor if the operating pressure reaches approximately 30 bar. The pressure sensor switches the compressor on again when the pressure drops below the normal operating state of approximately 26 bar. The pressure sensor also switches on the auxiliary fan, depending on outside temperature and coolant temperature.

Evaporator surface temperature sensor

The ECC system is equipped with a device to prevent icing of the evaporator surface. Instead of operating as a switch and interrupting the A/C command from the control panel, the Vectra-C uses a temperature sensor as an input of the ECC control unit. When the evaporator temperature drops below the threshold of $-1.5°C$, the ECM (Engine Control Module) will receive a signal from the ECC control module via the CAN bus to deactivate the compressor clutch. At $1.2°C$, the compressor is activated again.

Outlet air temperature sensors

Four NTC resistors ($10\,k\Omega$ at $25°C$) are used to measure the temperature of the air coming out of the heater housing: two for the floor outlets (left/right) and two for the upper outlets (left/right). The ECC control unit determines the desired position of the air mix flaps through the input from these sensors.

B77 dual sun sensor with in-car temperature sensor

The function of the dual sun sensor is to determine the sunlight intensity. The sensor is equipped with two identical photodiodes for direction recognition. The signals are supplied to the ECC control module via two separate terminals. The sun sensor is located on the centre of the instrument panel on the defroster nozzle. The sensor does not only contain the photodiodes but also the in-car temperature sensor (NTC, $5\,k\Omega$ at $25°C$) (Fig. 3.163).

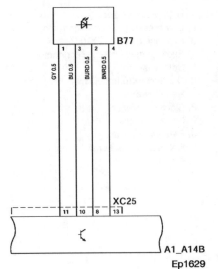

Figure 3.163 Sun load sensor wiring diagram

Pin	Designation
1	Sun sensor left
2	In-car temperature sensor
3	Sun sensor right
4	Ground

Component codes

B77	Sensor sun intensity
A1–A14B	Control module ECC

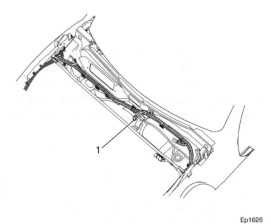

Figure 3.164 Air quality sensor position
(courtesy of Vauxhall Motor Corporation)

Technical data	
Step angle	15°
Number of steps	24
Restraint torque	0.001
Gear ratio	300:1
Diameter	48 mm
Height	27 mm
Door angle 60°	1200 steps

Figure 3.165 Stepper motor specification

Air quality sensor

The air quality sensor (see also Fig. 3.52) is located under the water deflector, left of the pollen filter (Fig. 3.164). This forms an input signal for the automatic recirculation. Like the pollen filter and the activated charcoal filter, it helps improve the on-board climate. By chemical reactions on its surface, the air quality sensor is able to detect localised ground level contamination, such as harmful diesel or petrol fumes. At traffic lights, e.g. when directly behind a truck or if driving through a tunnel, exhaust gas peaks occur which can be up to 1000 times higher than the exhaust gas concentration in the general environment.

If the vehicle is in such an exhaust gas cloud, the air intake process will always be stopped and the system will switch to recirculation mode. To prevent the air quality in the passenger compartment from becoming worse than the air outside, which could happen in some cases if recirculation mode were on permanently and the air in the passenger compartment was not being exchanged at all, the system works dynamically. In exceptional cases such as these, this mode of operation ensures that an adequate supply of fresh air is fed into the system. For technical data see section 3.2.

Stepper motors

The ECC system operates five stepper motors, two for the temperature blend function (one motor for the driver temperature zone and one for passenger temperature zone), and one each for the following, defrost flap (windscreen), face flap ventilation and floor flap. For motor specification see Fig. 3.165.

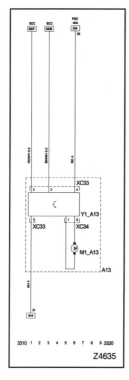

Code	Component
A13	Assembly – blower, passenger compartment
M1-A13	Motor – blower, passenger compartment
XC33	Instrument panel and voltage regulator
XC34	Voltage regulator and motor – blower
Y1-A13	Voltage regulator

Figure 3.166 The blower motor and regulator wiring schematic
(courtesy of Vauxhall Motor Corporation)

Blower motor with voltage regulator

The blower voltage regulator is connected to the ECC control module via a command line and a diagnostic line. Using a PWM (see section 3.2) signal the ECC control unit reports the desired blower speed to the blower voltage regulator. The blower control module controls the blower motor through a 2 KHz PWM signal. Figure 3.166 provides a wiring diagram.

ECC special features

The following special features apply to the ECC system.

Rest-climatisation

It is possible to make use of the residual heat/cold of the HVAC system after the ignition is switched OFF. If the Auto button on the control panel is pressed, after ignition OFF, the ECC is activated and the display indicates the ambient, driver and passenger temperature and the rest-climatisation symbol ('Residual Air-conditioning On').

It is now possible to adjust the desired temperature. The blower speed is fixed and cannot be adjusted manually in the rest-climatisation mode.

With an increasing deviation between the desired and actual outlet air temperature, the blower voltage is gradually reduced to 0 V. Below a battery voltage of 10.7 V, the rest-climatisation mode is disabled.

Rest-climatisation and anti-theft warning system

It is possible to activate the rest-climatisation, leave and lock the vehicle. When equipped with an Anti-Theft Warning System (ATWS), the circulation of air may cause problems when arming the interior monitoring (ultrasonic module).

To prevent this, the interior monitoring is deactivated/disabled with a request/message from the ECC control module to the BCM (Body Control Module). Once the rest-climatisation stops, a message from the ECC control module to the BCM ensures that the interior monitoring is armed.

Blower delay

Similar to previous ECC systems, in the Auto mode the blower speed is gradually increased after an engine start. This is done to cool down the evaporator first, before a large amount of air flows through it. The delay depends on the ambient temperature and can extend up to 8 seconds. This delay can be overruled manually by increasing the blower speed with the centre rotary control.

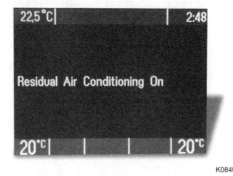

K0840

Figure 3.167 Residual A/C system
(courtesy of Vauxhall Motor Corporation)

Defroster delay

If the defrost button is pressed after an engine start, the defrost flap remains closed up to a maximum of 30 seconds. During this delay, the air is directed to the footwell to get rid of the moisture in the heating and ventilation housing. This way, fogging up of the windows is prevented in defrost mode. Like the blower delay, the defroster delay depends on the ambient temperature.

Logistic mode

In the Logistic mode, the ECC cannot be disabled completely, because the defrost function for the windscreen is required by law. All after-run functions (ignition OFF) are disabled because their only function is comfort and not safety. Furthermore the ECC control module is not allowed to 'wake up' any of the CAN buses, or activate the parking heater (future option) or rest-climatisation. The flaps remain in the last position used. No overstepping is performed after ignition OFF.

Electrical information

Fuse box information illustrated using Vauxhall's TIS (Technical Information System which is CD-ROM based).

Fuse box/relay plate

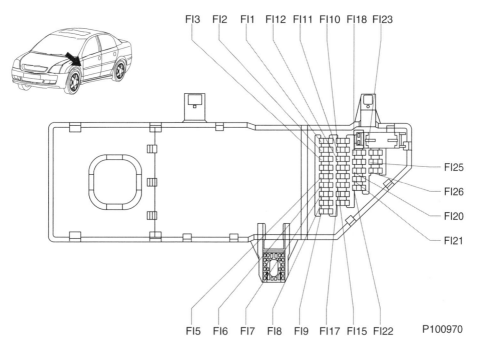

Figure 3.168 Central junction box
(courtesy of Vauxhall Motor Corporation)

Fuse (amps)	Circuit
F123 (40 A)	Blower motor regulator
F12 (7.5 A)	ECC module
F15 (7.5 A)	ECC module
F125 (7.5 A)	ECC module

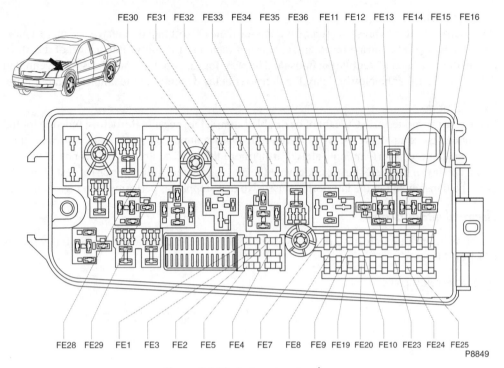

Figure 3.169 Battery junction box
(courtesy of Vauxhall Motor Corporation)

Fuse (amps)	Circuit
FE8 (10 A)	Air quality sensor
FE4 (10 A)	Compressor Relay K8

Relay	Circuit
K4	Blower motor
K5	Blower motor
K8	Compressor relay

ECC block diagram

The block diagram in Figure 3.170 illustrates the input and output relationships between the A/C system module, A/C components and other modules that share information via a multiplexed network. The use of the component information chart is used to match the codes with the

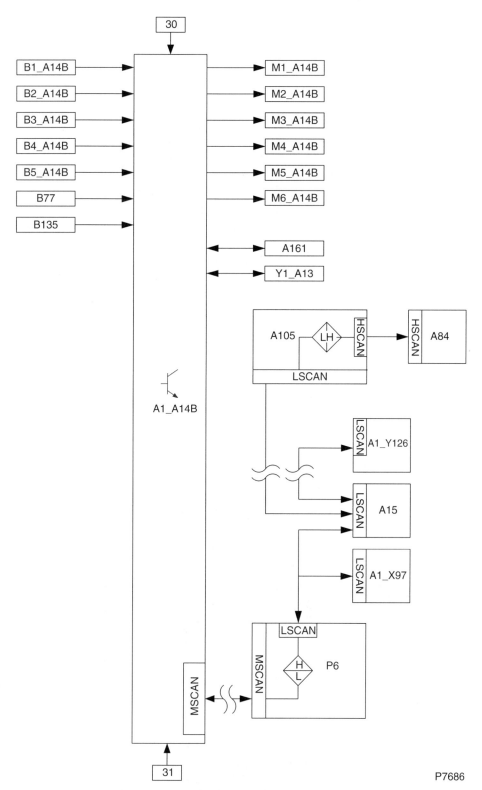

Figure 3.170 Block diagram of the ECC system
(courtesy of Vauxhall Motor Corporation)

Components

A1_A14B	Control unit – Electronic climate control
A1_X97	Control unit – Rear electronic module
A13	Assembly – Blower, passenger compartment
A14B	Adjustment unit – Electronic climate control
A161	Remote control – Independent heater
A15	Body control module
B1_A14B	Sensor – Outlet temperature footwell, right
B2_A14B	Sensor – Outlet temperature footwell, left
B3_A14B	Sensor – Outlet temperature passenger compartment, left
B4_A14B	Sensor – Outlet temperature passenger compartment, right
B5_A14B	Sensor – Temperature, evaporator
B18	Sensor – Pressure, air-conditioning
B135	Sensor – Air quality
B77	Sensor – Sun intensity and Compartment temperature
A84	Control unit – Engine
A105	Column integrated module
A1_Y126	Control unit – Independent heater
L7	Clutch – Compressor, air-conditioning
K7_X97	Relay – Heated back window
K8_X125	Relay – Compressor, air-conditioning
M1_A13	Motor – Blower, passenger compartment
M1_A14B	Stepper motor – Mixed air flap, left
M2_A14B	Stepper motor – Mixed air flap, right
M3_A14B	Stepper motor – Passenger compartment flap
M4_A14B	Stepper motor – Footwell flap
M5_A14B	Stepper motor – Defrost flap
M6_A14B	Control motor – Recirculation flap
P6	Info display
Y1_A13	Blower regulator

Connectors

X1	Body and Instrument panel
X4	Battery positive and Body
X1	Body and Instrument panel
XC25	Instrument panel and Adjustment unit – Electronic climate control
XC26	Adjustment unit – Electronic climate control and Actuators – Electronic climate control
XC33	Instrument panel and Blower regulator

| X26 | Body and Body front | | XC34 | Blower regulator and Motor – Blower, passenger compartment |
| XE3 | Battery positive and Underhood electrical centre | | X19 | Battery positive and Engine |

Abbreviations

7.5A	7.5 ampere		FI2	Fuse, instrument panel body electrical FI2
30	Constant voltage		IH	Independent heater
31	Ground		MSCAN-H	Mid-speed CAN bus – High
ECC	Electronic climate control		MSCAN-L	Mid-speed CAN bus – Low
EMP	Radio		15	Ignition voltage

Colour codes

BK	Black		GN/BN	Green/Brown
BK/BU	Black/Blue		GY	Grey
BK/GY	Black/Grey		GY/BN	Grey/Brown
BK/WH	Black/White		GY/BU	Grey/Blue
BN	Brown		RD	Red
BN/BK	Brown/Black		RD/WH	Red/White
BN/RD	Brown/Red		WH	White
BN/VT	Brown/Violet		WH/GN	White/Green
BN/WH	Brown/White		WH/RD	White/Red
BU	Blue		WH/VT	White/Violet
BU/BK	Blue/Black		YE	Yellow
BU/RD	Blue/Red		YE/BK	Yellow/Black
BU/WH	Blue/White		YE/GN	Yellow/Green
GN	Green			

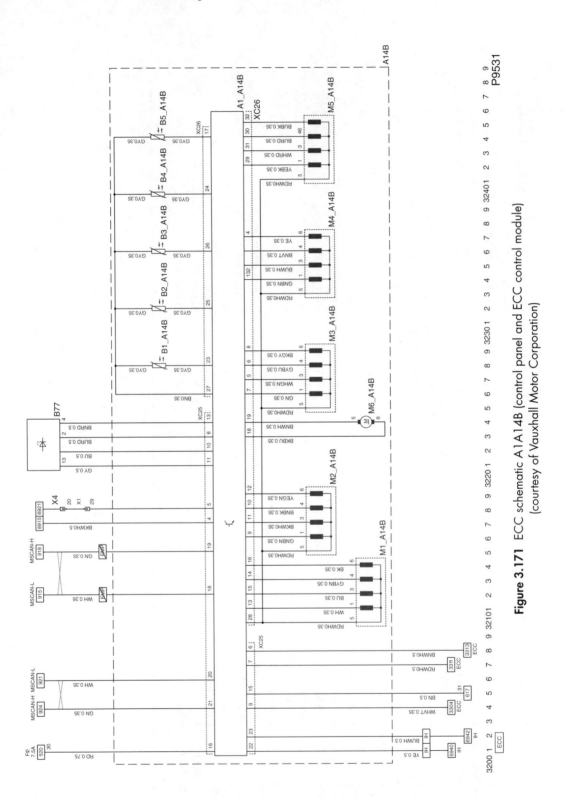

Figure 3.171 ECC schematic A1A14B (control panel and ECC control module) (courtesy of Vauxhall Motor Corporation)

components (DIN Standard Codes). Once an understanding of the block diagram is achieved make use of the wiring schematic.

Component information

ECC system wiring schematic

The wiring schematic and pin layout in Figure 3.171 is used to aid the understanding of the operation of the system.

Explanation of wiring schematics

The ECC is the central control unit for controlling the interior vehicle temperature. This module controls all the heating and ventilation controls and acts as a user interface by including selection knobs and switches. The module controls HVAC door motor positions and blower speed based on information provided from sensor inputs on temperature, sun intensity and air quality. The ECC module also controls the independent heater (see Chapter 1). The ECC has two connections to the mid-speed CAN bus. They serve to loop the mid-speed CAN bus through the control unit. The control unit electronics is then internally connected to the looped-through mid-speed CAN bus. If the ECC control unit is not connected, the mid-speed CAN bus is interrupted.

In order to prevent overloading the engine (in particular during idling), the ECC exchanges the ECC operating conditions and requests signals with the engine control unit via the CAN bus. These messages are transmitted to and from the HSCAN and MSCAN via CIM interface (HSCAN to LSCAN) and the DIS interface (LSCAN to MSCAN). The activation of the A/C compressor is not carried out by the ECC module, it is activated upon a request signal generated by the ECC module which is sent to the engine control module (A84) via the CAN bus system.

P6 (information display, Graphics Information Display (GID), Colour Information Display (CID)) shares information on a medium speed CAN data bus. Information will consist of signals for display purposes (temperature and blower speed etc.). The communication between the control panel and display takes place via the mid-speed CAN bus. Among other tasks, the GID/CID is responsible for the display of all ECC menus. Furthermore, information that needs to be transmitted from the ECC to other control units in the CAN network is transmitted to the low speed CAN bus by the GID/CID (interface function).

Important

If this information is destined for control units connected to the high-speed CAN bus, the CIM effects the transmission from the low-speed CAN bus to the high-speed CAN bus. For example, if the evaporator temperature sensor (B5-A14) output voltage equated to a temperature approaching ice formation on the evaporator housing then this signal would be sent via the MSCAN to P6. The signal would then be sent from P6 (display) to BCM (A15 body control unit) via the LSCAN. The BCM is connected to the CIM (A105 column integrated module) via the LSCAN. The CIM has the facility to interface with LSCAN and HSCAN systems so the signal will travel via the HSCAN to the ECM (A84 engine control module). Upon receiving the signal the ECM will de-energise the compressor coil.

Pin layout

Component	Pin	Test	Result
Power supply F12 (7.5A)	16	Power probe/Multimeter/ Oscilloscope/Serial data list	Battery supply (battery voltage) (power-to-power check <500 mV)
MSCAN Low/Hi	20/21	Oscilloscope/Serial data list	See multiplex waveform analysis (section 3.4)
B77 Sunlight sensor (photo diode)	11,10, 8,13	Multimeter/ Oscilloscope/Serial data list	5 V supply Variable output voltage between 0.1 and 4.9 V for left and right sunlight sensor. Compare volt drop with temperature chart or resistance check
Power supply to B1-5	27	Power probe/ Multimeter/Oscilloscope/ Serial data list	Battery voltage (power-to-power check <400 mV)
B1 – Outlet foot right temperature sensor	23	Multimeter/ Oscilloscope direct temperature measurement using thermometer/Serial data list	5 V supply Compare volt drop with temperature chart or resistance check
B2 – Outlet foot left temperature sensor	25	Multimeter/Oscilloscope direct temperature measurement using thermometer/Serial data list	5 V supply Compare volt drop with temperature chart or resistance check
B3 – Outlet pass comp left temperature sensor	26	Multimeter/Oscilloscope direct temperature measurement using thermometer/Serial data list	5 V supply Compare volt drop with temperature chart or resistance check
B4 – Outlet pass comp right temperature sensor	24	Multimeter/Oscilloscope direct temperature measurement using thermometer/Serial data list	5 V supply Compare volt drop with temperature chart or resistance check
B5 – Evaporator temperature sensor	17	Multimeter/Oscilloscope direct temperature measurement using thermometer/Serial data list	5 V supply Compare volt drop with temperature chart or resistance check
Power supply to M1–M5	28	Power probe/Multimeter/ Oscilloscope/Serial data list	Supply battery voltage (power-to-power check <400 mV)

Component	Pin	Test	Result
M1 – Temperature blend right (stepper)	13,14, 15,16	Power probe/Multimeter/ Oscilloscope, temperature test/Serial data list	Flashing LED, waveform pulse signal check (see section 3.4 on waveform analysis). Motor coil resistance test (see spec). Temperature probe test – measure the temperature exiting blend motor vent and then increase temperature incrementally and check change in temperature. The motor steps are 10% hence 10 blend door positions and possibly 10 temperature changes
M2 – Temperature blend left (stepper)	10,11, 12	Power probe/Multimeter/ Oscilloscope Serial data list	Flashing LED, waveform pulse signal check (see section 3.4 on waveform analysis). Motor coil resistance test (see spec). Temperature probe test – measure the temperature exiting blend motor vent and then increase temperature incrementally and check change in temperature. The motor steps are 10% hence 10 blend door positions and possibly 10 temperature changes
M3 – Passenger compartment door (stepper)	5,6,7,8	Power probe/Multimeter/ Oscilloscope/Serial data list	Flashing LED, waveform pulse signal. Coil resistance test (see spec). Temperature probe test – measuring vent temperature exiting and check air flow (with hand)
M4 – Footwell door (stepper)	1,2,3,4	Power probe/Multimeter/ Oscilloscope/Serial data list	Flashing LED, waveform pulse signal. Coil resistance test (see spec). Temperature probe test – measuring vent temperature exiting and check air flow (with hand)
M5 – Defrost door (stepper)	29,30, 31,32	Power probe/Multimeter/ Oscilloscope/Serial data list	Flashing LED, waveform pulse signal. Coil resistance test (see spec). Temperature probe test – measuring vent temperature exiting and check air flow (with hand)
M6 – Recirculation door (DC motor)	18,19	Power probe/Multimeter/ Oscilloscope/Serial data list and actuator test	On/off DC supply voltage measure-ment or signal. Check air entering pollen filter housing when recirculation is on

(continued)

Component	Pin	Test	Result
Compressor signal (activation signal)	4	Power probe/Multimeter/ Oscilloscope/Serial data list	Battery voltage when A/C compressor is activated. Measure current flow or waveform signal (see section 3.4)
Compressor signal (ground signal)	5	Power probe	Ground (green LED lights up) signal when A/C compressor is activated
Blower communication wire (PWM demand signal)	6	Multimeter/Oscilloscope/ Serial data list	See section 3.4 – blower control waveform
Blower communication wire PWM feedback signal)	7	Multimeter/Oscilloscope/ Serial data list	See section 3.4 – blower control waveform
ECC ground	15	Power probe/Multimeter/ Oscilloscope	Ground-to-ground test – no more than 400 mV difference
Air quality PWM signal	9	Power probe/Multimeter/ Oscilloscope/Serial data list	PWM signal output. Battery voltage input. Ground-to-ground test sensor earth
Independent heater communication wire	23	Power probe/Multimeter/ Oscilloscope/Serial data list	PWM signal
Independent heater communication wire	22	Power probe/Multimeter/ Oscilloscope/Serial data list	PWM signal

Engine Control Module (ECM) – compressor control (Fig. 3.172)

Compressor clutch L7 is controlled by the ECM. The ECM provides a ground signal to K8 compressor relay. This allows current to flow from relay K8 to compressor coil L7 and to ground. Current will also flow to pin 4 of the ECC module. This acts as confirmation that the ECM has activated or deactivated the A/C compressor clutch.

The compressor has a control pressure sensor to protect the system in case of overpressure or loss of pressure. This signal is sent to the ECM enabling the clutch to be de-energised in the event of a system failure.

Diagnostic information

These are the procedures to be followed to obtain the required test results.

Delivery temperature
Preparatory conditions:

1. All windows and doors closed.
2. All ventilation outlets fully open.

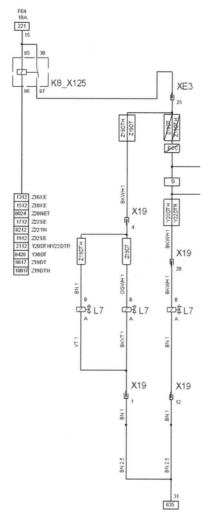

Figure 3.172 Compressor control schematic (courtesy of Vauxhall Motor Corporation)

3. Engine idling.
4. Select 'LO' on the A/C control module.
5. A/C/heater blower motor set to maximum speed.
6. Recirculation mode selected on the A/C control module.
7. A/C control module set to vent.

Checking:

1. Run engine at 1500–2000 rpm.
2. Position temperature probe 100 mm into fascia ventilation centre outlet.
3. Measure temperature after 5 minutes.

Ambient temperature	Humidity	Delivery temperature
21.5°C	Low	6–12°C

Refrigerant pressures
Preparatory conditions:

1. All windows and doors closed.
2. All ventilation outlets fully open.
3. Engine idling.
4. Select 'LO' on the A/C control module.
5. A/C/heater blower motor set to maximum speed.
6. Recirculation mode selected on the A/C control module.
7. A/C control module set to vent.
8. Run air-conditioning for 5 minutes prior to testing.

Checking:
1. Run engine at 1500–2000 rpm.

Ambient temperature	High pressure	Low pressure
20°C	12–16.5 bar	1–3 bar

Self-diagnosis
The A/C control module fault memory can only be checked using diagnostic equipment connected to the Data Link Connector (DLC).

Trouble codes
Suitable diagnostic equipment is required to obtain and erase data from A/C control module memory. See Table 3.11 for a list of codes.

Data logger using Tech 2 – ECC system

Table 3.12 provides a data list that was obtained from a vehicle at the Vauxhall Training Centre by the author. The data list was accessed using the scan tool Tech 2. Note this data is not live but simulated by the module (interfaced) to enable the technician to understand what operating conditions and input/output information is being received and sent by the ECC module.

Table 3.11 Fault code identification

DTC	Diagnostic trouble code
B0165	Compartment temperature sensor circuit low voltage
B0166	Compartment temperature sensor circuit high voltage
B0175	Left air outlet temperature sensor interior circuit low voltage
B0176	Left air outlet temperature sensor interior circuit high voltage
B0180	Left air outlet temperature sensor footwell circuit low voltage
B0181	Left air outlet temperature sensor footwell circuit high voltage
B0185	Left solar sensor circuit low voltage
B0186	Left solar sensor circuit high voltage
B0190	Right solar sensor circuit low voltage
B0191	Right solar sensor circuit high voltage
B0193	Fan controller (PWM) circuit no feedback
B0194	Fan controller (PWM) circuit high temperature or malfunction
B0195	Blower motor circuit low voltage
B0197	Blower motor circuit open
B0225	Recirculation flap motor circuit malfunction
B0226	Recirculation flap motor circuit high voltage or shorted
B0240	Defrost flap motor circuit open or low voltage
B0241	Defrost flap motor circuit high voltage or high current
B0250	Footwell flap motor circuit open or low voltage
B0251	Footwell flap motor circuit high voltage or high current
B0260	Interior flap motor circuit open or low voltage
B0261	Interior flap motor circuit high voltage or high current
B0283	Rear window heater high voltage
B0283	Rear window heater low voltage or circuit open
B0368	Key stuck
B0410	Left blending flap motor circuit open or low voltage
B0411	Left blending flap motor circuit high voltage or high current
B0420	Right blending flap motor circuit open or low voltage
B0421	Right blending flap motor circuit high voltage or high current
B0511	Right air outlet temperature sensor interior circuit low voltage
B0512	Right air outlet temperature sensor interior circuit high voltage
B0516	Right air outlet temperature sensor footwell circuit low voltage
B0517	Right air outlet temperature sensor footwell circuit high voltage
B1445	Stepper motors power supply circuit low voltage or high current
B1445	Stepper motors power supply circuit malfunction
B3843	Air quality sensor circuit low voltage
B3843	Air quality sensor circuit open or high voltage
P0530	A/C pressure sensor circuit high input
P0530	A/C pressure sensor circuit low input
P0530	A/C pressure sensor circuit open
P0530	A/C pressure sensor intermittent
P0532	A/C refrigerant pressure sensor circuit low input
P0533	A/C refrigerant pressure sensor circuit high input
P0646	A/C relay voltage low
P0647	A/C relay voltage high
P1530	A/C relay voltage high
P1530	A/C relay voltage low
P1530	A/C relay circuit open

(continued)

DTC	Diagnostic trouble code
P1532	Evaporator temperature sensor circuit low voltage
P1533	Evaporator temperature sensor circuit high voltage
P1545	A/C clutch relay control circuit

Table 3.12 ECC data list – obtained from vehicle

Sensor	Output	Range
System voltage	13.8 V	
Key in status	Active	Active/Inactive
Ignition status	Active	Active/Inactive
Engine crank signal	No	Yes/No
Engine running signal	Yes	Yes/No
Ambient air temperature	22°C	
Engine speed	850 rpm	
Under bonnet temperature	24°C	
Under bonnet voltage	1.76 V	0.1–4.9 V
Solar sensor left	4.61 V	0.1–4.9 V
Solar sensor right	4.61 V	0.1–4.9 V
Air quality sensor	100%	0–100%
Evaporator temperature	1.5°C	
Evaporator voltage	2.82 V	0.1–4.9 V
Parking heater status (fuel heater)	Inactive	Inactive/Active
Blower motor feedback	Okay	Okay/Not okay
Left blend flap (door)	13%	0–100%
Right blend flap (door)	13%	0–100%
Floor flap position	0%	0–100%
Defrost flap position	0%	0–100%
Interior flap position (face vent)	100%	0–100%
Recirculation position	Fresh	Fresh/Recirculation
Left floor outlet temperature	18.2°C	
Left floor outlet voltage	2.06 V	0.1–4.9 V
Right floor outlet temperature	19.6°C	
Right floor outlet voltage	1.98 V	0.1–4.9 V
Left interior (face vent) outlet temperature	8.3°C	
Left interior (face vent) outlet voltage	2.61 V	0.1–4.9 V
Right interior (face vent) outlet temperature	7.7°C	
Right interior (face vent) outlet voltage	2.69°C	0.1–4.9 V
Engine coolant	91°C	
A/C switch	Active	Active/Inactive
A/C pressure	1080 kPa	
A/C pressure	1.5 V	0.1–4.9 V
Compressor clutch	Active	Active/Inactive
A/C cut-off mode	Off	On/Off
Vehicle speed	0 km/h	

Note 0% is closed, 100% is open.

Actuator tests
The following actuator tests are available using a Tech 2 tester:

- Recirculation motor test results: recirculation – moving – fresh.
- LED test on display results: dim/bright.

Technical data

Technical data

Refrigerant
Type	R134a
Quantity	730 grams

Refrigerant oil
Type	PAG oil
Viscosity	ISO 46

Quantities
Compressor	Replace quantity drained
Condenser	Replace quantity drained
Evaporator	Replace quantity drained
Expansion valve	Replace quantity drained
Hose/lines	Replace quantity drained
System	120 ml

Compressor clutch
Adjustment type	Shim
Clearance	0.3 mm–0.7 mm
Resistance	

Note – A/C system is filled with tracer dye from manufacture.

Service note
The dashboard does not need to be removed to gain access to the evaporator.

4 Diagnostics and troubleshooting

The aim of this chapter is to:

- Enable the reader to understand the range of techniques that can be used in diagnosing faults which affect system performance.

4.1 Initial vehicle inspection

The initial vehicle inspection is not a checklist. Information from the customer on the symptoms, vehicle history and conditions upon which the fault occurs will allow the technician to be selective. The technician should first try to gather as much information as possible and assess if the symptom is normal behaviour (water dripping from underneath the vehicle) or not. The technician should then assess if the environment in which the fault occurs can be replicated. For example, a fault which occurs when the vehicle has been idle for 2 days cannot be replicated the same afternoon the vehicle has been delivered. The correct conditions (temperature, load conditions) must be available to enable accurate fault detection. If conditions are not right then the customer must be aware that an initial diagnostic period will be allocated to the vehicle to carry out a range of tests allowing a number of possible causes to be verified.

The technician should then ensure that they have access to all information required from the customer and for the vehicle. This includes fault finding charts, wiring diagrams, technical service data, diagnostic procedures, technical service bulletins etc. This information may be as simple as a radio code in case the power to the vehicle is interrupted to ensuring the customer has access to a fault code pod (card) which allows access to any fault codes held within the system (see Chapter 3, sections 3.8, 3.9 and 3.10 for examples of information). Manufacturers also have software-based fault diagnostic procedures which direct the technician through guided procedures. Technical helplines are also available.

Note – if the technician is inexperienced, then use the inspection as a checklist.

Simple inspection routine

CHECK CONDENSER FINS FOR BLOCKAGE OR DAMAGE

- If the fins are clogged, wash them with water.

Note – be careful not to damage the fins.

CHECK THE POLLEN FILTER FOR SERVICE CONDITION

- If dirty remove and replace.

MAKE SURE THAT DRIVE BELT IS INSTALLED CORRECTLY

- Check that the drive belt fins fit properly in the ribbed grooves.

CHECK DRIVE BELT TENSION

- Check the drive belt tension.

CHECK CONDENSER FAN FREELY ROTATES

Note – after installing the drive belt, check that it fits properly in the ribbed grooves.

CHECK ENGINE COOLANT LEVEL

- Check coolant level. If unsatisfactory then test coolant system.

START ENGINE AND TURN ON A/C SWITCH

- Check that the A/C operates at each position of the blower switch. If blower does not operate, check electrical circuits.

CHECK MAGNETIC CLUTCH OPERATION

- If magnetic clutch does not engage, check system pressure with gauges and power supply and operation of A/C control, e.g. electrical operation of low pressure switch.

CHECK THAT IDLE INCREASES

- When the magnetic clutch engages, engine rpm should increase.
- Standard idle-up rpm: 900–1000 rpm.

CHECK THAT CONDENSER FAN MOTOR CUTS IN

CHECK THAT THE HEATING PIPES LEADING TO THE HEAT EXCHANGER ARE HOT

CHECK THE PERFORMANCE OF THE A/C CONTROLS

- Check the air distribution control, vary the direction of the air distribution and check air flow. Vary air temperature to test blend operation. Use a temperature probe to verify temperature range (4–60°C) and air direction (panel, floor, face).

The initial vehicle inspection should direct the technician to one of the following:

1. A performance diagnostic test on the A/C operation:
   - A/C performance test.
   - Pressure gauge analysis.
   - Temperature measurement on A/C components.
   - Refrigerant identification test.
   - Level of refrigerant charge.
   - Recovery.
   - Leak testing – OFN, bubble, vacuum, UV dye.
   - Recharge and retest.

2. A/C electrical tests:
   - Self-test checking for fault codes via control panel LCD/graphics display.
   - Serial test using a handheld tester – wiggle test, actuator, DTC, data logger.
   - In-depth 'pin-by-pin' electrical test using a break-out box or directly from the module connector.

Note – systems with a fixed orifice valve and cycle switch (CCOT) are controlled mainly by pressure measurement. This means that pressure type tests like cycle tests are well suited to diagnosing system faults. Systems like TXV which are controlled by measuring temperature are well suited to all gauge and temperature tests.

4.2 Temperature measurements

Measuring the temperature at various points on the A/C system and making comparisons provide the technician with valuable information on system performance.

Pinpoint temperature measurements

Measuring the temperature of the refrigeration components at certain points around the A/C system allows the technician to verify the changes occurring within the system. Table 4.1 provides a guide to the temperature of the refrigerant flowing through the components within the A/C system.

Measuring the temperature of the air flowing inside the vehicle at certain points allows the technician to ensure the blend and air distribution system is functioning correctly. Placing temperature probes and varying the blend door position allow the technician to verify the available temperature range the system is capable of delivering and how quickly the range can be delivered. Measuring the temperature and rate of air flowing at different ventilation points tests the air distribution positions.

Temperature comparisons

Some important temperature comparisons:

1. Ambient temperature and condenser temperature.
2. Centre vent temperature and the ambient temperature (minimum difference of 20°C).
3. Temperature of the high and low pressure side of the A/C system.
4. Inlet and outlet of the condenser (difference of 15–30°C). Excessive difference indicates a blockage similar to the action of an orifice tube. A small difference indicates that the condenser efficiency is low. Parallel condensers are measured from left to right and serpentine condensers from top to bottom. The temperature difference must be progressive.
5. Inlet and outlet of the evaporator (maximum difference of 4°C). This is also referred to as the 'Delta T (ΔT)' check which is mainly used on FOV systems where access to the inlet of

Table 4.1 Surface temperature of A/C components

No.	Description	Temperature
1	Compressor	Up to 80°C
2	High pressure connection	Up to 80°C
3	Condenser	Up to 70°C
4	Dehydrator	Up to 60°C
5	Relief valve	60°C reduced to −4°C
6	Evaporator	Warmer than −4°C
7	Low pressure connection	Warmer than −4°C

the evaporator is available. Record the temperature of the inlet and outlet of the evaporator and compare the results with a chart that indicates the amount of refrigerant which is required to be added to the system. A large difference indicates the inability to transfer a large quantity of heat. This is due to low refrigerant charge. The most accurate method of determining the charge level is to recover the refrigerant and check the weight. Only this method is recommended.

A system under a small cooling load may still be able to produce a low temperature out of the centre vents but when placed under a high load may fail to provide adequate cooling performance. Measuring temperatures around the system and making comparisons allow the technician to evaluate how much load the A/C system is under and how it performs under that load.

4.3 Pressure gauge readings and cycle testing

Introduction

The gauge readings within this section are indications of possible pressures related to a range of faults regularly found on A/C systems. The reading will vary depending on the following:

1. Refrigerant in system R12, R134a.
2. Type of control system – FOV, TXV, EPR.
3. Type of compressor – fixed displacement, variable displacement.
4. A combination of the above.

Low pressure side of the system represents the amount of refrigerant metered and flowing through the evaporator and back to the compressor. The following information is examples of low pressure readings for some of the different systems available:

1. Fixed orifice valve (CCOT). Low pressure has a range between a lower and upper control point which the cycling switch operates at, e.g. 1.5–2.9 bar.
2. Expansion valve system regulates the flow of refrigerant by throttling. Generally normal system pressure is about 2 bar. A thermostatic expansion valve system will go as low as 0.7 bar.
3. EPR (Evaporator Pressure Regulator). The EPR normally allows the system to operate around a control point, e.g. Toyota EPR valve 2 bar.
4. Variable displacement compressors. Generally control the low side pressure to 2 bar.

The high pressure side of the system has a greater pressure range and represents system load. The high side pressure reflects the amount of heat which needs to be removed via the condenser. Ambient air temperature and humidity play an important part in determining the high pressure value.

CCOT system testing

Fault finding chart FOV system
The chart in Table 4.2 assists in diagnosing system faults. Use the compressor cycling time test and pressure gauge readings to identify possible system faults.

Three values are used for the fault diagnosis on an FOV system with low pressure cycling switch:

● Low pressure.
● High pressure.
● Compressor switching cycles (on/off).

Table 4.2 FOV fault finding chart

High pressure	Low pressure	Operating cycle time			Possible cause
		Interval	On	Off	
high	high	switched on continuously			poor cooling of condenser
high	normal to high				engine overheating
normal to high	normal				too much refrigerant (a); air in refrigerant
normal	high				O-rings at fixed orifice tube leaking or missing
normal	normal	slow or off continuously	long or on continuously	normal or off continuously	moisture in refrigerant; too much refrigerant oil
normal	low	slow	long	long	low pressure switch reacting too late
normal to low	high	switched on continuously			compressor output insufficient
normal to low	normal to high				suction (low pressure) line to compressor blocked or constricted (b)
normal to low	normal	fast	short	normal	evaporator blocked or air throughput too low
			short to very short	normal to long	condenser, fixed orifice tube or refrigerant line blocked or constricted
				short to very short	insufficient refrigerant
				long	evaporator blocked or constricted
normal to low	low	switched on continuously			suction (low pressure) line to compressor blocked or constricted (c): low pressure switch sticking
–		compressor running unevenly or not at all			low pressure switch opened continuously or contacts dirty, electrical connection faulty; electrical system faulty

(reproduced with the kind permission of Ford Motor Company Limited)

The following requirements must be met in order to carry out an accurate test:

1. Close both of the manual valves on the pressure gauges. Connect the pressure gauges to the high pressure and the low pressure side of the air-conditioning system.
2. Start the engine.
3. Set the air-conditioning to maximum cooling.
4. Air recirculation on.
5. Set the blower to maximum speed.
6. Run the engine at 1500 rev/min.
7. Engine at normal operating temperature.
8. All windows closed.
9. All vents closed except centre face vent.
10. Ventilation switched to face.

The measured values (R134a) for high and low pressure depend on the outside temperature. This is shown in Figures 4.1 and 4.2 and the chart below. The area between the two curves corresponds to the tolerance range. The measured value must lie in this range.

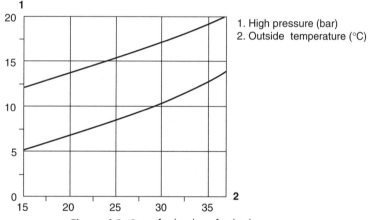

Figure 4.1 Specified values for high pressure

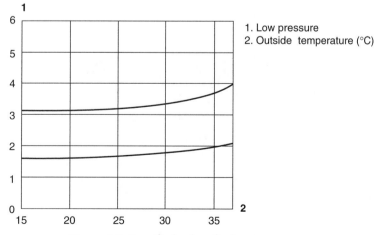

Figure 4.2 Specified values for low temperature

FOV system with cycling switch (CCOT)

Engine Off (static pressure)
Action
Start engine and carry out a dynamic test.

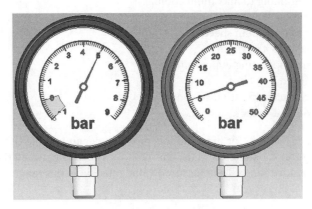

Figure 4.3 Low pressure side *normal,* High pressure side *normal*

Gauge reading R134a (CCOT).

Low pressure side						High pressure side				
Pressure	Bar	PSI	kPa	MPa	kgf/cm^2	Bar	PSI	kPa	MPa	kgf/cm^2
R12	5.00	72.52	500.01	0.50	5.10	5.00	72.52	500.01	0.50	5.10
R134a	5.00	72.52	500.01	0.50	5.10	5.00	72.52	500.01	0.50	5.10

This is no indication of whether the system has sufficient charge.

Engine running (dynamic test)
Action
Record the pressure in the system. Measure centre vent temperature and ambient temperature. Carry out cycling test.

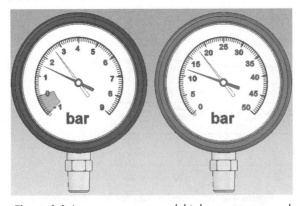

Figure 4.4 Low pressure *normal,* high pressure *normal*

Gauge reading R134a shows the low and high side will fluctuate between the upper and lower limits. The chart below is a snapshot of the pressures taken from CCOT system under light cooling load.

Humidity	low
Ambient air	10–15°C (50–59°F)

Low pressure side					High pressure side				
Pressure	Bar	PSI	kPa	MPa	kgf/cm^2	Bar	PSI	kPa	MPa	kgf/cm^2
R12	2.00	29.01	200	0.20	2.04	10.00	145.04	1000	1.00	10.20
R134a	1.80	26.11	180	0.18	1.84	9.50	137.79	950	0.95	9.69

FOV – moisture in the system

Poor cooling capacity of system. Example, outlet temperature 10°C under light load.

Action

Recover the refrigerant, weigh and recycle it (if available). Check the quantity of oil in the compressor (dipstick). Replace the accumulator. Adjust system oil quantity as required. Vacuum the system for minimum of 1 hour (longer if possible). Add tracer dye. Charge the system and check performance.

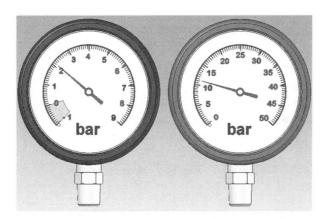

Figure 4.5 Low pressure side *normal*. High pressure side *normal*

Gauge reading R134a shows the low and high side.

Low pressure side					High pressure side				
Pressure	Bar	PSI	kPa	MPa	kgf/cm^2	Bar	PSI	kPa	MPa	kgf/cm^2
R12	2.20	31.91	220.00	0.22	2.24	12.00	174.05	1200.00	1.20	12.24
R134a	2.10	30.46	210.00	0.21	2.14	12.40	179.85	1240.00	1.24	12.64

The best method of testing a CCOT system is using a cycle test.

Cycle time testing

Figures 4.6–4.8 show the required values for the compressor switching cycles. Measure the cycles using a stopwatch and make a note of the result. If the measured value lies outside the tolerance range then there is an error in the system.

The total cycle time is obtained by adding the on-time to the off-time.

The following conditions must be met before checking the switching cycle:

1. Connect the pressure gauges to the high and low pressure side of the air-conditioning system.
2. Start the engine and allow it to run for approximately 5 min at 1500 rev/min.
3. Set the air-conditioning to maximum cooling and air recirculation.
4. Set the blower to maximum power.
5. Adjust the interior temperature to approximately 22°C (if automatic temperature controlled) measured between the front head rests.
6. Measure the switching cycles using a stopwatch and make a note of the results.
7. Read off the pressure from the pressure gauges, make a note of the values and compare them with the required values in the diagrams.

> Note – a serial tester can be used to monitor A/C compressor clutch activation. An oscilloscope can also plot a trend graph showing cycle operation. An LED tester can be placed across the cycling switch and used to monitor switch operation (LED will flash).

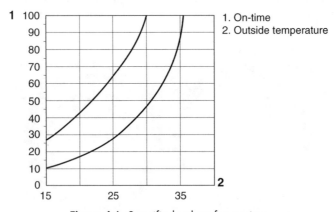

Figure 4.6 Specified values for on-time

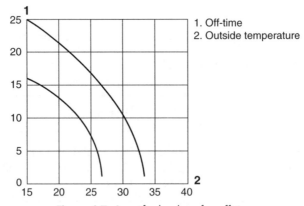

Figure 4.7 Specified values for off-time

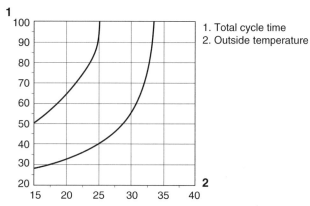

1. Total cycle time
2. Outside temperature

Figure 4.8 Specified values for total cycle time

Testing equipment and application – LED, power probe, multimeter, oscilloscope, OBD II and EOBD, break-out box.

Expansion valve system

Table 4.3 assists in diagnosing system faults. Use the pressure gauge readings to identify possible system faults.

Table 4.3 TXV fault finding chart

High pressure	Low pressure	Possible cause
High	High	Engine overheating; expansion valve open continuously; temperature in evaporator housing too high; coolant shut-off valve not closing correctly
High	Normal to high	Air in refrigerant circuit
High	Normal	Too much refrigerant (system overfilled)
Normal to high	High	Line from compressor to condenser constricted/blocked
Normal to high	Normal to high	Too much refrigerant oil; air humidity well above the normal value
Normal, but uneven	Normal, but uneven	Moisture in refrigerant circuit impairing operation of expansion valve
Fluctuating	Fluctuating	Temperature sensor of expansion valve faulty
Normal to low	Normal to low	Evaporator blocked; air throughput insufficient
High at compressor, low in high-pressure line	Low	Constriction/blockage in receiver/drier, condenser or high pressure line
Low	High	Suction line constricted; valves in compressor damaged, hence poor performance
Low	Low	Suction line or receiver/driver constricted; evaporator iced; condenser blocked; compressor clutch no longer disengaging; de-ice switch remaining closed; refrigerant leak or underfilling; temperature sensor of expansion valve faulty; blockage in high pressure line

Normal operation

Action

Record system pressure. Carry out performance test.

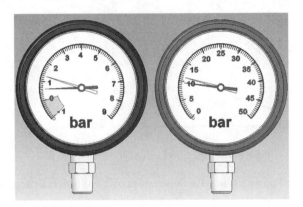

Figure 4.9 Low pressure side *normal*. High pressure side *normal*

Gauge reading R134a shows the low and high side will fluctuate between the upper and lower limits depending on the cooling load.

	Low pressure side					High pressure side				
Pressure	Bar	PSI	kPa	MPa	kgf/cm^2	Bar	PSI	kPa	MPa	kgf/cm^2
R12	1.20	17.40	120.00	0.12	1.22	11.50	166.79	1150.00	1.15	11.73
R134a	1.00	14.50	100.00	0.10	1.02	11.50	166.79	1150.00	1.15	11.73

Faulty compressor valve plate

Compressor operating temperature high, compressor is noisy.

Action

Check sight glass for any foreign matter. Remove the refrigerant from the A/C system and replace or overhaul the compressor (see section 5.7). Retest the system upon completion.

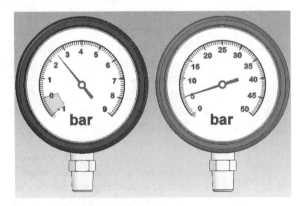

Figure 4.10 Low pressure side *too high*. High pressure side *too low*

Gauge reading R134a shows the low and high side pressure.

Low pressure side					High pressure side				
Pressure	Bar	PSI	kPa	MPa	kgf/cm^2	Bar	PSI	kPa	MPa	kgf/cm^2
R12	2.70	39.16	270.00	0.27	2.75	5.00	72.52	500.00	0.50	5.10
R134a	2.50	36.26	250.00	0.25	2.55	5.00	72.52	500.00	0.50	5.10

Insufficient refrigerant
Warm evaporator outlet. Bubbles in sight glass. Fluctuating temperature control.

Action
Remove the refrigerant from the A/C system and weigh it. Pressure test system: vacuum or OFN pressure test. Replace any faulty components and recharge the system. Clean A/C components to ensure all tracer dye stains are removed. Add tracer dye to system as a diagnostic aid against further loss. Retest the system operation upon completion.

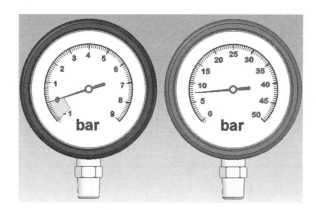

Figure 4.11 Low pressure side *too low*. High pressure side *too low*

Gauge reading R134a shows the low and high side pressure.

Low pressure side					High pressure side				
Pressure	Bar	PSI	kPa	MPa	kgf/cm^2	Bar	PSI	kPa	MPa	kgf/cm^2
R12	0	0	0	0	0	8.00	116.03	800.00	0.80	8.16
R134a	0	0	0	0	0	8.00	116.03	800.00	0.80	8.16

Insufficient condenser output
Only a small temperature difference across the condenser inlet and outlet pipe. No gradual reduction in temperature across the surface of the condenser. Possible internal blockage.

Action

Check sight glass if fitted. Check and clean condenser fins, check operation of condenser fans. Remove the condenser. Check the condenser for any foreign matter. Flush system if required (see section 5.4). Replace condenser and dehydrator if required. Recharge with refrigerant. Test system performance.

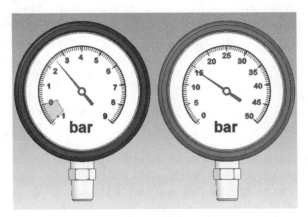

Figure 4.12 Low pressure side *too high*. High pressure side *too high*

Gauge Reading R134a shows the low and high side pressures.

Low pressure side					High pressure side				
Pressure Bar	PSI	kPa	MPa	kgf/cm² Bar	PSI	kPa	MPa	kgf/cm²
R12 2.70	39.16	270.00	0.27	2.75 14.00	203.05	1400.00	1.40	14.28
R134a 2.50	36.26	250.00	0.25	2.55 14.50	210.30	1450.00	1.45	14.79

Faulty expansion valve (stuck open)
Very small temperature drop across the expansion valve.

Action

Check the fitment and temperature of the thermal bulb. Test the bulb operation using cold spray and heat. Observe pressure changes. Cold spray – drop in low side pressure. Heat applied – increase in low side pressure. If the expansion valve fails to respond then replace the valve. Check the valve for any foreign matter. Flush the system if required (see section 5.4). Retest the valve operation and carry out a performance test.

Gauge reading R134a shows the low and high side pressures.

Faulty expansion valve (stuck closed)
Same symptom as a restriction in the high pressure side. Warm evaporator outlet. Frost on expansion valve. Very large temperature drop across the expansion valve.

Action

Check the sight glass for any foreign matter if fitted. Check for any sudden drops in temperature of the high side components which would indicate a partial blockage. Check the fitment and temperature of the thermal bulb. Test the bulb operation using cold spray and heat. Observe pressure changes. Cold spray – drop in low side pressure. Heat applied – increase in low side pressure. If the expansion valve fails to respond then replace the valve. Check the valve for any foreign matter. Flush the system if required. Carry out a performance test.

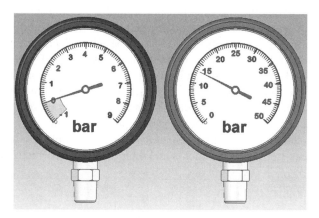

Figure 4.13 Low pressure side *too low*. High pressure side *too high*

Gauge reading R134a shows the low and high side pressures.

Low pressure side					High pressure side				
Pressure	Bar	PSI	kPa	MPa	kgf/cm^2	Bar	PSI	kPa	MPa	kgf/cm^2
R12	0	0	0	0	0	14.00	203.05	1400.00	1.40	14.28
R134a	0	0	0	0	0	14.50	210.30	1450.00	1.45	14.79

4.4 A/C system leak testing

A/C system leaks of up to 100 g per year are universally agreed to be normal. The greatest source of refrigerant leakage is the compressor seal. Other leaks include Schrader valves, connector seals and flexible hoses. The universal drive (industry and legislation) to reduce the leak rates will lead to sealed compressors (electric) and possibly an all metal pipe work system.

Under the EPA Act, section 33 states:

It is illegal to keep, treat or dispose of a controlled substance in a manner likely to cause pollution to the environment or harm to human health.

BS4434 section 3, subsection 6 includes:

it is an offence under section 33 and 34 of the EPA Act 1990, to deliberately discharge damaging refrigerant to the atmosphere.

If it is an offence to discharge refrigerant into the environment knowingly, then A/C technicians should not charge A/C systems with refrigerant if a leak is knowingly present.

Leak testing procedure

Different leak detection methods should be applied under the appropriate conditions. An oil stain test is only appropriate for R12 systems. If a system has no refrigerant in the system at all then OFN pressure testing with bubble spray should be selected. Often the leak will be quite large and easy to find. If the system has a low residual pressure then UV test to find an appropriate area where the leak may have occurred. Run the A/C system for a short period if possible and place the electronic leak detection (sniffer) tester around the system concentrating on areas where UV dye was found. Vacuum testing is particularly useful during servicing and applying a deep vacuum for moisture removal is important. Vacuum testing should never be used to test the correct fitment of components. OFN should be applied to the system to ensure that the system is leak free and components and seals have been correctly applied during the repair procedure.

UV tracer dye

A leak detecting agent that mixes with the refrigerant is placed inside the A/C system. Because the refrigerant evaporates under atmospheric pressure, if a leak occurs the dye is left behind. The dye is difficult to remove and is only visible under a UV (ultra violet lamp). The lamp is used in conjunction with PPE (Personal and Protective Equipment) and is a very useful method for detecting leaks. The dye is often placed in the system from manufacture so does not need to be initially added. The more service operations carried out on the system will dilute the dye eventually requiring a fresh charge. A fresh charge is generally injected under vacuum into the low side of the system allowing it to be induced into the compressor where most of the A/C lubricant is stored.

Problems with using this method include old dye traces that have not been removed which give false indications of a leak. System component replacement can also cause dye to spread around the outside of an A/C system. Once a leak has been repaired the system must be cleaned using dye removal fluid, removing all traces of the dye on the external surface of the A/C system.

Electronic leak detector (sniffer)

Electronic leak detectors are very useful in a system that still has refrigerant charge (e.g. 150 grams). When operating the detector the probe must be positioned at the highest point of the A/C circuit in an environment which is not drafty. Because refrigerant is heavier than air the probe is then placed below connectors and across components to detect a leak working towards the lowest point. Some detectors have audible and visual signal output. Once the detector has been switched on the sensitivity can be adjusted. While the detector is on a constant frequency audible bleeping can be heard. If a refrigerant leak is detected, and the gas concentration increases, this is signalled by a rise in the pitch and frequency of the audible bleep. There are two types of electric leak detector, one for use only with the R12 system, and one that can be used with both the R134a and R12 systems. Note, though, that the sensitivity level of the leak detector designed only for the R12 system is too low to be used for detecting leaks in the R134a system.

Figure 4.14 UV leak detection kit
(courtesy of Autoclimate)

Figure 4.15 Tracer dye and injector assembly
(courtesy of Autoclimate)

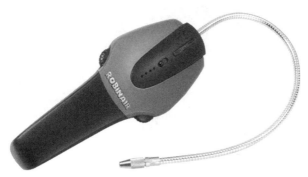

Figure 4.16 Shows a leak detector used for all halogenated refrigerants
(courtesy of Autoclimate)

Specification

Refrigerant/sensitivity	R134a	R12
HI-SENS.	15 g (0.5 oz) to 30 g (1 oz)/year	6 g/year (0.2 oz/year)
R134a	40 g/year (1.4 oz/year)	
R12		15 g/year (0.5 oz/year)
Sensor type	Coroner discharge	
Detection feature	Audible and visual leak indicators using LED bar graph and threshold balance control to eliminate background contamination	

Oxygen-free nitrogen testing

If the A/C system is empty then OFN (Oxygen-Free Nitrogen) is a useful method of pressurising an A/C system without damaging the environment. OFN is cheap and very easy to use and has a small molecular structure enabling easy leakage within an A/C system. The OFN is delivered via an A/C hose connected to a regulator and gauge. The system is pressurised up to 15 bar. While the system is being pressurised it is often useful to check the output of pressure switches and sensors to ensure they are operating correctly. An oscilloscope on waveform record across the cycling switch allows hands-free measurement while filling a system with OFN. A sniffer tester (electronic leak detector) cannot sense OFN. Often the A/C system will have a small quantity of refrigerant trapped in the refrigerant lubricant (PAG oil/mineral) which under pressure is released allowing the sniffer to alert the technician of a potential leak. A bubble solution (or soapy water) is available to spray around system components, connectors, compressor seal etc. When the testing is complete the OFN can be safely vented to the atmosphere.

Figure 4.17 OFN pressure regulator
(courtesy of Autoclimate)

Note – when a repair has been made to an A/C system it is important to OFN pressure-test the system before filling it with refrigerant. This is important when checking the correct fitment of parts like evaporators where long labour times are included when removing dash panels.

Vacuum testing

After the refrigerant has been recovered, to aid moisture removal or as a system pressure-test, an A/C system can be placed under vacuum. In a vacuum moisture boils and the pumping action of the vacuum pump helps to remove the moisture in the form of a vapour. A good vacuum pump is capable of creating a vacuum in a system of up to 1.006 bar. At this pressure water boils at $-1.1°C$.

If an A/C system is adequately sealed the vacuum should be held for a minimum of 10 minutes and the pressure drop should not exceed 20 mbar. A pressure rise is sometimes experienced due to trapped refrigerant within the compressor oil which boils off and creates an increase in pressure.

It is possible that an A/C system may seem leak free after being evacuated. This is often due to seals being pulled into leaking locations providing a temporary seal. Once the system is charged the leak reappears.

Oil stains

An oil stain on a connection or joint indicates that refrigerant is leaking from that place. This is because the compressor oil mixed with the refrigerant escapes when refrigerant gas leaks out from the refrigeration circuit, causing an oil stain to form at the place where the refrigerant gas is leaking out.

If such an oil stain is found, parts should be retightened or replaced as necessary to stop the gas leakage. Gasketed compressor joints and pipe connections are the places where oil stains are most likely to be found and the condenser due to its position is prone to leaks so it is important to check these places. R12 mineral oil leaves a clear oil stain but R134a PAG oil evaporates so this test will not be visual without the aid of a UV lamp. The UV lamp will highlight tracer fluid inside the system. Most manufacturers now place tracer dye inside the system from manufacture.

Figure 4.18 Vacuum pump with exhaust filter (reduce oil mist)

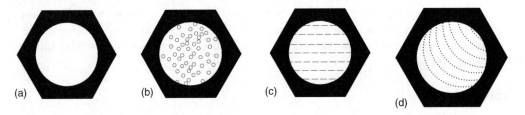

Figure 4.19 Sight glass: (a) clear; (b) foamy; (c) streaky; (d) cloudy

4.5 Sight glass

The sight glass is fitted into the top of the receiver-drier or built into the manifold gauge assembly. To obtain the maximum efficiency from the air-conditioning system, it is very important that it is charged with the correct amount of refrigerant. The sight glass can be used, by the experienced technician, to check the amount of refrigerant in the system. The main purpose of the sight glass is to visually check the condition of the refrigerant passing through the system. There are several 'indicators' that help the service technician to diagnose possible problems. The sight glass should only be used to gain a quick response to a problem and should be supported using a charging/reclaiming station.

> Note – because R134a refrigerant shows a milky colour when viewed with a sight glass it is not used a great deal for system diagnosis. R12 systems generally use the sight glass for additional diagnostic information.

Sight glass clear

A clear sight glass indicates the system has a correct charge of refrigerant. It may also indicate that the system has a complete lack of refrigerant (this will also be accompanied by a lack of any cooling action by the evaporator).

> Note – the sight glass may be clear but the system might be overcharged (too much refrigerant).

This must be verified by connecting the charging trolley and checking the gauge readings.

Sight glass foamy

A 'foamy' or 'bubbly-looking' sight glass indicates the system is low on refrigerant, and air has probably entered the system.
   However, if only occasional bubbles are noticed during clutch cycling or system start-up, this may be a normal condition.

Sight glass streaky

If oil or streaks appear on the sight glass a lack of refrigerant may be indicated.

Sight glass cloudy

A cloudy sight glass indicates that the desiccant contained in the receiver-drier has broken down and is being circulated through the system.

5 Service and repair

The aim of this chapter is to:

- Enable the reader to understand the need for safe working practice.
- Enable the reader to understand the correct procedures for A/C service and repair.

5.1 Servicing precautions

SAE standards

The Society of Automotive Engineers (SAE) has clear standards covering the safe handling and use of refrigerants.

CFC-12 SAE documents:

- SAE J1989: Service procedures
- SAE J1990: Specifications for recycling equipment
- SAE J1991: Standard of purity
- SAE J2209: CFC-12 Extraction equipment

HFC134a SAE Documents:

- SAE J2211: Service procedures
- SAE J2210: Specifications for recycling equipment
- SAE J2099: Standard of purity
- SAE J1732: HFC134a Extraction equipment

Handling refrigerant

Technicians often recover refrigerants from an A/C system during service. Depending on how these refrigerants are processed after removal, they can be classified as recycled, reclaimed, or extracted.

Before an A/C service machine is connected to an A/C system the refrigerant analyser must be used to sample the refrigerant. If the results indicate an NCG (Non-Condensable Gas) of no less than 98% then the refrigerant can be internally recycled to remove any service contaminants. If the analyser shows an NCG of less than 98% then the refrigerant should be reclaimed or disposed off using the correct procedures.

When handling refrigerant the following precautions must be observed:

1. Do not handle refrigerant in an enclosed area or near an open flame.
2. Always wear PPE (Personal Protective Equipment).
3. Be careful that refrigerant does not get in your eyes or on your skin.

If liquid refrigerant gets in your eyes or on your skin:

1. Do not rub the area.
2. Wash the area with a lot of cool water.
3. Apply clean petroleum jelly to the skin.
4. Go immediately to a physician or hospital for professional treatment.

When replacing parts on refrigerant line:

1. Discharge refrigerant slowly before replacement.
2. Insert a plug immediately in disconnected parts to prevent entry of moisture and dust.
3. Do not leave a new condenser or receiver-drier etc., lying around with the plug removed.
4. Discharge nitrogen gas from the charging valve before removing the plug from the new compressor.
5. Do not use a burner for bending or lengthening tubes.

If the nitrogen gas is not discharged first, compressor oil will spray out with the nitrogen gas when the plug is removed.

When tightening connecting parts:

1. Apply a few drops of compressor oil to O-ring fittings for easy tightening and to prevent leaking of refrigerant gas.
2. Tighten the nut using two wrenches to avoid twisting the tube.

Identifying refrigerants

Every vehicle with an air-conditioning system has a sticker located on the bonnet lock panel indicating the refrigerant used:

> Filled with R12 = Black sticker.
> Filled with R134a = Gold and yellow sticker.
> Converted to R134a = Pale blue sticker.

All refrigerant containers are clearly labelled showing the grade of refrigerant they contain.

R12	R134a
Container colour: White	Container colour: Light blue
Container markings: R12	Container markings: R134a
Container fitting size: 7/16″–20	Container fitting size: 1/4″ flare 1/2″–16 ACME
Chemical name: Dichlorodifluoromethane	Chemical name: Tetrafluoroethane
Boiling point: −29.70°C (−21.62°F)	Boiling point: −25.15°C (−15.07°F)
Latent heat of vaporisation: 9.071 calories or 38 007 J or 36 BTUs at 0°C (32°F)	Latent heat of vaporisation: 11.843 calories or 49 622 J or 47.19 BTUs at 0°C (32°F)

Recycled refrigerant (remove containments during normal operation)

Recycled refrigerant is cleaned to remove contaminants produced during normal operation of the A/C system. This is in the form of air, water and oil. The air is vented to the outside, the water is absorbed by a desiccant and the oil is separated. Most modern A/C machines include recycle facilities which are automatically built into the recovery process. Recycled refrigerant must meet the same standards as virgin refrigerant as stated in SAE J1991 and J2099. Standards of purity must not affect the performance or warranty of the system. Refrigerants sent off site for processing and from other sources must meet ARI 700 standard or EN12205 to ensure that the refrigerant is not contaminated and is in compliance with the law.

The SAE J2099 standard of purity for recycled HFC134a refrigerant for use in mobile A/C systems, which has been directly removed from automotive A/C systems, shall not exceed the following levels of contaminants:

- Moisture: 50 ppm (parts per million) by weight.
- Refrigerant oil: 500 ppm by weight.
- Non-condensable gases (air): 150 ppm by weight.

Single-pass recovery system

In single-pass systems refrigerant is drawn from the vehicle A/C system and then passes through an oil separator. This removes any oil. The filter/drier assembly removes moisture and particle contamination. After a single cycle, the contaminant-free, recycled refrigerant, is sent to a storage container.

Multi-pass system

In multi-pass systems refrigerant is drawn from the vehicle, passed through an oil separator, which removes any oil, and a filter/drier assembly, which removes moisture and particle contamination, and is then stored in a tank. When recycling is desired, the recycle solenoid valve is opened, allowing a continuous loop-filtering process in which the refrigerant passes through a desiccant (drier) cartridge several times, until the moisture is fully removed. The station has an indicator to alert the service technician or will automatically vent the recovery tank to remove air. An indicator will show when the refrigerant is ready for use.

Reclaimed refrigerant (contaminated with foreign refrigerant)

Reclaimed refrigerant is processed to the same standards and purity as new refrigerant. This process requires expensive equipment not ordinarily found in service departments. An A/C technician may send a refrigerant to be reclaimed if it contains a foreign refrigerant. The recovery of a refrigerant to be reclaimed must not be carried out using the same recovery equipment for R12 or R134a. Each refrigerant must use a separate machine and this includes contaminated and blends.

Extracted refrigerant (recovered but not recycled only stored)

Extracted refrigerant is simply removed and stored in an approved container. This process is used when servicing the refrigeration system and the refrigerant must be removed and stored from the A/C system.

Key dangers in using extracted refrigerants

The main sources of contamination in recovered refrigerants are:

1. Moisture – possible icing up of the expansion valve and reduced heat transfer of the evaporator could increase the formation of acid oil sludges.

2. Non-Condensable Gas (NCG) – chemically inert gases in a refrigerant system can cause the following:
   - reduced cooling efficiency;
   - higher than normal head pressures;
   - higher discharge temperatures.
3. Organic contaminants – these result from decomposition of various organic materials such as oil, insulation, varnish, gaskets and adhesives. This may cause problems by plugging small orifices resulting in restricted or plugged capillary tubes or sticky expansion valves.
4. Metallic contamination:
   - scoring of metallic components within compressors and bearings;
   - lodging in the motor insulation of a hermetic or semi-hermetic system causing shorts;
   - plugging oil holes in compressor parts thus leading to improper lubrication;
   - serving as a catalyst to increase the rate of chemical breakdown in the system.

   Note – this is why it is very important to use a refrigerant analyser and an A/C machine with recycle capabilities.

Storing refrigerant

   Note – refrigerant cylinders must be stored according to BS4434 recommendations.

Both R12 and R134a are gases at normal room temperature, and they can be hazardous if stored improperly. New refrigerant stored in its original, properly filled container usually poses no safety hazard. However, recycled refrigerant can be dangerous if it is stored in the wrong type of container or in an overfilled container. To prevent accidents when handling recycled refrigerant, never save disposable refrigerant containers for reuse. Remove all refrigerant and dispose of the containers properly. Only use containers approved for refrigerant. Never fill a container to more than 60% of container capacity. Never store refrigerant containers in direct sun or heat. High temperature causes the gas to expand, which increases the pressure in the container and may cause the container to burst.

To monitor the container pressure, install a calibrated pressure gauge with 6.9 kPa or 1 psi divisions and a thermometer to monitor the temperature within 10 cm of the container. Ensure that the pressure at the measured temperature does not exceed the limit given in Table 5.1.

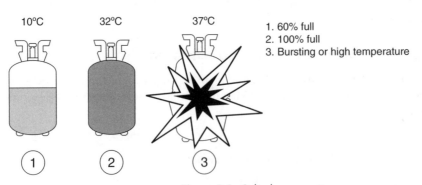

Figure 5.1 Cylinder capacity
(reproduced with the kind permission of Ford Motor Company Limited)

Refrigerant transfer

In the event that a refrigerant is required to be transferred from one storage container to another, the recipient (container to be filled) must be evacuated of air first. The container must be fitted to a vacuum pump and evacuated to at least 635 mm (25 inHg) below atmospheric pressure (vacuum) before transfer.

Refrigerant disposal

If a refrigerant cannot be recycled and contains hydrocarbons or other contaminants which prevent further use then it must be disposed of. This is achieved through following local government requirements. If the procedures are unknown then contact the local council office.

Table 5.1 Maximum allowable container pressure – recycled HFC134a

(a) Metric

Temp °C(F)	kPa	Temp °C(F)	kPa	Temp °C(F)	kPa	Temp °C(F)	kPa
18(65)	476	26(79)	621	34(93)	793	42(108)	1007
19(66)	483	27(81)	642	35(95)	814	43(109)	1027
20(68)	503	28(82)	655	36(97)	841	44(111)	1055
21(70)	524	29(84)	676	37(99)	876	45(113)	1089
22(72)	545	30(86)	703	38(100)	889	46(115)	1124
23(73)	552	31(88)	724	39(102)	917	47(117)	1158
24(75)	572	32(90)	752	40(104)	945	48(118)	1179
27(77)	593	33(91)	765	41(106)	979	49(120)	1214

(b) English

Temp °F	Psig	Temp °F	Psig	Temp °F	Psig	Temp °F	Psig
65	69	79	90	93	115	107	144
66	70	80	91	94	117	108	146
67	71	81	93	95	118	109	149
68	73	82	95	96	120	110	151
69	74	83	96	97	122	111	153
70	76	84	98	98	125	112	156
71	77	85	100	99	127	113	158
72	79	86	102	100	129	114	160
73	80	87	103	101	131	115	163
74	82	88	105	102	133	116	165
75	83	89	107	103	135	117	168
76	85	90	109	104	137	118	171
77	86	91	111	105	139	119	173
78	88	92	113	106	142	120	176

5.2 Refrigerant recovery, recycle and charging

A/C service units

Most A/C machines are separate units combined into a Refrigerant Management System (RMS). An RMS contains a recovery machine, vacuum pump, electronic scales and an LCD control panel. The unit often includes very simple programmes allowing the technician to select automatic recovery, recycle, evacuation and charging procedures. Some units include databases for vehicle information and flushing capabilities.

The unit's pipe work, connectors, switches and gauges are generally colour coded:

Red for the High Pressure (HP) side, also referred as 'liquid side' of the A/C system.
Blue for the Low Pressure (LP) side, also referred as the 'vapour side' of the A/C system.

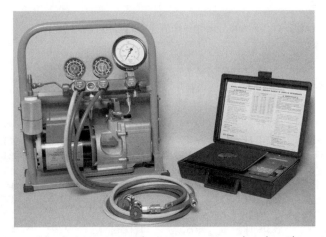

Figure 5.2 A portable charging unit with vacuum pump and scales. This unit can vapour charge or liquid charge using a heat belt or charging cylinder

Figure 5.3 RMS (Refrigerant Management Station) (courtesy of Autoclimate)

Other colours are used to assist in service operations directing the technician to carry out certain procedures, e.g. yellow service hose. When the unit is not in use all valves should be closed.

An A/C system can be charged with refrigerant in liquid or vapour form. The charging process is carried out by measuring the refrigerant charge weight. This is the only approved method of charging an A/C system.

The RMS (Refrigerant Management Station) unit is capable of automatic refrigerant recovery, recycle, evacuation and recharging. The unit also has a database on the oil quantities for different manufactured vehicles, which is updatable. The unit can also flush A/C systems using appropriate adapters.

Some RMS units can be used with multiple refrigerants.

Main components of a service station

Low pressure gauge

This gauge is a compound gauge and measures the low pressure side of the air-conditioning system and is coloured blue. It will also give an indication of the vacuum depression when the

Figure 5.4 Low pressure compound gauge
(courtesy of Autoclimate)

Figure 5.5 High pressure gauge
(courtesy of Autoclimate)

system is being evacuated. If the vehicle has a suspected leak that is very slight, this gauge may not be able to detect it.

High pressure gauge

This gauge measures the high pressure side of the air-conditioning system, and is coloured red. When the air-conditioning system operation is being checked, the various modes can be seen to operate on this gauge with the different pressure alterations being indicated as they are switched. The relationship between the pressure readings on the two gauges provides a reliable guide to the functioning of the system and an indication of when problems exist.

> Note – on R12 systems the couplings for connection to the service connectors are the same. Therefore, it is imperative to note the colour coding. Incorrect connection would damage components of the servicing unit. In R134a systems different couplings are used. The connectors on the high pressure and low pressure sides are of different diameters. This prevents incorrect connection or different refrigerants being mixed with one another.

Low pressure valve

This should be coloured blue to signify it is part of the low pressure side of the charging station system. It must not be overtightened. It should be turned using the fingertips until an impedance is felt (signifying it is closed) and then gently pinched to lock it. When the valve is reopened the operator should be able to 'crack' the locking torque without using undue pressure (with the fingertips).

High pressure valve

This should be coloured red to signify it is part of the high pressure side of the charging station. This valve must be tightened and unlocked using the same method as the other valves.

High pressure pipe

Colour coded red, it is usually of a suitable length so that a clearance can be maintained between the charging station and the vehicle. Minimising the possibility of the charging station damaging the vehicle paint work. The end connection to the vehicle system has a Schrader type valve depressor built into it.

> Note – at no time should the pipe be uncoupled at the charging station while it is connected to the vehicle.

There is a hose gasket fitted to provide a good seal between the vehicle system and the charging trolley. This should be inspected for wear or damage at frequent intervals. The sealing properties of this gasket can be prolonged with the application of a small amount of refrigerant oil before use.

Low pressure pipe

Coloured blue to signify that it is part of the low pressure circuit. It is usually the same length as the high pressure pipe with the same type of end connector fitted.

The hose gasket fitted into the connector requires the same attention and maintenance as for the high pressure pipe gaskets.

Figure 5.6 Manifold assembly
(courtesy of Autoclimate)

Note – to meet SAE requirements the connections into the vehicle system are dissimilar for the high and low pressure charging points – 7/16 Nf for high pressure and 3/8 Nf for low pressure.

Manifold assembly

This unit allows the charging station to evacuate, recharge and test the vehicle air-conditioning system in situ without disconnecting any hoses. For discharge purposes the pipe connecting the manifold to the charging station must be disconnected so that the refrigerant 12 can be discharged into a recovery station. This enables the operator to measure the amount of refrigerant oil lost from the air-conditioning system during the discharge operation.

To avoid the possibility of refrigerant entering the vacuum pump during the system discharging operation, ensure the vacuum pump valve is closed. The manifold is designed so that when the high and low pressure pipes are connected to the vehicle air-conditioning system, with both of the valves in the 'off' position, the gauges will read the vehicle system pressure. When the manifold is set to this position, the charging station is isolated from the air-conditioning system. The valves can then be opened, connecting the charging station to the air-conditioning system as required.

Vacuum pump valve

The vacuum pump valve's function is to switch the vacuum pump depression to the manifold, and to isolate the vacuum pump when refrigerant is being circulated through the charging system.

Refrigerant control valve to the charging cylinder

This valve, which is normally coloured red, connects the refrigerant bottle (fitted on the rear of the charging station) to the charging cylinder. It can be finely controlled so that refrigerant can be slowly measured into the charging cylinder.

Refrigerant control valve to the manifold assembly

This red coloured valve must be treated with the same consideration as the other valves. Its function is to control the refrigerant flow to the vehicle, via the manifold. If refrigerant is allowed to flow too quickly it will boil and vaporise, reducing the vacuum depression, which is required to draw the refrigerant into the vehicle system.

Charging cylinder gauge

The gauge is connected to the top of the charging cylinder sight glass. Its function is to measure the pressure variations of the refrigerant in the charging cylinder that can then be calibrated to a reading on the plastic shroud that surrounds the cylinder. This can be done by taking a reading from the charging cylinder pressure gauge and comparing it to the table that is marked around the top of the plastic shroud. Line this reading up with the charging cylinder to indicate the pressure of the refrigerant in the measuring cylinder. There are different types of refrigerant, with the plastic shroud depicting different tables for each of the most common types.

Charging cylinder

This charging cylinder can store and deliver a specific amount of refrigerant to the vehicle air-conditioning system with an accuracy of plus or minus 7 grams (4 oz). Dial-a-charge charging cylinders allow the operator to compensate for refrigerant volume fluctuations resulting from temperature variations. This allows the purchase of more economical drums of refrigerant. It also provides a greater degree of accuracy when charging the air-conditioning system, eliminating over- and undercharging, which can lead to problems when the vehicle has gone back into service.

These cylinders are provided with heating elements that allow the operator to overcome equalisation of pressure between the air-conditioning system and the charging cylinder, which reduces the time required to recharge the system. Simply 'dial' the calibrated plastic shroud so that the pressure reading at the top of the shroud corresponds to the gauge reading. The calibrated column that is in line with the sight glass will then show the amount of refrigerant that is in the sight glass at that pressure. The charging cylinder is now ready for the transfer of an accurately measured amount of refrigerant 12 to the vehicle air-conditioning system.

Vacuum pump

This is a very robust pump that enables the vehicle air-conditioning system to be evacuated quickly and effectively.

It requires very little routine maintenance, having its own oil circulation system that requires periodic checking according to the manufacturer's instructions.

Vacuum pump to manifold connecting pipe

This pipe connects the vacuum pump to the vacuum pump valve on the manifold. It requires a good seal so that there are no leaks between the vacuum pump and the manifold. The connections should be periodically checked for tightness.

System evacuation

A vacuum pump may be purchased and used as a separate item. The centre or yellow service hose connects to the vacuum pump and the two hoses on the manifold gauge connect to the low pressure (LP) and high pressure (HP) sides of the system. The vacuum pump should only be attached when system gauge pressure is zero or showing a vacuum. Before using the vacuum pump check the oil via the pump sight glass is satisfactory. Start the pump and open the

LP and HP service connectors to apply a vacuum. The reading on the LP gauge should steadily go into vacuum. If the vacuum pump fails to draw a vacuum then the system may have a leak. Periodically check the gauge reading and if unsatisfactory carry out a leak test. If satisfactory progress is made then evacuate for 30 minutes to a pressure of −1.006 bar. Upon satisfactory evacuation of the system shut the manifold gauge valve to the pump and check that the pressure is maintained for 10 minutes minimum, overnight if required. A slight pressure increase may be experienced if trapped refrigerant in the oil boils off.

System vapour charging

Vapour charging is carried out on the low side (blue hose) of the A/C system where during normal A/C operation vapour is flowing into the compressor. When vapour charging the compressor is running to draw vapour in. The refrigerant cylinder must be positioned so only vapour can leave which is generally the upright position with the blue valve open (vapour valve). A small amount of air is bled (purged) off from the service hoses at the manifold gauge end to ensure that air present in the hoses is removed. Some automated machines do not require air purging. The refrigerant cylinder is placed on electronic scales and weighed at the start and throughout the process of charging the system to monitor that the correct quantity of refrigerant is delivered. The A/C system is initially evacuated to achieve a deep vacuum; the refrigerant vapour will fill the system producing an initial charge level. This produces enough pressure to shut the low pressure switch/sensor, which allows the A/C compressor to engage and the A/C system to operate and draw the rest of the vapour into the system.

System liquid charging

Liquid charging is a much more dangerous process and is carried out on the high pressure (HP) side of the A/C. The HP connector is generally positioned as follows: on the compressor (early models), on the condenser, between the condenser and receiver-drier (TXV) or between the condenser and FOV. Systems that have the HP valve fitted to the compressors should not be liquid charged. This is because there is a danger of the compressor being internally damaged. The engine is not running during liquid charging. A liquid charging process uses high pressure to charge the A/C system which is obtained using either a charging unit or a charging cylinder.

System charging using a charging cylinder

The refrigerant is transferred from its cylinder to a charging cylinder. A charging cylinder as previously discussed has a calibrated shroud used for temperature compensation which is rotated until the graduated number for that refrigerant volume at that pressure is next to the sight column. The refrigerant is transferred to the charging unit by warming the source refrigerant bottle (max. 45°C) and by applying a vacuum to the charging cylinder. The heated source bottle is connected to the charging cylinder. The red 'liquid' valve on the source bottle is opened and then air is carefully purged out of the hoses. The valve on the charging cylinder is opened and liquid refrigerant can flow into the unit. Once the correct level is obtained the valve is closed. When the charging cylinder is filled to the required level plug the unit into a 240 V supply and heat the cylinder to about 35°C which is noted on the thermometer. The system can now be filled via the high pressure connection/liquid line. The refrigerant will flow from the red liquid valve on the charging cylinder straight into the A/C system on the high pressure side (normally into the condenser).

Note – never liquid charge into the compressor because of damage to the valve plate.

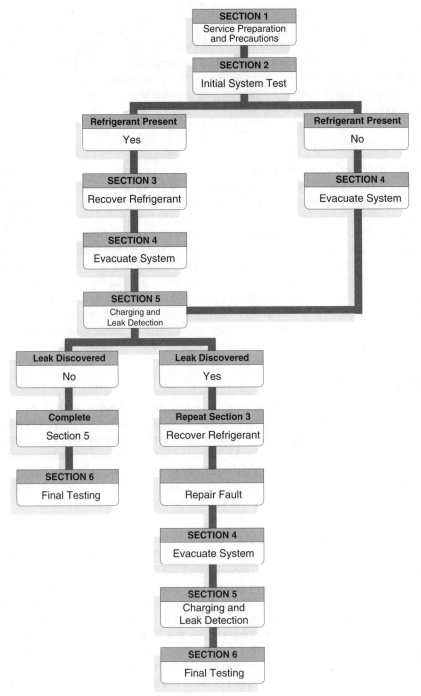

Figure 5.7 The stepped diagnostic approach

Figure 5.8 Personal protective equipment

Section 1 Service preparation and precautions

When handling refrigerant or carrying out repairs to an automotive air-conditioning system it is recommended that eye protection and gloves are worn. Extreme caution must be taken not to allow any refrigerant to come into contact with the skin and especially the eyes. Liquid refrigerant (R134a) evaporates at approximately $-26.3°C$, and because it evaporates quickly, it freezes anything it comes into contact with. Care must also be taken to avoid breathing refrigerant or system lubricant vapour. Exposure may irritate eyes, nose and throat.

Use only approved automotive air-conditioning service equipment. Avoid carrying out air-conditioning service repair work in any small unventilated area to avoid asphyxiation. We strongly recommend that all servicing technicians refer to their appropriate COSHH file for more detailed information. Ensure that protective covers are applied to the vehicle before commencing any work. The battery must be disconnected to prevent accidental starting of the engine and the possibility of personal injury if access to service connectors are close to fans, belts etc. Make sure that tools, measuring equipment and parts to be fitted are clean and dry.

Keep all necessary equipment and tools within easy reach so that the system is not left open any longer than is absolutely necessary. Before undoing any refrigerant lines, joints or connectors, clean off any dirt, moisture, oil etc. in order to prevent contamination of the system. All open connections should be capped or plugged (air tight) immediately to stop dirt, air or moisture getting into the system. Air inside the circuit will damage the system and reduce the cooling effect as it contains moisture. Any O-rings disturbed by undoing unions must always be renewed after lubricating with refrigerant oil prior to fitment. When removing O-rings from couplings, care must be taken not to scratch the sealing face. It is recommended that the receiver-drier/accumulator is replaced if the system has been open to the atmosphere for more than 4 hours (depending on the manufacturer), is physically damaged or has been in service for more than 2 years. Do not remove plugs from new components until each component is ready to be installed into the A/C system, this will limit the amount of air and moisture entering the system. When adding refrigerant oil, ensure that any filling equipment (hose, container etc.) is clean and dry. The oil container must be sealed immediately after use. To ensure the system works correctly after servicing, the system must be evacuated (vacuumed) for a minimum of 30 minutes before recharging. This will remove (by dehydration) any moisture from the system. One of the most important requirements when filling the air-conditioning system is to use clean refrigerant. Any foreign matter including air, moisture, dirt etc. in the air-conditioning circuit will have an adverse effect on refrigerant pressures and impair the system performance. After every repair or service procedure, the system must be leak checked to identify any leaks that may be present. If any leaks are found, the refrigerant within the system must immediately be recovered and the leak repaired.

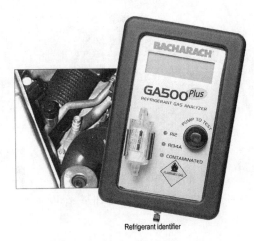

Refrigerant identifier

Figure 5.9 Refrigerant identifier

Section 2 Initial system test
Step 1

Position car in workshop bay.
Switch off engine.
Open bonnet.
Remove charge port caps.

Note – it is recommended that a refrigerant identifier is used to ascertain system refrigerant type, percentage of air and whether the refrigerant is contaminated prior to commencing.

Refrigerant identifier
A refrigerant vapour sample flows via the low side of the A/C system to an infrared sensor capable of sensing a range of refrigerants and blends. Audible and visual alarms indicate the presence of hydrocarbon-based refrigerants. The percentage of purity as a level of contamination is provided on an LCD screen. A printer port allows the connection to a printer to provide the customer with a detailed report.

Note – if the refrigerant is unknown, established not to be R12 or R134a, then the customer should be informed and advised of appropriate action. The refrigerant should be recovered into a waste bottle and the system oil and receiver/accumulator must be replaced. A system flush could also be carried out before recharging. When sampling a blend a reading will appear showing the different refrigerants and percentages e.g. the percentage of R12, percentage of R134a and % of hydrocarbons. OEMs in the UK do not recommend the use of blends but the US have a range of blends on the market. If blends are used then a fingerprint (percentage of each refrigerant) should be sampled before charging the system so the correct blend percentage of each refrigerant is known.

With the introduction of new CO_2 refrigerants the analysers will eventually be updated or superseded.

Refrigerant labels
A simple method of identifying a refrigerant is to look for the A/C system manufacturer's original label or retrofit label. These are often lost due to the vehicle engine bays being stream cleaned or body repair work being carried out. Service ports often provide evidence of whether the system is R12 or R134a but this cannot be treated as conclusive.

Figure 5.10 RMS connection to A/C system

A comparator
A comparator and thermometer with a set of pressure gauges can be used to identify a refrigerant. The comparator or slide rule uses the pressure and temperature relationship of the refrigerant in a saturated liquid/vapour window. This method is not always accurate due to some refrigerants' performance characteristics being closely related.

With the use of a refrigerant identifier record the refrigerant type and percentage of air in the system. Use a printer if available to connect to the analyser providing the customer with a detailed report.

Reminder
If the refrigerant NCG is 98% or greater then recover and recycle. If below then reclaim using a separate A/C machine (avoid contamination of A/C machine).

Step 2
Once the refrigerant has been identified *the correct service equipment* should be used.

> Note – technicians must ensure that there is sufficient space for the refrigerant in the recovery or waste bottle. A simple calculation must be made. The amount of refrigerant in the system must be added to the space available inside the bottle without exceeding 80% capacity.

Disconnect the refrigerant analyser from the low side of the A/C system. Make sure the low and high side connectors are shut off.

Connect the low pressure and high pressure hoses to the A/C system's low pressure and high pressure service ports. Open hose connectors to allow refrigerant to flow to the RMS (Refrigerant Management System) (Fig. 5.10).

Record system pressures (A/C Off)

> Note – if system pressure is at 0 bar go to Section 4 Evacuation.

Step 3
Using a digital thermometer take the ambient temperature reading (Fig. 5.11) near the front of the vehicle and record for reference during system performance tests.

Figure 5.11 Ambient temperature check

Figure 5.12 Operate A/C system

Lower the bonnet, do not close fully.

Note – refer to Performance Test/Recharge and Service Report.

Step 4
Start the engine.

Step 5
When normal engine operating temperature is achieved:

- check HVAC control's function, operation and air distribution;
- evaluate evaporator odours.

Note – if a musty smell is present the evaporator will need to be treated with evaporator anti-bacterial treatment.

Step 6
Fully open all windows front and rear.

Figure 5.13 Antibacterial treatment

Figure 5.14 Engine RPM check

Step 7
Stabilise engine speed between 1500 and 2000 rpm.

Note – use a throttle prop if required.

Step 8
Set HVAC controls to:

● air-conditioning system on.
● fan speed set to maximum.
● heater controls set to maximum cold.
● vents set to full face position.
● recirculation set to the off position.

Insert thermometer into centre face vent (Fig. 5.15). Record centre face vent temperature for reference during system performance test.

Figure 5.15 Centre vent temperature check

Figure 5.16 Low pressure and high pressure readings

Step 9
Record the system low and high pressure gauge readings on the RMS while the compressor is engaged and the system has stabilised.

Step 10
Switch off engine. Raise bonnet.

Section 3 Refrigerant recovery

> Note – Technicians must ensure that there is sufficient space for the refrigerant in the recovery or waste bottle. A simple calculation must be made. The amount of refrigerant in the system must be added to the space available inside the bottle without exceeding 80% capacity.

Step 1
Using the RMS recover the refrigerant from the vehicle's air-conditioning system.

Figure 5.17 Record the weight of refrigerant recovered

Figure 5.18 Visual inspection

Note – we recommend that to minimise system oil carry-out, recovery of refrigerant should be undertaken on the low side of the A/C system.

Step 2

Note – during the recovery process it is possible to begin inspecting overall system condition. The following items may be inspected and completed.

During the recovery ensure that the low pressure gauge progressively reduces and eventually goes into a slight vacuum. This is in case the system has an air leak and the machine starts to recover excess air from the system.

- Visually inspect the compressor drive belt.
- Check the compressor drive belt tension (where applicable).
- Visually inspect all refrigerant pipes and hoses.
- Visually inspect condenser.
- Visually inspect A/C electrical components and connections.

Figure 5.19 Oil recovery

Note – refer to Performance Test/Recharge and Service Report.

Refrigerant may be trapped in the oil which will boil off during the recovery process. This may cause the machine to recover the main refrigerant charge and then after approximately 5 minutes recover a small additional charge.

Note – if the recovery machine cannot draw the system into a slight vacuum then a leak may will be present.

Step 3
After the recovery process is completed observe the RMS lubricant separator and check the amount of oil recovered, if any, from the vehicle's A/C system.
   Record the amount of oil recovered.

Note – the A/C system is now safe to work on.

If excessive oil is drained from the system then the oil level within the system must be checked. This can be done by using a dipstick to measure the quantity within the compressor. Most modern compressors do not have sumps so the only way of checking the system quantity is by removing the compressor and draining it or flushing the whole system. The receiver-drier/ accumulator can be removed and drained into a measuring cylinder as a representative amount of oil within the system.

Section 4 Evacuation
Step 1

Note – we recommend the dehydrator is renewed if the A/C system has been open to atmosphere or has not been regularly maintained.

Visually inspect the dehydrator for leaks, damage, corrosion and security (excludes subcooled systems).

Note – if required system components should repaired/replaced.

Figure 5.20 Receiver-drier/dehydrator

Step 2

The air-conditioning system should now be evacuated (vacuumed) for at least 30 minutes (recommended).

After about 5 minutes, stop the vacuum process, close control panel valves and wait a few moments to see if the system begins to lose vacuum. If so this could indicate the A/C system has a leak. A small rise is expected due to refrigerant trapped within the oil boiling off.

> Note – system evacuation is vital for the removal of air and moisture from the vehicle's A/C system. It is a necessary procedure for continued reliability, performance and operation. See Figure 5.17.

Step 3

> Note – some of the items listed below may have been inspected during the recovery process, if not, please continue the inspection.

- Visually inspect the compressor drive belt.
- Check the compressor drive belt tension (where applicable).
- Visually inspect all refrigerant pipes and hoses.
- Visually inspect condenser.
- Visually inspect A/C electrical components and connections.
- Visually inspect evaporator drain tube.
- Visually inspect pollen and particulate filter (if fitted).
- Check compressor mountings.

Step 4

On successful completion of the evacuation process it will be necessary to add new lubricant equivalent to the amount removed during the recovery process plus any additional amount for components replaced. Record A/C system lubricant replaced.

> Note – if the evacuation process system has shown that a leak is present then carry out leak detection, see Chapter 4, section 4.4. The correct viscosity/type of lubricant must be identified. When replacing lubricant into an A/C system new lubricant must always be used.

Figure 5.21 Oil being added while the system is under vacuum

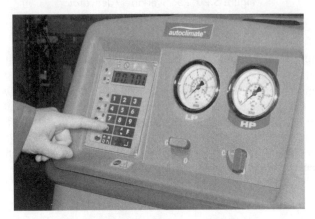

Figure 5.22 Charging with refrigerant

Note – if the system requires tracer dye to be added then do so while the system is under vacuum.

Section 5 Charging and final leak detection

Step 1
Confirm specified quantity of refrigerant to be charged. Assuming the vehicle's A/C system has no visible indication of refrigerant leakage, partially recharge the system with approximately 20% of the total refrigerant.

Note – we recommend that to minimise irreparable damage to the compressor, liquid refrigerant should only be charged through the *high* pressure connection.

Step 2

Note – the air-conditioning system should now be leak checked. If a leak is detected the refrigerant charge must be recovered immediately and the faulty component or seal replaced. See section 3 Refrigerant recovery and perform these steps again.

Figure 5.23 Check all refrigerant pipes, hoses and seals for leaks

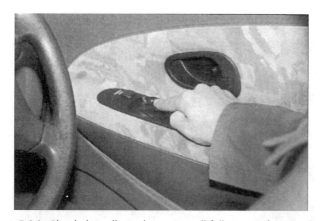

Figure 5.24 Check that all windows are still fully open (front and rear)

Step 3
If no leaks are found the remaining 80% of refrigerant can be charged into the system.

   Note – refrigerant weights (quantities) are critical for optimum performance.

Record the refrigerant quantity replaced (see Figure 5.23).

Section 6 Final testing
Step 1
Record ambient temperature. Lower bonnet, do not close fully.
   Start engine.
   Check HVAC controls have not been altered:

● Air-conditioning system on.
● Fan speed set to maximum.
● Heater controls set to maximum cold.

- Vents set to full face position.
- Recirculation set to the off position.

Step 2
Stabilise engine speed between 1500 and 2000 rpm.

Note – use throttle prop if required.

Step 3
Insert the temperature probe into centre face vent. When normal engine operating temperature is achieved record centre face vent temperature (see Figure 5.15).

Step 4
Record system pressure gauge readings on the RMS machine. Check compressor for excessive operational noise. Check condenser/radiator fan operation (see Figure 5.16).

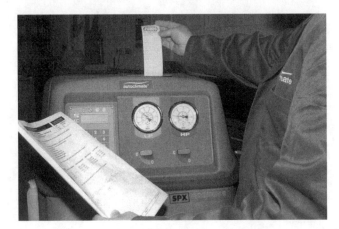

Figure 5.25 Print report

Figure 5.26 Replace caps

Estimated labour times

Section	Function	Comments	Labour	Machine	Total
Section 1	Preparation	Position Car in Workshop	02:00	02:00	
Section 2	Initial System Testing	Identify system refrigerant type and % of air*		01.20	
		Connect RMS		00.30	
		Record ambient temperature reading		00.05	
		Record system pressures (A/C Off)			
		Start engine			
		Check HVAC controls			
		Evaluate evaporator odours			
		Open windows			
		Stabilise engine – temp/RPM			
		Set HVAC controls			
		Record system pressures (A/C On)			
		Record face vent temperature			
		Switch off engine		04.40	06.35
Section 3	Refrigerant Recovery	Set RMS to recover		00:05	05.25
		Check system condition. (See section 3 Refrigerant Recovery)		04:30	
		Check oil recovered		00:05	00.05 05.30
Section 4	Evacuation	Set RMS to evacuate mode	00:05	30:00	
		Check system condition. (See section 4 Evacuation)		23.00	
		Add system oil (if required)		00:05	00.05 30.05
Section 5	Charging/ Leak Detection	Set RMS to charge mode		00:05	
		Partial charge			00:30
		Leak checking (Electronic)		02:00	
		Charge completion			01:00 03:35
Section 6	Final Testing	Lower bonnet			
		Note ambient temp			
		Stabilise engine – temp/RPM			

(*continued*)

Section	Function	Comments	Labour	Machine	Total
		Set HVAC controls Record centre face vent temperature Record system pressure readings Switch off engine		05:00	05.00
Total Times			**37.45**	**37.05**	**52.45**
Times	Measured in minutes				
Labour	Physical hands-on time			37.45	
Machine	Machine operation time				37.05
Estimated	Total times including cross-over between labour/ machine				52.45

Notes:

Upon identifying the vehicle refrigerant, if the refrigerant is unknown, established not to be R12 or R134a the vehicle cannot be checked any further. The customer should be informed and the appropriate action taken to repair the vehicle. If any trace of air is found in the air-conditioning system it is recommended that the air-conditioning system is serviced to ascertain full air-conditioning efficiency and performance.

It is important to note that if a vehicle has a low refrigerant charge this can have a direct effect on vehicle performance, efficiency and premature failure of components due to excessive loads and wear due to lack of lubrication.

Step 5

Print service/test results and attach to Performance Test/Recharge and Service Report.

Compare previously noted ambient temperature, gauge pressures and centre vent temperature against manufacturer's or aftermarket specifications to ascertain system performance.

Step 6

Check HVAC controls – operation and air distribution. Switch off engine.

Step 7

Raise bonnet. Disconnect low pressure and high pressure hoses from the A/C system's high pressure and low pressure service ports. Remove the RMS from the vehicle. Leak check the service ports. Replace service port caps. Complete the Performance Test/Recharge and Service Report. The service is now complete.

Example service sheet

A/C Performance Test ☐   Recharge ☐   Service Report ☐ (Please tick appropriate box)

Mr/Mrs: _____   Vehicle Make: _____
Address: _____   Vehicle Model: _____
_____   Registration No.: _____
_____   Odometer Reading: _____
                                                          VIN: _____
Post Code: _____
Telephone: _____   Job No.:

PLEASE COMPLETE ALL BOXES

✓	– Inspected OK
X	– Attention required (See below)
n/a	– Not applicable

T = TEST R = RECHARGE S = SERVICE

	OPERATION	COMMENTS	T	R	S		OPERATION	COMMENTS	T	R	S
1	Record	Refrigerant Type				20	Visually inspect	Inspect Refrigerant Pipes and Hoses	n/a	n/a	
2	Record	% of Air in System				21	Visually inspect	Inspect Condenser for Leaks, Damage, Air Flow etc.	n/a	n/a	
3	Record	System Pressures*				22	Visually inspect	Inspect A/C Electrical Components & Connections	n/a	n/a	
4	Record	Ambient Temperature				23	Visually inspect	Inspect Evaporator Drain Tube (Where Possible)	n/a	n/a	
5	Check	HVAC Controls – Function, Operation & Air Distribution				24	Visually inspect	Inspect Pollen & Particulate Filter (If Fitted)	n/a	n/a	
6	Evaluate	Evaporator Odour				25	Check	Compressor Mounting	n/a	n/a	
7	Record	Centre Face Vent Temperature				26	Check	Condenser / Radiator Fan Operation (If Possible)	n/a	n/a	
8	Record	Low Side System Pressure				27	Add	A/C System Lubricant (If Necessary)	n/a		
9	Record	High Side System Pressure				28	Record	A/C System Lubricant Replaced	n/a		
10	Record	System Refrigerant Quantity Recovered				29	Recharge	A/C System with Specified Quantity of Refrigerant	n/a		
11	Record	Lubricant Quantity Recovered**				30	Record	Amount of Refrigerant Replaced	n/a		
12	Add	A/C System Lubricant (If Necessary)		n/a	n/a	31	Check	All Refrigerant Pipes, Hoses and Seals for Leaks	n/a		
13	Record	A/C System Lubricant Replaced		n/a	n/a	32	Record	Ambient Temperature	n/a		
14	Recharge	System with Refrigerant Quantity Recovered		n/a	n/a	33	Record	Centre Face Vent Temperature	n/a		
15	Record	Amount of Refrigerant Replaced		n/a	n/a	34	Record	Low Side System Pressure	n/a		
16	Visually inspect	Dehydrator for Leaks, Damage, Corrosion & Security	n/a	n/a		35	Record	High Side System Pressure	n/a		
	Note: Replace Dehydrator if Required					36	Check	Compressor – Operational Noise	n/a	n/a	
17	Evacuate	A/C System	n/a			37	Check	HVAC Controls – Function, Operation & Distribution	n/a		
18	Visually inspect	Compressor Drive Belt	n/a	n/a		38	Add	UV Leak Detection Dye (If Required)	n/a		
19	Check	Compressor Drive Belt Tension	n/a	n/a		39	Check	Service Ports for Leaks	n/a		

A/C Performance Test Results/Record	RESULTS	
	Initial	Final
Refrigerant Type		n/a
% of Air		n/a
System Pressure (A/C Off) Low Pressure		n/a
High Pressure		n/a
Ambient Temperature (˚C)		
Centre Face Vent Temperature (˚C)		
System Pressure (A/C On) Low Pressure		
High Pressure		
Refrigerant Quantity Recovered		n/a
Lubricant Quantity Recovered		n/a
Lubricant Quantity Replaced	n/a	
Refrigerant Quantity Replaced	n/a	

Attach RMS
Printout Here

* If pressure is below 0.5 bar advise customer that system requires attention/service ** After completing Point 11 please advise customer of Results and how to proceed. **Please Note:** The performance test does not in any way indicate the overall condition of the air conditioning system and any of its associated components. The condition of the air conditioning system and its associated components are inspected and checked as part of an A/C Service. For further information please contact our Service Reception.

Attention Required:

CHECKED BY: _____ DATE: _____

Please provide the customer with a copy of this A/C Performance Test / Recharge / Service Report

5.3 System oil

Lubricant

Refrigeration oil lubricates the moving parts and seals of an A/C system. The oil flows with the refrigerant throughout the system.

Mineral and PAG oil

The type of refrigeration oil used in an A/C system depends on the type of refrigerant. When engineers develop a refrigerant, they simultaneously develop the lubrication oil to use with it. R12 A/C systems use mineral oil as a lubricant. R134a systems use oil made of polyalkylene glycols, commonly called PAG oil. PAG oil and mineral oil are completely incompatible and should never be mixed. Refrigeration oil, is highly refined and free of the additives and detergents found in conventional motor oil. Refrigeration oil flows freely at temperatures well below freezing, and it includes an additive to prevent foaming in the A/C system. Refrigeration oil readily absorbs moisture. If stored improperly, the oil becomes unusable. If you use saturated oil in an A/C system, acids form, damaging seals and other components. Always seal refrigerant oil properly after use, and never reuse oil removed from an operating A/C system. In an A/C system, the components hold the refrigerant oil. The compressor helps to mix the oil with the refrigerant and circulates it throughout the system. Both oils are available in different viscosities. Always ensure the correct oil type and viscosity is placed into an A/C system.

Polyol ester oil

Polyol Ester (POE) has good miscibility with HFCs. It is an alcohol-based oil. Its ability to change viscosity with a change in temperature makes it an attractive alternative to PAG oil although some controversy exists which states the POE breaks down into alcohol and acid if in contact with moisture. It is compatible with R12, R134a and CO_2-based A/C systems. POE is not an OEM oil and should only be used if specified by the A/C manufacturer. Aftermarket retrofitting is generally associated with the use of POE.

Oil viscosity

The higher the viscosity number the greater the viscosity and the higher its resistance to flow. This means the liquid is thicker.

The following is a list of recommended lubricants for R134a compressor applications:

Behr/Bosch rotary compressors – Ester 100
Behr/Bosch piston compressors – PAG 46
Calsonic V5 – PAG 150
Calsonic V6 – PAG 46
Chrysler RV2 – Ester 100

Chrysler C171, A590 & 6C17 – PAG 46

Diesel/Kiki (Zexel) DKS, DKV & DCW – PAG 46
Ford FS6, FX15, FS10, 10P & 10PA – PAG 46

Matsushita (all) – Ester 100
Mitsubishi FX80 – PAG 100
Mitsubishi FX105 – PAG 46
Nihon (all) – Ester 100
Nippondenso 6P, 10P, 10PA, 10P08E – PAG 46
Nippondenso SP127, SP134 & 6E171 – PAG 46
Nippondenso TV series – PAG 125

Panasonic (all) – PAG 46

GM A6, R4, DA6, HR6, HT, V5 & V7 – PAG 150

Sanden SD500 & SD700 – PAG 100

GM V5 retrofit – PAG/FLR-118

Sanden SD710, SDB, TV & TRS – PAG 46

Hitachi (all) – PAG 46

Seik-Seiki (all) – Ester 100

Keihin (all) – PAG 46

York/Tecumseh – PAG 46

When replacing an A/C component, the oil trapped in the component that is being replaced must also be replaced. Service manuals contain charts describing how much oil to add for various component replacements. Charts that refer to replacing the quantity drain should be ignored. The correct amount of oil must be replaced if the system is to function correctly. If the system has a severe oil shortage then replacing the quantity drained will not help the system operation and will not stop the system from eventual failure. This can cause a great deal of customer dissatisfaction and responsibility may be aimed at the last technician who worked on the system. This is another reason why it is important as a technician to record all work carried out on a vehicle's system especially A/C related. Oil removed and oil added must be carefully recorded including refrigerant and components.

Table 5.2 ISO grades used to identify viscosity

Viscosity grade	Container colour	Viscosity equivalent
ISO viscosity grade 46	Yellow container	KLHOO-PAGSO Sanden SP10 Nippondenso ND8 PAG 1 PAG 3 UC RL244
ISO viscosity grade 100	Green container	KLHOO-PAGRO Sanden SP20 Nippondenso ND9 PAG 2
ISO viscosity grade 150	Red container	UC RL488 DH G25X

Table 5.3 Example oil replacement quantities (always use manufacturers' information)

Component	Refrigerant oil
Compressor	90 ml or check compressor code and match against technical data for exact quantity and viscosity. Use a dipstick to measure the quantity of oil in the compressor.
Evaporator	90 ml
Condenser	30 ml
Accumulator	90 ml
Lines	60 ml
Every recovery and recharge	20 ml

Oil replacement

If the oil quantity within the system is unknown or in doubt then the use of a dipstick placed inside the compressor will give the best indication of system oil quantity. Most modern A/C compressors do not have a sump. This means that the compressor must be removed and drained. This may take considerable time depending on its position. The accumulator/receiver can be removed and drained to give an indication of oil quantity within the system. Upon the recovery of refrigerant from an A/C system the collected oil from the separator within the machine can act as an indication of the amount of oil separated to the amount of refrigerant recovered, e.g. 20 ml is the normal amount of oil during one recovery procedure if the full quantity of refrigerant within the system is available and recovered.

Oil stains

An oil stain on a connection or joint indicates that refrigerant is leaking from that place. This is because the compressor oil mixed in with the refrigerant escapes with the refrigerant when refrigerant gas leaks out from the refrigeration circuit, causing an oil stain to form at the place where the refrigerant gas is leaking out.

If such an oil stain is found, parts should be retightened or replaced as necessary to stop the gas leakage. Gasketed compressor joints and pipe connections are the places where oil stains are most likely to be found, and the condenser, due to its position, is prone to leaks so it is important to check all these places. R12 mineral oil leaves a clear oil stain but R134a PAG oil evaporates so this test will not be visual without the aid of a UV lamp. The UV lamp will highlight tracer fluid inside the system. Most manufacturers now place tracer dye inside the system from manufacture. The amount of oil lost depends on the size of the leak and the length of time it was leaking. After you repair a leak, replace the amount of lost oil. Carefully measure the oil removed during evacuation and replace it.

> Note – overfilling the system with refrigerant oil reduces the cooling effect. Follow the instructions provided by the manufacturer of the servicing unit when topping up the refrigerant oil. When topping up the refrigerant oil, make sure that the filling equipment (hose, container) is clean and dry. The oil container must be sealed immediately after use.

5.4 System flushing

Automotive air-conditioning system flushing has become in recent months a very controversial subject. A lack of guidance and commitment from OEMs regarding flushing and the varying opinions from industrial bodies have left the aftermarket with no conclusive flushing method or flushing procedure. Consequently, the end user is left bewildered and/or unprepared to make a decision on the best flushing method available and service to offer his customers.

Autoclimate, an aftermarket supplier, has considered all OEM methods and procedures, taken into account numerous industry opinions and subsequently developed a liquid refrigerant flushing procedure, which minimises valuable labour time and at minimal cost to the end user.

An automotive air-conditioning system may require flushing for several reasons:

1. Retrofit (conversion from R12 to R134a).
2. System contamination.
3. System component failure.

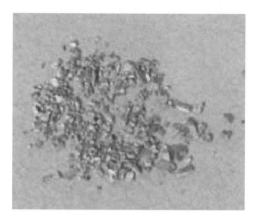

Figure 5.27 Debris from an A/C system
(courtesy of Autoclimate)

Retrofitting is the name given to a procedure that involves converting an R12 automotive air-conditioning system, which uses refrigerant R12, into a system that uses refrigerant R134a. Consequently any R12 system which needs repair must to be retrofitted to R134a, if system longevity is to be considered. System contamination is a phrase that covers a wide scope of potential issues. System contamination is considered when any other substance outside design parameters has been introduced into the A/C system. Debris, burnt lubricant, contaminated lubricant and incorrect lubricant are some of the issues that need to be considered when system contamination has occurred. A low system refrigerant charge will result in poor lubricant circulation, thus increasing friction and heat causing premature system failure. Increased friction and heat will also begin to burn the system lubricant causing it to carbonise inside components, on pipes, seals etc.

Debris is usually the result of a system component failure, this can vary from fine metal particles to desiccant, a result of a receiver-drier/accumulator failure. Once this debris has begun to circulate in the system, it must be removed. Irreparable damage and expensive repairs are a direct consequence of debris in the A/C system. Debris passing over 'burnt' lubricant will begin to adhere to it, and slowly a blockage may form. In these instances the system must be flushed to remove all debris.

The lubricant must also be flushed from the system and the correct amount and viscosity added; if the flushing procedure and lubricant renewal are not carried out correctly it might result in a repeat failure. The build-up of lubricant/debris can also affect the heat transfer characteristics of both the condenser and the evaporator. Modern condensers and evaporators have multipath parallel circuits, so the possibility of clearing every path that is blocked is unlikely, thus affecting performance. In these instances it may be necessary to remove and replace the component. A similar effect may be experienced if the system has too much lubricant, although flushing will remove excess lubricant.

Flushing method

The flushing procedure may be carried out in a closed loop configuration (with the system split in two halves) or each individual component separately. Flushing each individual component can be labour intensive and expensive for the customer. Closed loop flushing with liquid refrigerant is a quick, effective method of flushing, which does not introduce any foreign chemicals to the system that may lead to further problems. The flushing process can be improved by

Figure 5.28 Flushing fluid
(courtesy of Autoclimate)

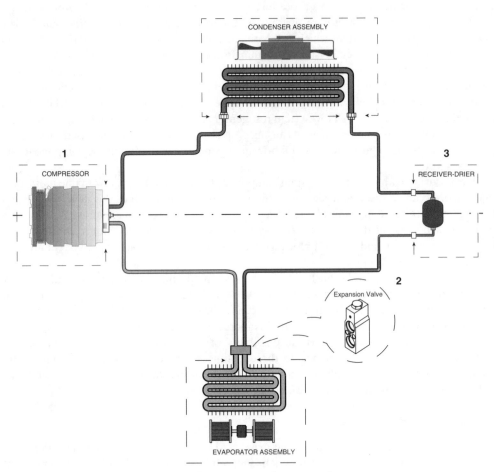

Figure 5.29 A typical expansion valve air-conditioning system
split into two halves, low side and high side
(courtesy of Autoclimate)

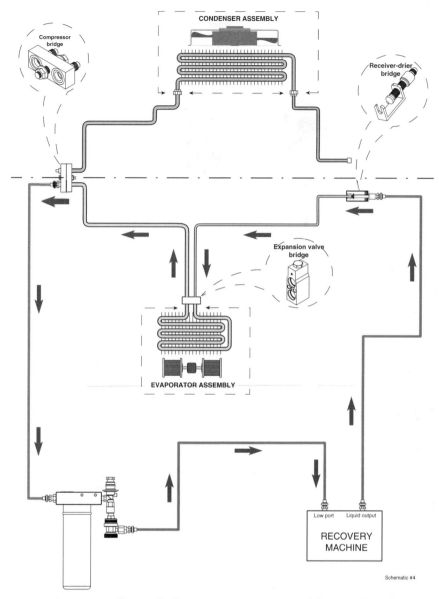

Figure 5.30 Shows which components are removed/bypassed, and
where the flushing hoses are connected
(courtesy of Autoclimate)

the introduction of an oil-based cleaning agent. The introduction of oil-based cleaning agents greatly improves the first time removal of oil and debris from the A/C system. By allowing the A/C system to be split into two halves, high side and low side, an effective forward/backward flush can be achieved. Any restrictions in the system, such as the receiver-drier, expansion valve, accumulator or orifice tube, are bypassed. Adaptors have been developed which allow for most restrictions to be bypassed; a set of universal adaptors may also be used for less common vehicles.

Figure 5.31 Examples of flushed material
(courtesy of Autoclimate)

Results

Once the system has been successfully flushed and all contamination/debris removed the system may be reassembled. It is recommended that the receiver-drier/accumulator is always replaced after flushing an A/C system. With the A/C system cleaned, increased heat transfer should result in improved system performance. No contamination, the correct viscosity, and the correct amount of lubricant in the system will ensure that the compressor also operates to system performance expectations.

5.5 Odour removal

Musty smell in interior

Occasionally, the air-conditioning system can cause a musty damp smell in the interior of the vehicle. This is due to particles of dirt, which are carried into the evaporator housing with the induced fresh air, sticking to the evaporator fins covered with condensation. When the air-conditioning system is switched off, a moist warm environment is created in which the partially organic deposits of micro-organisms and fungi are broken down. This process produces unpleasant smells which are conveyed into the interior of the vehicle when the air-conditioning system is switched on again. The smell rapidly disappears due to the continuous change of air when the system is running. However, it recurs after every break in operation. Vehicles used in moist warm climates are more frequently affected by this type of smell. Often it also occurs in vehicles with horizontal evaporator fins. The droplets of condensation are unable to drain away completely from these fins and a constantly damp environment is produced which promotes the multiplication of micro-organisms and fungi. The interior of the evaporator housing must not be allowed to remain continuously moist if the smell is to be avoided. Therefore, the drainage lines must be cleaned regularly. In addition, it is important to ensure that as little dirt as possible enters the evaporator housing. Vehicles parked regularly under trees pick up particularly large amounts of dirt.

Action to cure smells

If smells ever occur, the interior of the evaporator housing must be treated with a chemical which has a disinfecting and deodorising action (see Figure 5.13). This chemical is distributed throughout the evaporator housing with the aid of a special spray gun. Before using the chemical, locate the precise cause of the smell. This chemical can only be used to cure smells emanating from the evaporator housing. Other sources of smells such as damp carpets or the like must be treated differently. Before treatment with the chemical, the evaporator must be completely dry. This is achieved by operating the system with maximum heating at the highest blower speed in fresh air mode with the vents opened for about 15 minutes. The compressor must not be allowed to operate (unplug the low pressure switch or the magnetic clutch). Moisture in the air would condense on the cold evaporator again immediately. The blower motor or the blower resistor can be removed to gain access to the evaporator housing, depending on the type of vehicle. If possible, the housing must be cleaned by hand from inside. If a particularly large accumulation of dirt is found, it is imperative to check whether the screens and filters are intact. The chemical is distributed over the entire surface of the evaporator with the spray gun through the opening. The blower motor or resistor can be refitted while the chemical is allowed to work for 10 minutes. To ensure that the chemical cannot be carried away by condensation, it must be allowed to dry on the evaporator. For this, the vehicle heating must be operated for a further 15 minutes without the compressor running. Then the interior of the vehicle must be aired thoroughly. The low pressure switch or the magnetic clutch can be reconnected. On completion of the work, the vehicle interior must be aired for 30 minutes.

5.6 Retrofitting

Conversion from R12 to R134a

1. If the R12 vehicle air-conditioning system is optional, run it at idle with the A/C blower on high speed for 5 minutes to maximise the amount of oil in the compressor.
2. Recover all R12 refrigerant from the vehicle's A/C system.
3. Remove the compressor from the vehicle.
4. Remove the compressor oil plug and then drain as much mineral oil as possible from the compressor body.
5. Drain mineral oil from the cylinder head suction and discharge ports while turning the shaft with a socket wrench on the clutch armature retaining nut.
6. Remove the existing R12 receiver-drier or accumulator-drier from the vehicle and discard. Allow as much oil as possible to drain from the A/C hoses.
7. Change any O-rings on the receiver-drier or accumulator-drier joints to approved HNBR O-rings; replace any other O-rings that have been disturbed.
8. Replace the receiver-drier or accumulator-drier with a new R134a compatible one containing XH7 or XH9 desiccant.
9. If a CCOT system is being repaired due to compressor damage, or foreign material is found in the oil drained from the system, this foreign material must be removed from the system. At this time an inline filter should be installed in the liquid line. Allow as much oil as possible to drain from the A/C lines when installing the filter. Change any O-rings disturbed in the installation of the filter to approved HBNR O-rings.
10. Perform any necessary repairs to the compressor or A/C system.
11. Using the original refrigerant oil quantity specification, add SP-20 or SP-10 oil to the compressor (SP-10 for TR, SDV-710, SDB-705, SDB-706 and SDB-709; SP-20 for all other SD compressors).
12. Replace the compressor oil plug O-ring with an HNBR O-ring.

13. Reinstall the compressor oil plug. The plug seat and O-ring must be clean and free of damage. Torque the plug to 11–15 ft lb (15–20 N m, 150–200 kgf cm).

14. Change any seals at the compressor ports to approved HNBR seals.

15. Reinstall the compressor to the A/C system. Evacuate the A/C system for at least 45 minutes to a vacuum of 29 inHg, using R12 equipment, to remove as much R12 as possible from the residual mineral oil.

16. Remove all R12 service equipment and disable the R12 service fittings to prevent any refrigerant other than R134a from being used. Permanently install R134a quick connect service fittings to the A/C system.

17. Connect R134a service hoses and other equipment. Re-evacuate the system for 30 minutes using the R134a equipment.

18. Charge the A/C system with R134a. Generally, about 5% (by weight) less than the R12 charge amount is required. Leak check the system per SAE J1628 procedure.

19. If the A/C system is a CCOT type, which has been repaired due to damage or the discovery of foreign material in the oil drained from the system, run the system for 60 minutes to capture this material in the filter installed in step 9. Recover the refrigerant, remove and dispose of the filter, reconnect the lines, evacuate for at least 45 minutes, and recharge the A/C system. This step should not be necessary for TXV systems, since the drier is fitted with an internal filter.

20. Check the A/C system operating parameters. The system should function correctly within acceptable limits of temperatures and pressures. This will ensure that the correct amount of R134a has been charged.

21. In extreme circumstances when expected cooling performance cannot be achieved and high discharge pressures are experienced, it may be necessary to add more condensing capacity to the A/C system. An electric fan(s) and/or larger capacity condenser can be used.

22. Replace all R12 compressor labels with retrofit labels per SAE J1660 in order to provide information on the R134a retrofit which has been performed.

5.7 Replacement and adjustment of compressor components

This section is reproduced from the *Compressor Service Manual Sanden SD6V12* (courtesy of Sanden International (Europe) Ltd).

Figure 5.32 Sanden compressor (SD6V12)

To the reader

Service shall be given at risk of owner, user, operator or service personnel of the A/C system and/or the compressor for which this Service Manual is destined. Sanden International (Europe) Ltd shall neither assume responsibility nor be kept liable for any loss or damage to the human life or body and/or the property which occurs or has occurred in conducting or in relation to services carried out in accordance with or in reference to this Service Manual.

1.0 Prechecks

Unusual noise not due to compressor

Unusual noises may be caused by components other than the compressor.

1. Compressor mounting – check for:
   - Loose belt – see belt tension specifications.
   - Broken bracket or compressor mounting ear. Replace broken component.
   - Missing, broken, or loose mounting bolts. Replace, reinstall, or tighten.
   - Flush fit of compressor to bracket and vehicle engine. Replace any part not properly fitted.
   - Loose or wobbling crankshaft pulley. Check for damage to pulley, incorrect center bolt torque or center bolt bottoming. Repair to vehicle manufacturer's specifications.
   - Bad idler pulley bearing. Replace if necessary.
2. Other engine components – check for noise in:
   - Alternator bearing
   - Air pump (if present)
   - Water pump bearing
   - Valves
   - Timing belt or chain
   - Power steering pump (if present)
   - Loose engine mount bolts.

Unusual noises due to compressor

1. Suction pressure less than about 5 psig can cause unusual noise. Charge refrigerant to proper amount and test by applying heat to evaporator to increase suction pressure.
2. Clutch bearing – see clutch check.
3. Oil level – insufficient oil can cause unusual noise. See oil level check.
4. Valve Noise – test for valve plate assembly failure per valve plate check.

System pressure release

Before disconnecting any lines, always make sure that the refrigerant has been removed from the A/C system by recovering it with the appropriate equipment. When working on compressors, always be sure to relieve internal pressure first. Internal compressor pressure can be relieved by removing the oil plug (if necessary) or by removing shipping caps/pads from both ports.

Recovery of refrigerant

Never discharge refrigerant to the atmosphere. Always use approved refrigerant recovery equipment.

Handling of refrigerant

Always wear eye and hand protection when working on an A/C system or compressor.

Ventilation
Keep refrigerants and oils away from open flames. Refrigerant can produce poisonous gases in the presence of a flame. Work in a well-ventilated area.

Avoid use of compressed air
Do not introduce compressed air into an A/C system due to the danger of contamination.

2.0 Compressor specifications

SD6V12 assembly torques

Item	N·m
Armature retaining nut	19.6 + 1.9/−0.9
Cylinder head bolts	14.7 ± 0.9
Oil filler plug	14.7 ± 4.9
Pad fitting bolt	4.9 ± 1.9
Clutch lead wire clamp screw	1.30 ± 0.3
High pressure relief valve	9.8 ± 1.9

SD6V12 PAG oil
The SD6V12 compressors leave the factory production line with SP10 PAG oil. The quantity of oil in the SD6V12 compressors supplied to Opel is 120 cc. When an existing compressor is to be installed on a vehicle it is necessary to add the correct amount of compressor oil. This should be done via the oil plughole.

3.0 Service operations – clutch

3.1 Armature assembly removal

1. Insert pins of armature plate spanner into holes of armature assembly.
2. Hold armature assembly stationary while removing retaining nut with 14 mm socket wrench (Figure 5.33).

Figure 5.33 Removal of armature retaining nut

3. The armature can be removed by pulling it manually upwards off the splined shaft (Figure 5.34).

Service operations – clutch

3.2 Rotor assembly removal

1. Remove rotor retaining snap ring using snap ring pliers (Figure 5.35).
2. Remove rotor using Sanden removing tool (Figure 5.36).
3. The rotor bearing is not changeable as it is staked into position (Figure 5.37).

Service operations – clutch

3.3 Field coil assembly removal

1. Remove lead wire clamp screw with Phillips screwdriver so that coil wires are free. Take care not to round off retaining screw head (Figure 5.38).
2. Remove coil retaining snap ring using snap ring pliers (Figure 5.39).
3. Remove the field coil assembly.

Figure 5.34 Slide armature up and off shaft

Figure 5.35 Remove snap ring

Figure 5.36 Remove rotor

Figure 5.37 Remove rotor
(note – bearing staked into place)

Figure 5.38 Remove clamp screw

Figure 5.39 Remove snap ring

Figure 5.40 Remove shims

4. Remove the shims from the shaft. Use a pointed tool and a small screwdriver to prevent the shims from binding on the shaft (Figure 5.40).

Service operations – clutch

3.4 Field coil assembly installation

1. Reverse the steps of section 3. The protrusion on the underside of the coil ring must match hole in the front housing to prevent the movement and rotation of the coil and to correctly locate the lead wire(s).

3.5 Rotor assembly installation

1. Place the compressor on support stand, supported at rear end of compressor.
2. Set rotor squarely over the front housing boss.
3. Place the rotor installer ring into the bearing bore. Ensure that the edge rests only on the inner race of the bearing, not on the seal, pulley, or outer race of the bearing.

Figure 5.41 Installation of rotor using arbor press

Figure 5.42 Reinstall snap ring

4. Place the driver into the ring and drive the rotor down onto the front housing with a hammer or arbor press. Drive the rotor against the front housing step. A distinct change of sound can be heard when using a hammer to install the rotor (Figure 5.41).
5. Reinstall the rotor retaining snap ring using snap ring pliers (Figure 5.42).

3.6 Armature assembly installation

1. Install clutch shims. Note – Clutch air gap is determined by shim thickness. When installing a clutch on a used compressor, try the original shims first. When installing a clutch on a compressor that has not had a clutch installed before, first try 1.0, 0.5 and 0.1 mm shims (Figure 5.43).
2. Reinstall the armature on the splined shaft. Manually push the armature down the shaft until it bottoms on the shims.
3. Replace retaining nut and torque to specification: 19.6 + 1.9/−0.9 N·m (Figure 5.44).
4. Check air gap with feeler gauge. Specification is 0.4–0.8 mm. If air gap is out of specification remove armature and change shims as necessary (Figure 5.45).

Figure 5.43 Replace shims

Figure 5.44 Torque armature retaining nut

Figure 5.45 Check air gap with feeler gauges

Figure 5.46 Felt ring removal

Figure 5.47 Snap ring removal

4.0 Service operations – shaft seal

4.1 Replacement of lip type shaft seal

Note – Lip seal assembly and felt ring must never be reused. Always replace these components.

1. Be sure all gas pressure inside the compressor has been relieved.
2. Remove clutch assembly as detailed in Section 3 Service operation clutch assembly.
3. Remove the felt ring assembly using a small screwdriver to pry it out (Figure 5.46).
4. Remove seal snap ring with internal snap ring pliers (Figure 5.47).
5. Use lip seal removal and installation tool to remove lip seal assembly. Twist the tool until the two lips on the tool engage the slots in the lip seal housing and pull the seal out with a twisting motion (Figure 5.48). Clean out shaft seal cavity completely. Make sure all foreign material is completely removed.

Figure 5.48 Shaft seal removal

Figure 5.49 Protector sleeve fitting

Service operations – shaft seal

6. Place shaft seal protector sleeve over compressor shaft. Inspect the sleeve to ensure that it has no scratches and is smooth so that the lip seal will not be damaged. Make sure there is no gap between the end of the sleeve and the seal surface of the shaft (Figure 5.49).
7. Engage the lips of the seal removal and installation tool with the slots in the new lip seal housing. Make sure the lip seal assembly, especially the O-ring, is clean. Dip the entire lip seal assembly, on the tool, into clean refrigerant oil. Make sure the seal assembly is completely covered with oil.
8. Install the lip seal over shaft and press firmly to seat. Twist the tool in the opposite direction to disengage it from the seal and withdraw the tool (Figure 5.50).
9. Reinstall shaft seal snap ring with internal snap ring pliers. Bevelled side should face up (away from the compressor body). Ensure that the snap ring is completely seated in the groove.

Figure 5.50 Replacement of lip seal

Figure 5.51 Removal of cylinder head bolts

Figure 5.52 Removal of cylinder head

Figure 5.53 Removal of valve plate

10. Tap new felt ring assembly into place.
11. Reinstall clutch assembly as detailed in Section 3 Service operation clutch assembly.

5.0 Service operations – cylinder head

5.1 Cylinder head, valve plate removal and installation

1. Be sure all internal compressor pressure has been relieved. This can be achieved by undoing the oil plug slowly (refer to section 6).
2. Remove cylinder head bolts (Figure 5.51).
3. Use a small hammer and gasket scraper to separate the cylinder head from the valve plate. Be careful not to scratch the gasket surface of the cylinder head.
4. Carefully lift the cylinder head from the valve plate (Figure 5.52).
5. It is recommended that both the head gasket (between the cylinder head and the valve plate) and the block gasket (between the valve plate and the cylinder block) be replaced at any time the cylinder head is removed. However, if no service is required to the valve plate, it may be left in place. If the valve plate comes loose from the cylinder block, the block gasket must be replaced.
6. Carefully remove old head gasket from top of valve plate with gasket scraper. Be careful not to disturb the valve plate to cylinder block joint if valve plate is to be left in place. If valve plate comes loose from cylinder block, proceed to section 5.2 Valve plate removal.

5.2 Valve plate removal

1. Using a small hammer and gasket scraper carefully separate valve plate from cylinder block. Be careful not to damage sealing surface of cylinder block (Figure 5.53).
2. Inspect reed valves, retainer and MFCV. Replace valve plate assembly if any part is damaged (Figure 5.54).
3. Carefully remove any gasket material remaining on valve plate, cylinder block or cylinder head.

5.3 Valve plate and cylinder head installation

1. Coat new block gasket with clean refrigerant oil.
2. Install block gasket. Align new gasket to location pin holes and orifice(s). Notch (if present) should face same direction as oil plug or adaptor.

Figure 5.54 Valve plate removed showing MFCV

(a)

(b)

Figure 5.55 (a) Tightening cylinder head bolts with a torque wrench.
(b) Tighten opposite bolts to avoid distortion

3. Place valve plate on cylinder block with discharge valve, retainer and nut facing up (away from cylinder block) and location pins properly located in holes.
4. Ensure that there is no residual oil in each bolt hole. If oil is present it must be removed to prevent thread damage.
5. Coat head gasket with clean refrigerant oil.
6. Install head gasket over location pins on the cylinder block, checking for correct orientation.
7. Install cylinder head on the locating pins on the cylinder block.
8. Install cylinder head bolts and tighten opposite bolts alternately to avoid distortion of cylinder head (Figure 5.55).

6.0 Service operations – replacement of oil plug

6.1 Remove the oil plug using a 17 mm socket or spanner. To replace, coat the sealing O-ring with oil and reinstall using the torque wrench. Tightening torque 14.7 ± 4.9 N·m.

Note – the compressor crank chamber is pressurised. The oil plug must never be removed while the compressor is attached to a pressurised system. The correct oil must be used on plug installation. SD6V compressors use Sanden SP-10 PAG oil.

Figure 5.56 Oil plug removal

Figure 5.57 Removal of high pressure relief valve

7.0 Service operations – replacement of high pressure relief valve (HPRV)

7.1 Remove the high pressure relief valve using a 16 mm socket or spanner (Figure 5.57).

7.2 To replace coat the sealing O-ring with oil and reinstall using a torque wrench.

7.3 Tightening torque of HPRV 9.8 ± 1.9 N·m.

Note – the correct oil must be used on HPRV installation. SD6V compressors use Sanden SP-10 PAG oil.

8.0 Service tools

8.1 Special tools for service requirements detailed in this manual.

Operation	Tool name	Sanden tool number
1. Clutch	Armature securing tool	2123
2. Clutch	Rotor remover	2130*
3. Clutch	Shim remover	3031
4. Shaft seal	Shaft seal removal	3111

(Continued)

Operation	Tool name	Sanden tool number
5. Shaft seal	Protection cap	3107
6. All	Compressor stand	1030

Note – General rotor puller available from any tool manufacturer.

9.0 Compressor replacement

9.1 Replacement with new compressor. New compressors are supplied with sufficient refrigerant oil for the air-conditioning system. If a new compressor is replacing an existing compressor, it is necessary to take some oil out of the new compressor, so that the oil amount in the new compressor is the same as the quantity of oil removed in the old compressor. Note – if a new compressor is installed without the excess oil being drained, the refrigerating capacity of the air-conditioning system will be reduced. Replace the oil plug as quickly as possible after draining oil and keep the A/C system sealed whenever possible to minimise moisture absorption of oil in the compressor.

9.2 Replacement with used or repaired compressor. When a repaired or previously used compressor, which does not contain oil, is to be installed in the vehicle the correct oil should be added via the oil filler plug hole. SD6V compressors use Sanden SP-10 PAG oil. The quantity of oil to be added is the same as the quantity of oil, which was contained, in the removed compressor. Therefore whenever a compressor is removed from a vehicle it is important to drain out the oil and measure its quantity in a measuring cylinder. Should inadequate oil be charged to the replacement compressor, seizure is likely due to lubrication failure. Replace the oil plug as quickly as possible after draining oil and keep the A/C system sealed whenever possible to minimise moisture absorption of oil in the compressor.

Oil level measurement (in vehicle)

Oil level in the compressor should be checked when a system component has been replaced, when an oil leak is suspected, or when it is specified as a diagnostic procedure.

1. Run the compressor for 10 minutes with the engine at idle.
2. Recover all refrigerant from the system, slowly so as not to lose any oil.
3. Determine the mounting angle of the compressor from horizontal (i.e. oil plug or adaptor on top). This is most readily done by using a machinist's universal level, if access to the compressor permits.
4. Remove the oil filler plug. Using a socket wrench on the armature retaining nut, turn the shaft clockwise until the counterweight is positioned as shown.
5. Insert oil dipstick up to the stop, as shown in Figure 5.58, with the angle pointing in the correct direction.
6. Remove dipstick and count number of notches covered by oil.
7. Add or subtract oil to meet the specifications shown in the table.
8. Reinstall oil plug. Seat and O-ring must be clean and not damaged. Torque to 11–15 ft·lb (15–20 N·m, 150–200 kgf·cm).

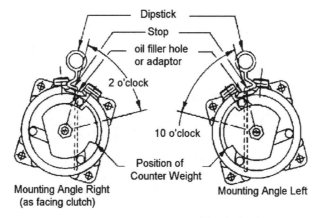

Figure 5.58 Compressor oil level check

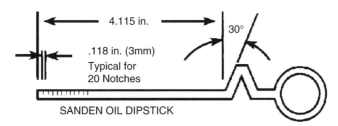

Figure 5.59 Oil measurement using a dipstick

	Acceptable oil level in increments	
Mounting angle (degrees)	SD5H14	SD7H15
0	3–5	5–7
10	4–6	6–8
20	5–7	7–9
30	6–8	8–10
40	7–9	9–11
50	8–10	10–12
60	8–10	11–13
90	8–10	16–18

Compressor repaired internally and reinstalled in the system

1. Before any internal repair is done, drain the oil from the compressor.
   - Remove the oil plug and drain as much oil as possible into a suitable container.
   - Remove the caps (if present) from suction and discharge ports.
   - Drain oil from the suction and discharge ports into a suitable container while turning the shaft clockwise only with a socket wrench on the armature retaining nut.

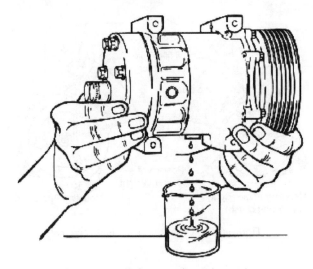

Figure 5.60 Oil removal via drain plug

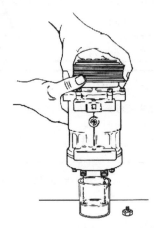

Figure 5.61 Oil removal via suction and discharge ports

2. Measure and record the amount of oil drained from the compressor.
3. Inspect the oil for signs of contamination such as discoloration or foreign material.
4. Perform repair to the compressor.
5. Add the same amount of new oil to the compressor as was measured in step 2. Be sure to use the correct oil for the compressor.
6. Reinstall oil plug. Seal and O-ring must be clean and not damaged. Torque to 11–15 ft·lb (15–20 N·m, 150–200 kgf·cm). Be careful not to cross thread the oil plug.
7. It is recommended that the oil quantity be confirmed after reinstallation of the compressor to the vehicle.

Sanden compressor replaced by a new Sanden compressor of the same type

1. Drain oil from the old compressor; measure and record the amount as per the procedure.
2. Drain oil from the new compressor.
3. Add new oil of the correct type to the new compressor. Use the same quantity as was removed from the old compressor in step 1.

Figure 5.62 Fixed orifice valve remover
(courtesy of Autoclimate)

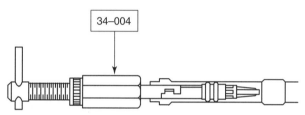

Figure 5.63 Withdrawing the FOV valve
(reproduced with the kind permission of Ford Motor Company Limited)

4. Reinstall oil plug. Seal and O-ring must be clean and not damaged. Torque to 11–15 ft·lb (15–20 N·m, 150–200 kgf·cm).
5. It is recommended that the oil quantity be confirmed after installation of the new compressor to the vehicle.

Sanden SP-20 refrigerant oil for R134a SD compressors

Sanden provides field service containers of SP-20 PAG oil for Sanden SD-series compressors in convenient 250 cc cans. These cans are designed to withstand moisture ingression. Always keep the cap of the can tightly closed when not handling the oil. Cans are packed in 'six-packs' and available through your Sanden representative. Material safety data is also available. Sanden limits the warranty of SD compressors for field service with the condition that only Sanden-approved SP-20 is utilized. 'Six-pack' of 250 cc cans of SP-20 oil – Sanden Number 7803–1997.

5.8 Fixed orifice valve remove and replace

Tools

The following tools (Figs 5.62–5.64) are available for removing and installing the fixed orifice tube:

1. Remover and installer for fixed orifice tube (with threaded sleeve) 34–004.
2. Remover for damaged fixed orifice tube (without threaded sleeve) 34–005.

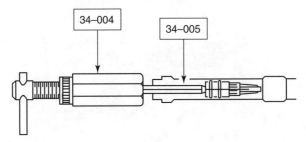

Figure 5.64 Removing the broken orifice valve
(reproduced with the kind permission of Ford Motor Company Limited)

The tool with the number 34–004 is hooked into the housing of the fixed orifice tube. The fixed orifice tube can then be withdrawn from the line by turning the threaded sleeve. This tool is also used to install the fixed orifice tube. This must be slid into the line as far as the shoulder.

If the housing of the fixed orifice tube breaks off during removal, the damaged fixed orifice tube can be removed with tool 34–005 in conjunction with the threaded sleeve of tool 34–004. For this purpose, the tool is screwed into the fixed orifice tube.

6 The environment

The aim of this chapter is to:

- Give the reader an understanding of the environmental research that underpins the global pressure on the A/C industry to reduce ozone depletion and global warming.

Climate change is one of the greatest environmental and economic challenges facing humanity. The increased frequency of extreme weather events and the fact that the 1990s was the warmest decade of the 20th century, which was the warmest of the millennium, demonstrates that climate change is not some distant threat but is happening right now.

6.1 Global Warming

Greenhouse effect

As a result of the use of large quantities of fossil fuel (such as oil, coal and spontaneous gas) and the depletion of forests, the concentration of carbonic acid, Freon, methane etc. in the atmosphere is increasing. The heat from the surface of the earth is being absorbed into the atmosphere. Under these conditions, it is thought that the earth is getting warmer, i.e. global warming.

If conditions remain as they are and greenhouse effect gas continues to be released into the atmosphere, it is estimated that by the year 2030 the temperature will have risen by 1.5 to 4.5°C (2.7 to 8.5°F). The surface of the ocean will also have risen by 0.2 metres to 1.4 metres. Quite simply, the atmospheric release of Freon is not only causing ozone depletion, it is also bringing about the greenhouse effect.

Greenhouse effect gas

Gases such as carbon dioxide (CO_2), Freon, dinitrogen monoxide (N_2O), and methane (CH_4) are the substances that are having an influence on the greenhouse effect. Automobile air-conditioning systems because of their design have real drawbacks in terms of greenhouse gas emission:

- They use refrigerants which have a high impact on the increase of the greenhouse effect. GWP (Global Warming Potential) of HFC 134a is 1300 which means that 1 kg of HFC emitted into the atmosphere has the same impact as 1.3 tons of CO_2.
- Flexible hosepipes' poor connections allow refrigerant loss.
- Compressor shaft seals are a major source of refrigerant loss.

Production, recovery, recycle, reclaim, and disposal all contribute to the greenhouse problem.

Table 6.1 GWP (Global Warming Potential)

Refrigerant type	GWP
CO_2	1
R12	8500
R134a	1300

6.2 The Ozone Layer

In recent years, an awareness about environmental protection has been growing throughout the world. Particular concerns about possible depletion of the ozone layer from CFCs was first raised in 1974 with the publication of research which theorised that chlorine released from CFCs could migrate to the stratosphere and destroy ozone molecules. Ozone layer protects life on earth by absorbing the greater part of the harmful ultraviolet rays emitted by the sun. When scientists began studying ozone depletion in the early 1970s, they investigated several natural phenomena, such as volcanoes and evaporation of seawater. Volcanoes can produce large quantities of hydrochloric acid. However, most volcanic discharges are not powerful enough to reach the stratosphere. Chlorine evaporation from seawater is dissolved in rain and does not reach the stratosphere. Chlorine produced by volcanoes or oceans does not leave the troposphere and poses no threat to the ozone layer. However, CFCs, being extremely stable, do not release chlorine until they reach the stratosphere.

CFCs released into the atmosphere are said to be destroying or damaging this ozone layer. It was discovered that CFC12 which has been used as a refrigerant in car air-conditioning system units until now is one of the substances that could cause depletion of the ozone layer. Hence an air-conditioning system that uses a different substance as a refrigerant type has been developed. Since this alternative refrigerant has characteristics that are different from the previous refrigerant, the conventional air-conditioning system must be modified to accept it.

Unusual phenomenon

Ultraviolet rays of a certain wavelength can be harmful to living organisms, and possibly cause skin cancer; it also exerts an influence on genes. The ozone layer absorbs these ultraviolet rays, thereby performing an extremely important role in preserving life on earth. In 1985, it was discovered that a phenomenon can be seen over the South Pole, whereby the ozone layer is reduced in the spring and restored to its normal level in the summer. A satellite using a special sensor captured this phenomenon, and the pictures that it transmitted revealed that the ozone layer over the continent of Antarctica was being depleted. Since this appears as a 'hole' in the ozone layer, it has come to be called the 'ozone hole'.

This 'ozone hole' attracted the concern of scientists. The fact that the ozone layer was being depleted by Freon and that there was a danger of harmful ultraviolet rays pouring onto the earth's surface had been pointed out more than 10 years previously. A decision was made to mount a large-scale observation in order to investigate the mechanism of this ozone hole and to clarify its relationship with CFC12.

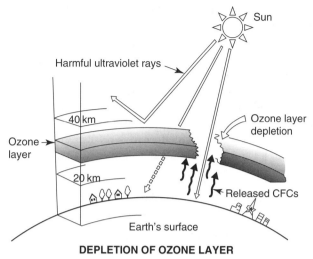

DEPLETION OF OZONE LAYER

Figure 6.1 Ozone depletion
(with the agreement of Toyota (GB) PLC)

Role of the ozone layer

The atmosphere which envelops the surface of the earth is divided into a number of layers. The layer closest to the earth is called the troposphere. In the troposphere, temperature decreases. For this reason, convection occurs in the atmosphere and is manifested as atmospheric phenomena.

In the layer above this, called the stratosphere, contrary to the troposphere, the temperature becomes higher as the altitude increases, and the air flows slowly. Between the 20 to 40 km altitude range of the stratosphere, the degree of concentration of ozone is high. This is called the ozone layer, which has a pungent smell and is slightly blue in colour. A certain wavelength of ultraviolet ray is damaging to living beings, is a source of skin cancer, and has an effect on genetic structure. The ozone layer, by absorbing these ultraviolet rays, plays a critical part in preserving life on the earth.

Ozone formation

Oxygen atoms absorb ultraviolet rays and are broken up into oxygen atoms. These oxygen atoms combine with oxygen molecules to form ozone. Ozone is formed near the equator where the amount of solar radiation is highest, and spreads in the direction of the Poles through slow atmospheric movement.

Theory of the ozone layer depletion

Freon is an extremely stable substance, hence it passes from the earth through the troposphere and reaches the stratosphere without breaking up. There, the diffused CFC12 is bombarded with strong ultraviolet rays and breaks up, releasing chlorine. With this chlorine as a catalyst, a reaction occurs, repeatedly combining with, and breaking apart, ozone molecules, and the ozone is depleted. Because chlorine is a very stable substance it gets into the stratosphere, where it remains for a long period of time (up to 120 years), causing ozone depletion to continue. *It is believed that one CFC molecule can destroy as many as 100 000 ozone molecules.*

In December 1994, NASA announced that three years of satellite data confirmed that CFCs are the primary source of stratospheric ozone chlorine.

Specified Freon (CFCs), chemically stable substances that are superior for heat resistance and non-combustibility, have the characteristics of being colourless and odourless without being flammable, corrosive, or toxic. For these reasons, they came to be used for a wide range of purposes such as refrigerants for air-conditioning and refrigeration systems, aerosol spray agents, cleaning agents for electronic systems, fire extinguisher material, foam agents, and raw material for synthetic resins.

The most important characteristic of an alternative refrigerant is that the ozone depletion potential is zero, and that it can be used safely in its required environment.

Table 6.2 Ozone depleting potential

Refrigerant type	ODP
R11	1
R12	1
R134a	0

7 Legislation

The aim of this chapter is to:

- Give the reader a historical perspective on how the A/C industry has been shaped by legislation over the past 20 years.
- Give the reader an understanding of the future of A/C technology.

7.1 Historical perspective

In June 1974, two chemists from the University of California in the United States, F. Sherwood Rowland and Mario Molina, published a report in the journal *Nature*. It indicated that Freon gas, which is the trade name for chlorofluorocarbons (CFC), may have an influence on the ozone layer and consequently may affect human life and the ecological system. This thesis provoked a great debate throughout the world, which continued to grow until the use of Freon came to be regulated.

In March 1985, in accordance with the United Nations Environmental Plan (UNEP), the 'Vienna convention for the Protection of the Ozone Layer' was adopted, and was published in September 1988.

In September 1987, the Montreal Protocol on substances that deplete the ozone layer was adopted, and many countries as well as organisations throughout the world signed it. The Montreal Protocol identified five kinds of Freon that were to be the subject of regulations (Freon 11, 12, 113, 114, 115 referred to as 'specified Freon'). A schedule regulating production and consumption was determined. At the same time it was decided to prohibit the import and export of products that include the target substances between countries that had not ratified the treaty. Along with this, it was agreed to promote mutual co-operation, research and development, and information exchange in regard to finding alternatives for the target substances.

During May of 1989, the 'Vienna Treaty, Montreal Protocol First Treaty Powers Meeting' was held. The proposal to strengthen regulations mandating the total abolition of 'specified Freon' by the year 2000 was examined in detail. At the same time, with the participation of countries who ratified the Treaty and the Protocol as well as observer countries, the 'Helsinki Declaration', which published international policy goals, was adopted. With this 'Helsinki Declaration', the countries that had not yet signed the Vienna Treaty and the Montreal Protocol were encouraged to do so. The participants agreed that the production and consumption of specified Freon should be completely phased out as quickly as possible. In June of 1990, the 'Montreal Protocol Second Treaty Powers Meeting' was held, and the participating powers agreed to strengthen the regulations of specified Freon. They agreed the use of the target Freon would be reduced to less than 100% (based on the actual results of Freon consumption in 1986) from July of 1989, to 50% or less from January of 1995, and to 15% or less from January of 1997. By the year 2000 it would be totally abolished.

1992 United Nations Framework Convention on Climate Change (UNFCCC)

1995 – EU halts CFC production for all except export and essential uses.

1997 – Kyoto Protocol requires industrialised countries to reduce their collective emissions of greenhouse gases by 5.2% below their 1990 levels for the period 2008 to 2012 (the first commitment period). The EU is fully committed to comply with its obligations under the UNFCCC and the Kyoto Protocol, namely to reduce its greenhouse gas emissions by 8% over 1990 levels by the period 2008–2012. Complying with its Kyoto emission reduction target of 8% within the first commitment period – an overall reduction of 336 million tonnes of carbon dioxide equivalent – is crucial to the EU's credibility. The major greenhouse gas is carbon dioxide (CO_2). However, the basket of greenhouse gases controlled by the Kyoto Protocol includes hydrofluorocarbons (HFCs) which have large Global Warming Potentials (GWP).

Established in June 2000 – European Climate Change Programme (ECCP)

Meeting the Kyoto Protocol target depends on putting in place effective policies and measures. Through the European Climate Change Programme (ECCP), established in June 2000, the European Commission has identified cost-effective measures that could be taken and so enable the European Community to meet its Kyoto Protocol target.

The report of the fluorinated gases (F gases) working group of June 2001 stated that the EU's fluorinated gas emissions in 1995 were around 65 million tonnes of CO_2 equivalent or 2% of its total greenhouse gas emissions. Assuming no additional measures were taken, emissions of these gases were forecast to increase to around 98 million tonnes of CO_2 equivalent by 2010, representing 2% to 4% of total projected EU greenhouse gas emissions.

On the basis of the work of the ECCP and a number of studies undertaken by the Directorate General for Environment of the European Commission on costs and benefits, the Commission made a proposal for a Regulation to control certain fluorinated greenhouse gases in August 2003. The proposal is expected to reduce projected emissions of fluorinated gases by around 23 million tonnes of CO_2 equivalent by 2010, and even greater reductions in the period after as some provisions will not have a significant impact until then.

The Commission proposal for a Regulation that covers the three industrial greenhouse gases in the Kyoto protocol has three main elements. First, there are provisions on containment and recovery which covers the operation, maintenance and end-of-life of certain equipment and includes minimum training and certification for servicing personnel. Second, there are reporting provisions that will require information on production, imports and exports of fluorinated gases to be obtained in order to help to validate the forecasts made by the emissions' models. Third, there are a number of marketing and use bans of these fluorinated gases including the phase out of mobile air-conditioning systems using fluorinated gases with a *global warming potential above 150.*

October 2004 – European Council Political Agreement

In its first reading on the Commission's proposal on 31 March 2004, the European Parliament proposed a number of amendments to the Commission's proposal. The Commission indicated its willingness to adopt a significant number of these amendments where they could reinforce and clarify the Commission's proposal. On 14 October 2004 the Council of Ministers meeting in Luxembourg adopted a Political Agreement, which also included a significant number of the Parliament's amendments, on the basis of the Commission's proposal.

The Political Agreement now consists of two elements, a Directive to phase out HFC134a from motor vehicle air-conditioning systems that will apply to all new vehicle models coming off the production lines as of 2011 and by 2017, every new vehicle produced will have to use alternatives. The other element is a Regulation that will apply containment, recovery and labelling provisions to so-called 'stationary' applications.

The Political Agreement reached by the Council of Ministers on 14 October 2004 will provide a solid base for the final shape of this Regulation. However, the legislative process is still not finished; the next step is for the European Parliament to consider the political agreement in a second reading. This will probably take place in the first half of 2005 or early 2006. If the Parliament agrees with the political agreement then this can go forward and the proposed Regulation can then become law. If, however, the Parliament wishes to make amendments to the Council's common position which the Council cannot accept then there will have to be negotiations between the two institutions in order to agree on a final outcome. The Commission with its technical knowledge can play an important role in facilitating agreement during this process.

Overview

The European Union is moving to phase out R134a as a refrigerant because it is on the hit list of gases that contribute to global warming. Most of the European nations signed the Kyoto Protocol which calls for cutbacks in carbon dioxide emissions and the release of other global warming gases. But within the next five to eight years, the Europeans will start introducing higher pressure CO_2 (R-744) A/C systems.

EU Council 14 October 2004:

- Article 95.
- No quota system.
- No 'low leak' system.
- GWP below 150.
- Ban R134a by 2011 for new vehicle types.
- Final ban by 2017 for all new vehicles.

During the second and third quarter of 2005 the EU Parliament will give a second reading for the legislation. The legislation should be approved by the end of 2005/beginning of 2006 and put into force by 2007/8. Manufacturers are looking to launch new HVAC modular CO_2-based systems to the mass market by 2008.

> Note – in the European Union the normal legislative procedure is Co-Decision by the European Parliament and the Council. The European Parliament consists of directly elected MEPs that adopt legislation, after consideration by formal committees, through formal opinions voted by the whole Parliament. The Council is the institution that represents the 25 Member State governments.
>
> The political agreement becomes a 'common position' once it is translated into all the official languages of the EU and is then formally adopted by the Council. The common position is then published in the *Official Journal of the EU.*

Europe and the view of the motor industry (provided by the Society of Motor Manufacturers and Traders (SMMT))

Most of the industry supports the EU Council's decision that gases whose GWP is over 150 times that of carbon dioxide should be banned. This would allow manufacturers to choose which technology to use to replace HFC134a, minimising costs while achieving the environmental objectives set out in the regulation. It is felt by the industry that the European Union should not act in isolation. Members of the Society of Motor Manufacturers and Traders (SMMT) welcome the proposal to properly contain and use fluorinated gases and accept the ultimate phase-out of 134a from mobile air-conditioning. However, SMMT urges the European Parliament to recognise that Europe should not act in isolation and manufacturers producing in the EU should be allowed to meet the demands of other markets.

Mobile air-conditioning systems using CO_2 have significant cost and safety implications (typical incremental production costs over and above the ones of HFC134a range in the region of 100 euros per vehicle). For many companies this may involve a complete rethink of investment plans, infrastructure, suppliers and, above all, aftersale servicing. SMMT is concerned that if a technology prescriptive approach is chosen, manufacturers exporting from the UK to countries such as the US where the decision has been made not to phase out HFC134a, but to move to enhanced HFC134a, will have to face disproportionate additional costs to meet their servicing and repair obligations.

A non-technology prescriptive approach is very important to industry as SMMT member companies, within both vehicle manufacturers and component supplier membership, have heavily invested in both CO_2 and HFC152a systems as possible alternatives to HFC134a. They both offer significant environmental benefits compared to HFC134a. The environmental benefits associated with the phasing out of HFC134a are gained in the quantum shift from HFC134a with a GWP of 1300 to an alternative with a GWP nearly ten times lower. The incremental environmental benefit of further reducing the legislative requirement to GWP 50 is negligible.

Leakage control

The agreed Council's text suggests the introduction of leakage requirements 12 months after adoption of a harmonised test procedure (or 1 January 2007, whichever is later) for new vehicle types and 24 months after the adoption of a harmonised test procedure for all new vehicles (or 1 January 2008, whichever is later). This effectively means a transition period of just 12 months. This would require manufacturers to re-engineer MAC systems within an unrealistically short timeframe.

While SMMT accepts the Council's position on the introduction of leakage rate requirements after adoption of a harmonised test procedure (or 1 January 2007, whichever is later) for new vehicle types, SMMT proposes the introduction of leakage rate requirements after the adoption of a harmonised test procedure for all new vehicles (or 1 January 2009, whichever is later). This would extend the transition period to 24 months – technologically challenging but achievable.

Ban dates

SMMT reinforces the need for sustainable lead times to move to alternative mobile air-conditioning systems. SMMT consider the earlier possible implementation dates to be 2012 and 2018. However, we understand the political imperative for the early phase-out of HFC 134a and take note of the compromise position agreed by the Council in October 2004.

We are committed to ban HFC134a from MAC, but this should be done within a realistic timeframe and ensuring that the EU remains a competitive and attractive place to do business, where new technologies can be developed, manufactured and exported outside the EU.

7.2 US perspective

In the US the Montreal Protocol is implemented by the Clean Air Act 1990.

The Clean Air Act (CAA 1970)

The Clean Air Act of 1970 set a national goal of clean and healthy air for all. It established the first specific responsibilities for government and private industry to reduce emissions from vehicles, factories, and other pollution sources.

The Clean Air Act Amendments (CAA 1990)

The 1990 Clean Air Act Amendments directed the EPA to develop regulations to comply with the Montreal Protocol and maximise recycling, ban non-essential use, develop labelling requirements and examine safe alternatives for ozone depleting substances. The Clean Air Act gives the EPA the authority to establish environmentally friendly procedures and codes of practice and standards which are often contributed to by the SAE (Society of Automotive Engineers). While the EPA may use standards set by the SAE they are only guidelines and not regulations. CFC-12 (also known by the brand name Freon) was widely used in air-conditioners for automobiles and trucks in the US for over 30 years. The Clean Air Act, and regulations issued under the Act, ended the production of CFC-12 for air-conditioning and refrigeration on 31 December 1995. While new vehicles no longer use CFC-12, most vehicles built before 1994 still require its use for servicing. As a result, 30 million cars or more needed conversions to use an alternative refrigerant. In the US approximately 5% of vehicles still use R12. Current supplies of recycled and reclaimed R12 are sufficient to meet the continual demand of these vehicles.

Section 608 of the Clean Air Act Amendments of 1990

Section 608 of the Clean Air Act prohibits releasing CF-12 into the atmosphere. The prohibition on venting CFC-12 has been in effect since 1992. Section 608 of the Clean Air Act prohibits releasing HFC134a into the atmosphere. The prohibition on venting HFC134a has been in effect since November 1995.

Section 609 of the Clean Air Act Amendments of 1990

Section 609 covers technician certification in the motor vehicle sector only. All technicians repairing and servicing the refrigerant circuit must now be certified in refrigerant recovery and recycling procedures to be in compliance with section 609 of the Clean Air Act Amendments of 1990. To make sure existing CFC-12 is used as much as possible, rather than being wasted and released to the atmosphere, EPA issued regulations under section 609 of the Clean Air Act to require that shop technicians use approved service machines to recycle refrigerants. Service shops must also certify to EPA that they have acquired and are properly using approved refrigerant recovery equipment. Becoming certified allows you to: (1) perform refrigerant servicing of vehicles with R12, R134a, or Significant New Alternatives Policy (SNAP)-approved refrigerants; and (2) purchase R12 and ozone-depleting blend substitutes for R12 (many of the blends approved for use as an R12 alternative in automotive air-conditioning are ozone depleting). Although you have to be certified to perform refrigerant servicing of vehicles equipped with R134a, currently, you do not have to be certified to purchase R134a.

Significant New Alternatives Policy (SNAP) 1994

In 1994, EPA established the SNAP Program to review alternatives to ozone-depleting substances like CFC-12. Under the authority of the 1990 Clean Air Act (CAA), EPA examines new substitutes for their ozone-depleting, global warming, flammability, and toxicity characteristics. EPA has determined that several refrigerants are acceptable for use as CFC-12 replacements in motor vehicle air-conditioning systems, subject to certain use conditions. It is important to understand the meaning of 'acceptable subject to use conditions'. EPA believes such refrigerants, when used in accordance with the conditions, are safer for human health and the environment than CFC-12. This designation does not mean that the refrigerant will work in any specific system, nor does it mean that the refrigerant is perfectly safe regardless of how it is used. The EPA does not approve or endorse any one refrigerant that is acceptable subject to use conditions over others also in that category. Note that the EPA does not test refrigerants under the SNAP process. Rather, they review information submitted to them by manufacturers and various independent testing laboratories.

Misleading use of 'drop-in' to describe refrigerants

Many companies use the term 'drop-in' to mean that a substitute refrigerant will perform identically to CFC-12, that no modifications need to be made to the system, and that the alternative can be used alone or mixed with CFC-12. The EPA believes the term confuses and obscures several important regulatory and technical points. First, charging one refrigerant into a system before extracting the old refrigerant is a violation of the SNAP use conditions and is, therefore, illegal. Second, certain components may be required by law, such as hoses and compressor shut-off switches. If these components are not present, they must be installed. Third, it is impossible to test a refrigerant in the thousands of air-conditioning systems in existence to demonstrate identical performance. In addition, system performance is strongly affected by outside temperature, humidity, driving conditions etc., and it is impossible to ensure equal performance under all of these conditions. Finally, it is very difficult to demonstrate that system components will last as long as they would have if CFC-12 were used. For all of these reasons and more, EPA does not use the term 'drop-in' to describe any alternative refrigerant.

System connections

The EPA states that each new refrigerant must be used with a unique set of fittings to prevent the accidental mixing of different refrigerants. These fittings are attachment points on the car itself, on all recovery and recycling equipment, and on all refrigerant containers. Adaptors to mate with those fittings may therefore be needed on charging equipment. If the car is being retrofitted, any service fittings not converted to the new refrigerant must be permanently disabled. Unique fittings help protect the consumer by ensuring that only one type of refrigerant is used in each car. They also help protect the purity of the recycled supply of CFC-12, which means it will last longer, so fewer retrofits will be necessary nationwide.

Applicability to manifold gauges and refrigerant identifiers (SAE standard)

Manifold gauges allow technicians to diagnose system problems and to charge, recover, and/or recycle refrigerant. A standard fitting may be used at the end of the hoses attached to the manifold gauges, but as stated above, unique fittings must be permanently attached at the ends of the hoses that attach to vehicle air-conditioning systems and recovery or recycling equipment. Similarly, refrigerant identifiers may be used with multiple refrigerants. The connection between the identifier or similar service equipment and the service hose may be standardised and work with multiple hoses. For each refrigerant, however, the user must attach a hose to the identifier that has a fitting unique to that refrigerant permanently attached to the end going to the vehicle. Adaptors for one refrigerant may not be attached to the end and then removed and replaced with the fitting for a different refrigerant. The guiding principle is that once attached to a hose, the fitting is permanent and is not removed.

Fitting sizes refrigerant	High side service port			Low side service port		
	Diameter (inches)	Pitch (threads/inch)	Thread direction	Diameter (inches)	Pitch (threads/inch)	Thread direction
CFC-12 (post-1987)	6/16	24	Right	7/16	20	Right
CFC-12 (pre-1987)	7/16	20	Right	7/16	20	Right
HFC-134a	quick-connect			quick-connect		
Freeze 12	7/16	14	Left	8/16	18	Right
Free Zone/ RB-276	8/16	13	Right	9/16	18	Right
Hot Shot	10/16	18	Left	10/16	18	Right
McCool Chill-It	6/16	24	Left	7/16	20	Left
GHG-X4/ Autofrost	.305	32	Right	.368	26	Right
GHG-X5	8/16	20	Left	9/16	18	Left
R-406A	.305	32	Left	.368	26	Left
GHG-HP	Not developed yet			Not developed yet		
Ikon-12/ Ikon A	Not developed yet			Not developed yet		
FRIGC FR-12	quick-connect, different from HFC134a			quick-connect, different from HFC134a		
SP34E	7/16	14	Right	8/16	18	Left
RS-24	Quick connect, different from HFC134a and FRIGC FR-12			Quick connect, different from HFC134a and FRIGC FR-12		

Figure 7.1 System connection requirements

Record keeping

Service shops must maintain records of the name and address of any facility to which refrigerant is sent. Service shops are also required to maintain records (on-site) showing that all service technicians are properly certified.

Labels

When a car air-conditioner originally designed to use R12 is retrofitted, the technician must apply a detailed label giving specific information about the alternative.

The label shows:

- the name and address of the technician and the company performing the retrofit;
- the date of the retrofit;
- the trade name, charge amount, and, when applicable, the ASHRAE numerical designation of the refrigerant;
- the type, manufacturer, and amount of lubricant used; and
- if the refrigerant is or contains an ozone-depleting substance, the phrase 'ozone depleter'
- if the refrigerant is flammable according to ASTM Standard E681, the sentence 'this refrigerant is FLAMMABLE. Take appropriate precautions'.

This label covers up information about the old refrigerant, and provides valuable details on the alternative and how it was used. It also tells the owner who performed the retrofit.

Refrigerant	Background
CFC-12	White
HFC134a	Sky blue
Freeze 12	Yellow
Free Zone/RB-276	Light green
Hot Shot	Medium blue
GHG-X4	Red
R-406A	Black
GHG-X5	Orange
GHG-HP	not yet developed*
Ikon-12/Ikon A	not yet developed*
FRIGC FR-12	Grey
SP34E	Tan
RS-24	Gold

Figure 7.2 Label colour requirements

Barrier hoses

HCFC-22, a component in some blends, can seep out through traditional hoses. Therefore, when using these blends, the technician must ensure that new, less permeable 'barrier' hoses are used. These hoses must be installed if the system currently uses old, non-barrier hoses. The table of refrigerants on the EPA's website notes this additional requirement where appropriate.

Compressor switch

Some systems have a device that automatically releases refrigerant to the atmosphere to prevent extremely high pressures. When retrofitting any system with such a device to use a new refrigerant, the technician must also install a high pressure shut-off switch. This switch will prevent the compressor from increasing the pressure to the point where the refrigerant is vented.

List of alternative refrigerants

Figure 7.3 summarises the following information about refrigerants reviewed under EPA's SNAP Program for use in motor vehicle air-conditioning systems. Note that 'air-conditioning' means cooling vehicle passenger compartments, not cargo areas, so refrigeration units on trucks and rail cars are not covered by this list.

Many refrigerants are sold under various names. All known trade names are listed, separated by slashes.
Status:

1. *acceptable subject to use conditions*: May be used in any car or truck air-conditioning system, provided the technician meets the conditions described above. Note that EPA cannot guarantee that any refrigerant will work in a specific system.
2. *unacceptable*: Illegal to use as a substitute for CFC-12 in motor vehicle air-conditioners.

Acceptable subject to use conditions (2)								
Name (1)	Date	Manufacturer	Components						
			HCFC-22	HCFC-124	HCFC-142b	HFC-134a	Butane (R-600) (3)	Isobutane (R-600a) (3)	HFC-227ea
HFC-134a	3/18/94	Several	–	–	–	100	–	–	–
FRIGC FR-12	6/13/95	InterCool Distribution 800-555-1442	–	39	–	59	2	–	–
Free Zone/ RB-276 (4)	5/22/96	Hi Tech Refrigerants, LLC 800-530-4805	–	–	19	79	–	–	–
Ikon-12	5/22/96	Ikon Corp. 601-868-0755	Composition claimed as confidential business information						
R-406A/ GHG(5)	10/16/96	People's Welding 800-382-9006	55	–	41	–	–	4	–
GHG-HP (5)	10/16/96	People's Welding 800-382-9006	65	–	31	–	–	4	–
GHG-X4/ Autofrost/ Chill-It (5)	10/16/96	People's Welding 800-382-9006 McMullen Oil Products 800-669-5730	51	28.5	16.5	–	–	4	–
Hot Shot/ Kar Kool (5)	10/16/96	ICOR 800-357-4062	50	39	9.5	–	–	1.5	–
Freeze 12	10/16/96	Technical Chemical 800-527-0885	–	–	20	80	–	–	–
GHG-X5 (5)	6/3/97	People's Welding 800-382-9006	41	–	15	–	–	4	40
SP34E	12/18/00	Solpower 888-289-8866	Composition claimed as confidential business information						
RS-24	12/20/02	Refrigerant Products Ltd	Composition claimed as confidential business information						

Figure 7.3 Table of alternative refrigerants from the EPA's website

3. *proposed acceptable subject to use conditions*: May be used legally. EPA will accept public comment on these refrigerants and then make a final ruling. There is no formal EPA position until then, and it is inappropriate for advertising to imply that EPA has found the product acceptable.
4. *not submitted*: Illegal to use *or sell* as a substitute for CFC-12 in motor vehicle air-conditioning systems.

Refrigerant handling

Recycle and reclaim
Recycling occurs in a service shop. Reclamation means the removal of all oil and impurities beyond that provided by on-site recycling equipment. Reclaimed refrigerant meets the identical

purity standards required of new and unused refrigerant. Reclamation cannot be performed in the service shop. Rather, the shop generally sends refrigerant either back to the manufacturer or directly to a reclamation facility.

Technicians who repair or service HFC134a MVACs must be trained and certified by an EPA-approved organisation. To be certified, technicians must pass a test demonstrating their knowledge in these areas. A technician already trained and certified to handle CFC-12 does not need to be recertified to handle HFC134a. Service shops must certify to EPA that they have acquired and are properly using approved refrigerant recovery equipment. Note that this certification is a one-time requirement, so that if a shop purchased a piece of CFC-12 recycling equipment in the past, and sent the certification to EPA, the shop does not need to send a second certification to EPA when it purchases a second piece of equipment no matter what refrigerant that equipment is designed to handle.

There is no restriction on the sale of HFC134a so anyone may purchase it.

Retrofitting vehicles

Although section 609 of the Clean Air Act does not govern retrofitting, section 612 of the Act requires that when retrofitting a CFC-12 vehicle for use with another refrigerant, the technician:

- must first extract the CFC-12. The original CFC-12 must be removed from the system prior to charging with the new refrigerant. This procedure will prevent the contamination of one refrigerant with another. Refrigerants mixed within a system probably will not work and could damage the system. This requirement means that no alternative can be used as a 'drop-in'.
- must cover the CFC-12 label with a label that indicates the new refrigerant in the system and other information and
- must affix new fittings unique to that refrigerant.
- if the system includes a pressure relief device, must install a high pressure compressor shut-off switch to prevent the compressor from increasing pressure until the refrigerant is vented;
- in addition, if retrofitting a vehicle to a refrigerant that contains R-22, must ensure that only barrier hoses are used in the A/C system.

CFC-12 equipment may be permanently converted for use with HFC134a under certain conditions. The retrofitted unit must meet SAE Standard J2210 and must have the capacity to purify used refrigerant to SAE Standard J2099 for safe and direct return to the air-conditioner following repairs.

Mobile Air Conditioning Climate Protection Partnership 2004

On Earth Day in 2004, the Mobile Air Conditioning Climate Protection Partnership initiated a global voluntary effort to reduce greenhouse gas emissions from vehicle air-conditioning systems. Members include environmental authorities from Australia, Canada, Europe, and Japan; environmental and industry non-governmental organisations; and global vehicle manufacturers and their suppliers. Partnership members are working to reduce greenhouse gas emissions from mobile air-conditioning systems in two ways. First, they are working to reduce fuel consumption from the operation of vehicle air-conditioning by at least 30%. This will reduce the greenhouse gas emissions from burning fuel. Refrigerant leakage also results in emissions, so partnership members are working to reduce direct refrigerant emissions by 50% as well.

The work of the Mobile Air Conditioning Climate Protection Partnership is important because improved mobile air-conditioning will avoid millions of tons of greenhouse gas emissions and will save billions of gallons of fuel each year. US drivers will save money by using less fuel and will benefit from increased reliability due to improved mobile air-conditioning technology.

Under the Montreal Protocol on Substances that Deplete the Ozone Layer, countries agreed to take steps to protect the ozone layer. As a result, vehicle air-conditioning systems worldwide were redesigned. Now, mobile air-conditioning systems use a refrigerant, HFC134a, that is much friendlier to the environment. HFC134a has no ozone depleting potential and just one-sixth the global warming potential of the former mobile air-conditioning refrigerant. Nonetheless, HFC134a is still part of 'the basket' of greenhouse gases.

2005 US update

Several things are going on to reduce automotive R134a emissions in the US. The Improved Mobile Air Conditioning Climate Protection Partnership is working towards cutting system leakage in half, improving the fuel efficiency of HFC motor vehicle air-conditioning equipment by 30%, and reducing servicing emissions by 50% through improved recovery equipment and training. The project is a collaborative effort funded in part by the US EPA.

The state of California includes auto air-conditioning refrigerant emissions in their new greenhouse gas regulation. Cars must reduce greenhouse gas emissions substantially; reducing refrigerant emissions is one way to do it.

When fully implemented, these efforts will save hundreds of millions of metric tons of carbon dioxide equivalent in the next few decades. Furthermore, the Improved Mobile Air Conditioning Climate Protection Partnership programme is international; benefits will not be limited to the US.

Many studies have compared the benefits of HFC134a systems to others such as carbon dioxide, hydrocarbon, and HFC152a. From an environmental and safety standpoint, carbon dioxide is mildly toxic, HFC152a is slightly flammable, and hydrocarbons are highly flammable. There is a non-environmental problem too: toxic and flammable refrigerants are banned under at least a dozen state laws. All systems would have to be designed to be leak-tight and very sturdy; adding weight to the car, increasing emissions due to fuel combustion. Entirely new training, systems and tools would have to be designed for the servicing industry if CO_2, and HFC152a are adopted.

Appendices

Chart 1 DIN 72 552 terminal numbers

Terminal designation enables technicians to correctly identify and correct connection of the conductors to various devices. The terminal numbers do not identify the conductor (wire), they identify the terminal at the device (component) and its state.

For example – 15 at a module connection indicates current is present when the ignition is switched to run position.

Terminal	Definition
Ignition system	
1	coil, distributor, low voltage
1a, 1b	distributor with two separate circuits
2	breaker points magneto ignition
4	coil, distributor, high voltage
4a, 4b	distributor with two separate circuits, high voltage
7	terminal on ballast resistor, to distributor
15	battery + from ignition switch
15a	from ballast resistor to coil and starter motor
Preheat (diesel engines)	
15	preheat in
17	start
19	preheat (glow)
50	starter control
Battery	
15	battery + through ignition switch
30	from battery + direct
30a	from 2nd battery and 12/24 V relay
31	return to battery − or direct to ground
31a	return to battery − 12/24 V relay
31b	return to battery − or ground through switch
31c	return to battery − 12/24 V relay
Electric motors	
32	return
33	main terminal (swap of 32 and 33 is possible)

Terminal	Definition
33a	limit
33b	field
33f	2. slow rpm
33g	3. slow rpm
33h	4. slow rpm
33L	rotation left
33R	rotation right
Turn indicators	
49	flasher unit in
49a	flasher unit out, indicator switch in
49b	out 2. flasher circuit
49c	out 3. flasher circuit
C	1st flasher indicator light
C2	2nd flasher indicator light
C3	3rd flasher indicator light
L	indicator lights left
R	indicator lights right
L54	lights out, left
R54	lights out, right
AC generator	
51	DC at rectifiers
51e	as 51, with choke coil
59	AC out, rectifier in, light switch
59a	charge, rotor out
64	generator control light
Generator, Generator voltage regulator	
61	charge control light
B+	Battery+
B−	Battery−
D+	Dynamo+
D−	Dynamo−
DF	Dynamo field
DF1	Dynamo field 1
DF2	Dynamo field 2
U, V, W	AC three phase terminals
Lights	
54	brake lights
55	fog light
56	spot light
56a	high beam and indicator light
56b	low beam
56d	signal flash
57	parking lights
57a	parking lights

(Continued)

Terminal	Definition
57L	parking lights left
57R	parking lights right
58	licence plate lights, instrument panel
58d	panel light dimmer
Window wiper/washer	
53	wiper motor + in
53a	limit stop+
53b	limit stop field
53c	washer pump
53e	stop field
53i	wiper motor with permanent magnet, third brush for high speed
Acoustic warning	
71	beeper in
71a	beeper out, low
71b	beeper out, high
72	hazard lights switch
85c	hazard sound on
Switches	
81	opener
81a	1 out
81b	2 out
82	lock in
82a	1st out
82b	2nd out
82z	1st in
82y	2nd in
83	multi-position switch, in
83a	out position 1
83b	out position 2
Relay	
85	relay coil−
86	relay coil+
Relay contacts	
87	common contact
87a	normally closed contact
87b	normally open contact
88	common contact 2
88a	normally closed contact 2
88b	normally open contact 2
Additional	
52	signal from trailer
54g	magnetic valves for trailer brakes
75	radio, cigarette lighter
77	door valves control

Conversion chart

Unit	Convert to	Formula	Input (unit)	Output (conversion)
Temperature				
Celsius (°C)	Fahrenheit (°F)	(°C × 1.8) + 32	10	50.0000
Fahrenheit (°F)	Celsius (°C)	(°F − 32)*0.556	40	4.4480
Pressure				
Bar	psi	bar × 14.50377	10	145.0377
psi	bar	psi × 0.0689	30	2.0670
psi	kPa	psi × 6.8947	30	206.8427
kPa	psi	kPa × 0.145	200	29.0000
kgf/cm^2	psi	kgf/cm^2 × 14.2233	21	298.6893
psi	kgf/cm^2	psi × 0.0703	30	2.1090
MPa	bar	MPa × 10	2	20.0000
Torque				
lbf-ft	N/m	lb-ft × 1.35582	30	40.6746
lbf-in	N/m	lbf-in × 0.1129	30	3.3870
N/m	lbf-ft	N/m × 0.737561	30	22.1268
N/m	lbf-in	N/m × 8.850	30	265.5000
Length				
inch	mm	inch × 25.4	1	25.4000
foot	metre	foot × 0.305	1	0.3050
mile	kilometre	mile × 1.609	1	1.6090
millimetre	inch	mm × 0.03936	1	0.0394
metre	foot	m × 3.280	1	3.2800
kilometre	mile	km × 0.62137	1	0.6214
Power				
watts	BTU/hour	W × 3.4151	100	341.51
BTU/hour	watts	BTU/h × 0.2930	100	29

Chart 2 DIN 40 719 destination letters of electrical devices

Electrical devices which are represented by symbols are identified by a destination code (letter code) which is positioned next to the symbol. For example, K16 is identified as a relay due to the letter code K and numbered 16 due to the number of relays fitted to the vehicle.

Letter code	Type	Examples
A	System, subassembly or parts group	Electronic control modules, CD players/radios
B	Sensor/transducer for the conversion of non-electrical variables	Speed, pressure, oxygen sensors, switches
C	Condenser, capacitor	General
D	Binary element, memory	Intergrated circuits
E	Various devices	Heating devices, air-conditioners, lights, lamps, spark plugs
F	Protective devices	Fuses, current protection
G	Power supply	Generators, batteries
H	Signalling device, alarm, monitor	Indicator lamps, warning lamps and buzzers
K	Relay, contactor	All general relays
L	Inductor	Choke coils, coil windings
M	Motor	Blower, fan, windscreen washer motors, starter motors
N	Regulators, amplifiers	Voltage stabilisers
P	Measuring equipment	Ammeters, diagnostic connectors, tachometers
R	Resistor	NTC resistors, variable resistors, glow plugs, PTC heating elements
S	Switch	General switches
T	Transformer	Ignition coils
U	Modulator, converters	DC transformers
V	Semiconductors	Darlington transistors, diodes, rectifiers, other semiconductor devices
W	Transmission path, antenna	Shielding components, cable harnesses, car antenna
X	Terminal plug and socket connection	All types of electrical connections
Y	Electrically operated mechanical device	Injector solenoids, idle valves, electro-magnetic clutches, fuel pumps
Z	Electrical filter	Suppressors, filters

Chart 3 DIN 40 900 electrical symbols

Symbols are the smallest component of a circuit diagram and are the simplest way to represent an electrical device. The symbols do not indicate the shape or dimensions of the device, just the connections. Often complex internal circuitry can be reduced to a symbol. An enormous range of symbols is available which are beyond the scope of this book. A small selection have been chosen to illustrate their application.

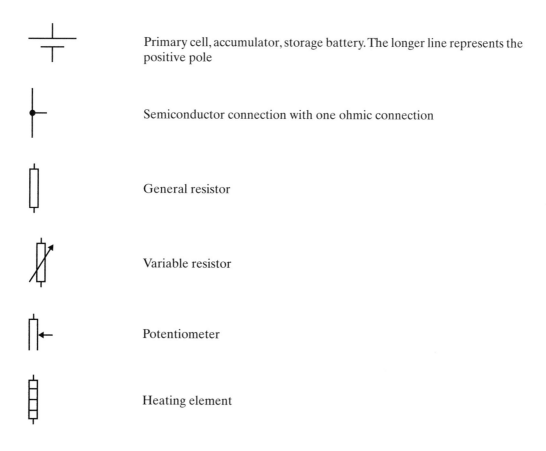

Primary cell, accumulator, storage battery. The longer line represents the positive pole

Semiconductor connection with one ohmic connection

General resistor

Variable resistor

Potentiometer

Heating element

Index